PUBLISH/SUBSCRIBE SYSTEMS

WILEY SERIES IN COMMUNICATIONS NETWORKING & DISTRIBUTED SYSTEMS

Series Editors: David Hutchison, *Lancaster University, Lancaster, UK*
Serge Fdida, *Université Pierre et Marie Curie, Paris, France*
Joe Sventek, *University of Glasgow, Glasgow, UK*

The 'Wiley Series in Communications Networking & Distributed Systems' is a series of expert-level, technically detailed books covering cutting-edge research, and brand new developments as well as tutorial-style treatments in networking, middleware and software technologies for communications and distributed systems. The books will provide timely and reliable information about the state-of-the-art to researchers, advanced students and development engineers in the Telecommunications and the Computing sectors.

Other titles in the series:

Wright: *Voice over Packet Networks* 0-471-49516-6 (February 2001)
Jepsen: *Java for Telecommunications* 0-471-49826-2 (July 2001)
Sutton: *Secure Communications* 0-471-49904-8 (December 2001)
Stajano: *Security for Ubiquitous Computing* 0-470-84493-0 (February 2002)
Martin-Flatin: *Web-Based Management of IP Networks and Systems* 0-471-48702-3 (September 2002)
Berman, Fox, Hey: *Grid Computing. Making the Global Infrastructure a Reality* 0-470-85319-0 (March 2003)
Turner, Magill, Marples: *Service Provision. Technologies for Next Generation Communications* 0-470-85066-3 (April 2004)
Welzl: *Network Congestion Control: Managing Internet Traffic* 0-470-02528-X (July 2005)
Raz, Juhola, Serrat-Fernandez, Galis: *Fast and Efficient Context-Aware Services* 0-470-01668-X (April 2006)
Heckmann: *The Competitive Internet Service Provider* 0-470-01293-5 (April 2006)
Dressler: *Self-Organization in Sensor and Actor Networks* 0-470-02820-3 (November 2007)
Berndt: *Towards 4G Technologies: Services with Initiative* 0-470-01031-2 (March 2008)
Jacquenet, Bourdon, Boucadair: *Service Automation and Dynamic Provisioning Techniques in IP/MPLS Environments* 0-470-01829-1 (March 2008)
Gurtov: *Host Identity Protocol (HIP): Towards the Secure Mobile Internet* 0-470-99790-7 (June 2008)
Boucadair: *Inter-Asterisk Exchange (IAX): Deployment Scenarios in SIP-enabled Networks* 0-470-77072-4 (January 2009)
Fitzek: *Mobile Peer to Peer (P2P): A Tutorial Guide* 0-470-69992-2 (June 2009)
Shelby: *6LoWPAN: The Wireless Embedded Internet* 0-470-74799-4 (November 2009)
Stavdas: *Core and Metro Networks* 0-470-51274-1 (February 2010)
Gómez Herrero, van der Ven, *Network Mergers and Migrations: Junos® Design and Implementation* 0-470-74237-2 (March 2010)
Jacobsson, Niemegeers, Heemstra de Groot, *Personal Networks: Wireless Networking for Personal Devices* 0-470-68173-X (June 2010)
Minei, Lucek: *MPLS-Enabled Applications: Emerging Developments and New Technologies, Third Edition*, 0-470-66545-9 (December 2011)
Barreiros: *QOS-Enabled Networks*, 0-470-68697-9 (December 2011)

PUBLISH/SUBSCRIBE SYSTEMS

DESIGN AND PRINCIPLES

Sasu Tarkoma
University of Helsinki, Finland

WILEY

A John Wiley & Sons, Ltd., Publication

Library of Congress Cataloging-in-Publication Data

Tarkoma, Sasu.
  Publish/subscribe systems : design and principles / Sasu Tarkoma.
      p. cm.
  Includes bibliographical references and index.
  ISBN 978-1-119-95154-4 (pbk.)
  1. Push technology (Computer networks) I. Title.
  TK5105.887.T37 2012
  006.7'876–dc23
                                                    2012010711

A catalogue record for this book is available from the British Library.

Paper ISBN: 9781119951544

Set in 10/12pt Times by Laserwords Private Limited, Chennai, India.

Printed in Singapore by Ho Printing Singapore Pte Ltd

Contents

About the Author

Sasu Tarkoma received his MSc and PhD degrees in Computer Science from the University of Helsinki, Department of Computer Science. He is full professor at University of Helsinki, Department of Computer Science and Head of the networking and services specialization line. He has managed and participated in national and international research projects at the University of Helsinki, Helsinki University of Technology, and Helsinki Institute for Information Technology (HIIT). He has worked in the IT industry as a consultant and chief system architect as well as principal researcher and laboratory expert at Nokia Research Center. He has over 100 publications, several patents in the area of distributed systems and mobile computing, and has also authored several books on distributed systems.

Notes on Contributors

Dr. Weixiong Rao contributed to the topic-based pub/sub part of Chapter 7, and contributed Chapter 10. Dr. Rao is a post-doctoral researcher at University of Helsinki.

Mr. Kari Visala contributed Chapter 13. He is a PhD student at Aalto University and researcher at Helsinki Institute for Information Technology.

Ms. Nelli Tarkoma produced most of the diagrams used in this book. She is a professional graphic artist and illustrator.

This work was supported by the Academy of Finland, grant numbers 255932, 139144, 135230.

Preface

The book offers a unified presentation of the publish/subscribe technology including the design, implementation, and evaluation of new systems based on the technology. Publish/subscribe is a frequently used paradigm for connecting information suppliers and consumers across time and space. The paradigm is extensively applied in modern distributed services, and it has a profound role in current and forthcoming enterprise, cloud, and mobile solutions. The book covers the basic design patterns and solutions, and discusses their application in practical application scenarios. The book examines current standards and industry best practices as well as recent research proposals in the area. The necessary content matching, filtering, and aggregation algorithms and data structures are extensively covered, and the mechanisms needed for realizing distributed publish/subscribe across the Internet.

1

Introduction

Publish/subscribe (pub/sub) technology encompasses a wide number of solutions that aim at solving a vital problem pertaining to timely *information dissemination and event delivery* from publishers to subscribers [1, 2]. In this chapter, we give an overview to pub/sub systems, examine their history, and motivate the contents and structure of this book.

1.1 Overview

The pub/sub paradigm is very useful in describing and monitoring the world around us. Any person meets a constant barrage of events in his waking hours. Most of these events are irrelevant and they should not be allowed to consume the decision maker's resources of awareness, watchfulness, processing and deciding upon actions. Some events are useful to notice and then there are others which are important, even critically important and create the need to muster all the tools and resources to hand. The ability to be aware of a rich stream of events with minimal exertion and to immediately detect critical events for further processing is central to any successful person or organization. The task of efficient event awareness is formidable.

There are a couple of mitigating factors, though. Typically we might know something about the probable sources of interesting events, although we are not actually interested in knowing who sends the notification of an event. Also we might know in advance something about the type of interesting events and can use this knowledge to preselect sources and also to recognize which are critical events. Thus we are interested in event streams of certain types and sources. One can say that we want to subscribe only such a subset of events streams that is enriched for our purposes.

For digital communication purposes this can be interpreted like this: we need a useful communication paradigm, a pub/sub, also called event notification, service that enables the communication components to dynamically detect and isolate particular events. Simultaneously the pub/sub service must allow introduction of new kinds of events. The participating components are generally unaware of each other, that is, an event may be sourceless from the viewpoint of the receiver.

Publish/Subscribe Systems: Design and Principles, First Edition. Sasu Tarkoma.
© 2012 John Wiley & Sons, Ltd. Published 2012 by John Wiley & Sons, Ltd.

> The pub/sub information dissemination and event delivery problem can be stated as follows: How to deliver information from its publishers to interested and active subscribers in an efficient and timely manner? Information is delivered in the form of asynchronous events, which are first detected, and then delivered by publishers to active subscribers in the form of notification messages.

The problem is vital, because many applications require timely data dissemination. To give some examples, stock market data updates, online advertising, asynchronous events in a *graphical user interface (GUI)*, purchase and delivery tracking, digital news delivery, online games, Web feeds (RSS), and in signalling in many embedded and industrial systems. Indeed, pub/sub is a general enabler for many different kinds of applications and it is especially useful in connecting distributed components together forming a basis for loosely coupled systems.

This problem is also challenging, because the information delivery and processing environments can be diverse and a single technological solution cannot address all these environments and the scenario specific requirements. Thus many different pub/sub systems have been developed. Research oriented systems have demonstrated algorithms, structures, and optimizations to pub/sub technology being applied in a certain operating environment. Industry standards have defined the conventions, interfaces, and *Application Programming Interfaces (APIs)* for creating interoperable pub/sub-based products and solution that use the technology. Thus academic research and industry standardization address two different but partially overlapping facets of the information dissemination problem.

Pub/sub and event-based systems are very different from database systems, because they enable data dissemination from publishers to subscribers in the present and future. This contrasts the traditional database model, in which queries are performed on existing data that is available in a database. The notions of database query and subscription are similar, but the query is about the past whereas the subscription is about the future when it is issued. Data tuples stored in a database and the published event, or notification, are also similar, but differ in that the event is forwarded from the publisher to the subscriber and is not stored by the pub/sub system other than for queuing purposes.

Pub/sub is a broad technology domain and consists of many solutions for different environments. Experiences in building pub/sub solutions and implementing them suggest that no single solution is able to meet the demands of the differing application environments and their requirements. This is evident in the number of pub/sub related standards, implementations, protocols, and algorithms. Yet, the goal of connecting diverse communicating entities through a substrate that supports asynchronous one-to-many communication is shared by these solutions.

Pub/sub is a potential candidate to become a key enabler for Web and mobile applications. On the Web, pub/sub enables the asynchronous communication of various Web components, such as web pages and web sites. Figure 1.1 presents a vision for content dissemination on the Internet that has inspired Google's Pubsubhubbub system.[1] In this vision, anyone can become content publisher and aggregator. Open

---

[1] http://code.google.com/p/pubsubhubbub/.

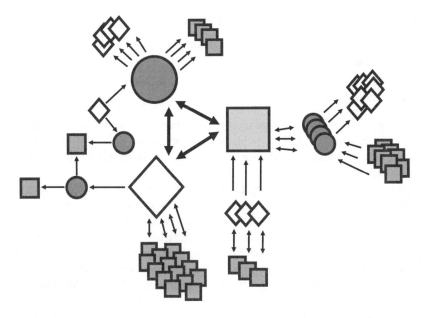

Figure 1.1   A vision of a self-organizing content dissemination system.

interfaces and protocols allow the integration of various content sources. Some publishers and sites become large and others remain small and topical.

Popular alert services, such as Google Alerts[2] and Microsoft Live Alerts[3] allow end users to input keywords and receive relevant dynamic Web content. They are examples of centralized pub/sub solutions for the Web. Their implementation details are not available, but it is believed that alert services are still based on batch processing through search engines. The search engines need to crawl and index live content. Except for a small number of frequently crawled selected sites, the crawling period is typically in the order of a week or tens of days. Thus, they offer a limited form of pub/sub. The next step would be a more decentralized, scalable, and real-time service with support for expressive content matching. Unfortunately, expressive matching semantics and scalability contrast each other making the design, implementation, and deployment of such a global pub/sub service challenging.

Architecture and protocol design should support self-organization and preferential attachment to content sources as well as efficient and timely content dissemination from content publishers through the intermediaries to the content subscribers. The mechanism, techniques, and algorithm are in the key focus of this book. We will address the different facets of the information dissemination problem, and present a collection of frequently employed pub/sub solutions as well as guidelines on how to apply them in practice.

---

[2] http://www.google.com/alerts.
[3] http://alert.live.com.

1.2   Components of a Pub/Sub System

Before going deeper into the topic, we first define the central terms and components, and the overall structure of a pub/sub system.

1.2.1   Basic System

The main entities in a pub/sub system are the publishers and subscribers of content. A publisher detects an event and then publishes the event in the form of a notification. A notification encapsulates information pertaining to the observed event. The notification can also be called the event message.

There are many terms for the entities in pub/sub or event systems; for example, the terms subscriber, consumer, and event sink are synonymous. Similarly, publisher, producer, supplier, and event source are synonymous. As mentioned above, the notification or event message denotes that an observed event has happened.

An event represents any discrete state transition that has occurred and is signalled from one entity to a number of other entities. For example, a successful login to a service, the firing of detection or monitoring hardware and the detection of a missile in a tactical system are all events.

Events may be categorized by their attributes, such as which physical property they are related to. For instance spatial events and temporal events note physical activity. Moreover, an event may be a combination of these, for example an event that contains both temporal and spatial information. Events can be categorized into taxonomies on their type and complexity. More complex events, called composite or compound events, can be built out of more specific simple events. Composite events are important in many applications. For example, a composite event may be fired

- in a hospital, when the reading of a sensor attached to a patient exceeds a given threshold and a new drug has been administered in a given time interval;
- in a location tracking service, where a set of users are in the same room or near the same location at the same time; or
- in an office building, where a motion detector fires and there has been a certain interval of time after the last security round.

After the notification has been published, it is the duty of the pub/sub system to deliver the message to interested recipients – the subscribers. A subscriber is an entity that has expressed prior interest to a set of events that meet certain requirements that the subscriber has set. The actual delivery depends on the pub/sub solution being used; for example, it could be based on the following:

- The message is broadcast on the network and devices on the same network will see the message. The pub/sub system running on a device can then process the message and deliver it to the subscriber if it is active on the device.
- The message is delivered via network supported multicast, in which a specific network primitive is used for delivering the message from one publisher to many subscribers.
- The message is sent directly by the publisher to subscribers that have informed the publisher that they are interested in receiving a notification. The publisher then utilizes

a one-to-one message delivery protocol on top of the communication primitives offered by the network, typically the TCP/IP protocol stack.

- The message is first sent to a broker server and then delivered by the broker to active subscribers. In this case, the subscribers have expressed their interest in receiving notifications with the broker.
- The message is delivered through a network of brokers. The scalability of a pub/sub system can be increased by deploying a network of pub/sub brokers.

The two first cases are based on communication primitives provided by the underlying network, namely broadcast and multicast. Typically these primitives are not usable with Internet applications, because they are supported only within specific regions of the Internet and thus cannot be used to deliver messages in the global environment. The third case is very typical and extensively used when the number of subscribers is known to be small. This strategy does not scale when the number of subscribers increases. The fourth and fifth case introduce the concept of a broker, also called pub/sub router, that mediates events and provides a routing and matching engine for the publishers and subscribers. This is a commonly used solution for the distributed environment. A well-known technique for deploying pub/sub systems is to create them as overlay networks that operate on top of the current Internet routing system [3].

1.2.2 Distribution and Overlay Networks

A pub/sub system may be centralized or distributed in nature. The notification processing and delivery responsibility may be provided by different entities:

- publishers;
- a centralized broker;
- a set of brokers in a routing configuration typically realized as an overlay network.

Event and notification processing can be easily implemented in publishers and with a centralized broker; however, as observed above, these approaches do not scale well when there are many entities and events in the system. Scalability can be improved by implementing the pub/sub system with a set of brokers as an overlay construct on top of the network layer.

An application layer overlay network is implemented on top of the network layer and it provides services such as resource lookup, overlay multicast, and distributed storage. An overlay network typically provides useful features such as easy deployment of new distributed functions, resilience to network failures, and fault-tolerance [3]. An overlay-routing algorithm is based on the underlying packet-routing primitives. A pub/sub overlay system is implemented as a network of application layer brokers or routers that communicate by using the lower layer primitives, typically TCP/IP.

Figure 1.2 illustrates a pub/sub overlay network. The two important parts of a distributed pub/sub network are the broker topology and how routing state is established and maintained by the brokers. By propagating routing state we mean how the interests of the subscribers are sent towards the publishers of that information. In essence, the routing state stored by a broker must enable it to forward event messages either to other brokers or to subscribers that have previously subscribed to the notifications.

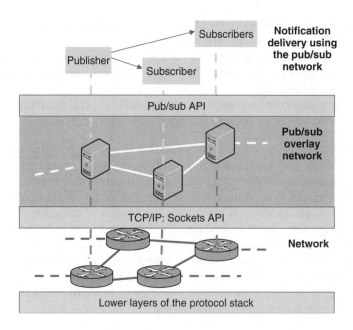

Figure 1.2   Example of a pub/sub overlay network.

In this book, we will investigate the above ways of realizing the notification as well as solutions for achieving high performance, expressiveness, availability, fault resilience, and security.

1.2.3   Agreements

The pub/sub system is used to facilitate the delivery of the messages; however, the meaning of the event is application and domain specific. In order to build a pub/sub system with many entities the following agreements need to be considered:

- Agreement of the notification message format and syntax. For example, many systems utilize a typed-tuple-based format or XML. This agreement may consist of additional details such as those pertaining to timestamps and content security.
- Agreement of the message protocol that is used to transfer the event between two entities. This can include many parameters, for example security, reliability, etc.
- Agreement of the notification filtering semantics. This specifies what elements of the message can be used to make a notification decision. For example, a notification is forwarded based on the publisher, observation time, and type of the event.
- Agreement on the visibility of the published event. It may be necessary to restrict the delivery and processing of the event in the operating environment.
- Agreement of the application and domain specific interpretation of the event. This agreement is outside the scope of a pub/sub system.

Thus many implicit or explicit agreements are needed to design and implement a pub/sub system for an environment that consists of many entities.

1.2.4 The Event Loop

The *event loop* is a key construct in creating event-based applications. The event loop is a frequently used approach in implementing applications that react to various events. For example, Microsoft Windows programs are based on events. The main thread of the application contains the event loop, which waits for new events to process. The event loop can use a blocking function call for receiving messages or a nonblocking peek message function. Typically when a message is received it is processed and delivered to callbacks for further processing.

The event loop is a crucial part of an application that needs to react to events in a timely manner, for example GUI events. The event loop naturally combines with a distributed pub/sub system and it is a key construct for implementing pub/sub engines. A simple pub/sub engine can be implemented as an event loop that reacts to incoming subscription and publishing requests.

1.2.5 Basic Properties

Pub/sub technology has evolved since its inception in the late 1980s to a promising technology for connecting software components across space, time, and synchronization [4]. These three properties summarize the salient features of the technology. We will examine each of the three properties presented in Figure 1.3 in detail in this section.

Space decoupling is illustrated by subfigure A, in which the event notification service decouples the publisher and the subscribers. The event message is transferred to the event service, and then it is transferred to the subscribers. Thus memory space is not shared by the entities. Subfigure B presents an example of time decoupling. The setting is the same as for the space decoupling case with the exception of message buffering at the service side. Time decoupling is achieved by storing the message in a message buffer at the event notification service for eventual delivery to subscribers. The synchronization decoupling is illustrated by subfigure C, which emphasizes the temporal aspect. The publish and notify phases of event delivery are decoupled and they do not require synchronization. The message is first delivered to the event notification service and then to the subscriber.

Figure 1.4 summarizes the decoupling properties of well-known communication techniques. As observed before, the communication techniques are not orthogonal but rather they are combined in order to implement more sophisticated systems. Message passing, *Remote Procedure calls (RPC)* and *Remote Method Invocation (RMI)*, and asynchronous RPC/RMI do not offer decoupling in space and time. They can offer decoupling in synchronization. Tuple spaces offer decoupling in space/time through the shared space; however, the reader of the tuple space is blocked and thus tuple spaces do not offer decoupling in synchronization [5]. Message queuing, on the other hand, offers decoupling of all three properties and it is a building block for the more sophisticated pub/sub systems.

Pub/sub is based on message queuing and message-oriented middleware. Message queuing is a communication method that employs message passing between a sender and a receiver with the help of a sender-side message queue. A message being sent is first stored in the local message queue. After the delivery has been made, the message can be removed from the queue. If the message cannot be delivered or the message is incorrectly received, the message can be resent.

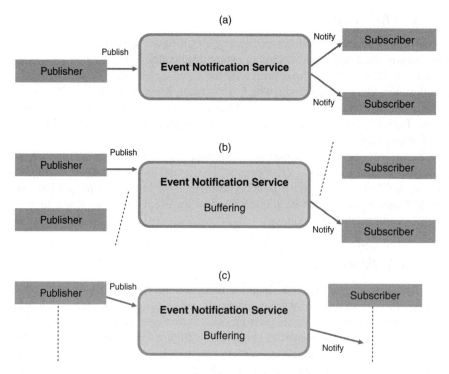

Figure 1.3 Decoupling properties in pub/sub. (a) Space decoupling; (b) Time decoupling; (c) Syschronization decoupling.

Abstraction	Space/time decoupling	Synchronization decoupling
Message passing	No	Varies
RPC/RMI	No	Invoker is blocked
Async RPC/RMI	No	Yes
Tuple spaces	Yes	Reader is blocked
Message queuing	Yes	Varies
Pub/sub	Yes	Yes

Figure 1.4 Summary of decoupling properties.

Queuing is a basic solution in achieving reliability in data communications. Queuing also supports disconnections during which the message cannot be sent. Message queuing is thus the basic ingredient for achieving decoupled communications.

One distinction between message queuing systems and pub/sub is that they typically offer one-to-one communications and require that the receivers are explicitly defined. Pub/sub on the other hand supports one-to-many and many-to-many communications and the subscribers can be defined implicitly by the event message being delivered and the a priori subscriptions that the subscribers have set.

The key properties of pub/sub systems are: decoupling in space, time and synchronization, many-to-many communications, and information filtering.

1.3 A Pub/Sub Service Model

Figure 1.5 illustrates a generic pub/sub service design. In the figure, the pub/sub service is a logically centralized service that provides the necessary functions and interfaces for supporting notification delivery from publishers to subscribers. The pub/sub service consists of the following key components:

- A notification engine that builds and maintain an index structure of the subscriptions, and uses the index table to forward notifications to subscribers. The engine offers the necessary interfaces for subscribers and publishers that allow them to subscribe, unsubscribe, and publish content.
- A subscription manager that accepts subscriptions from the engine and maintains those. The two mandatory operations are insert and remove a subscription.
- A subscription storage that stores subscriptions and data related to the subscriptions.
- An event storage is a facility for storing published events so that they can be retrieved later.
- A notification consumer that is an intermediary component in the notification process. A consumer receives notifications from the engine and then forwards those to the proper subscriber. The consumer can buffer, compress, and process notifications before the final delivery.

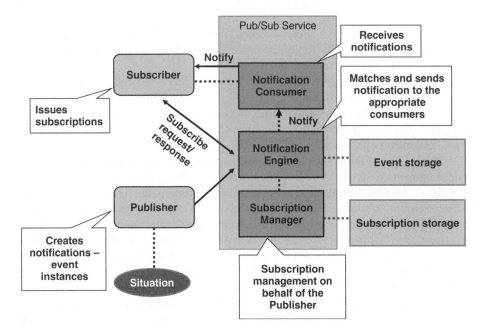

Figure 1.5 Components of a pub/sub system.

A publisher observes a situation and when an event of interest is observed, a notification is created and sent to the notification engine using its publication interface. The notification is then matched with the subscription index maintained by the engine with the help of the subscription manager. The notification is given by the engine to the notification consumers of subscribers that have expressed interest in the notification. In other words, the notification matches with the subscribers' subscriptions. The notification is then prepared by each consumer for delivery to the associated subscriber.

This model of a pub/sub service separates the management of the subscriptions, the matching process with the notification engine, and the final delivery to the subscribers. This separation allows, for example, changing of the notification consumer without changing the engine.

The design of Figure 1.5 is logically centralized and it hides the distribution of the components. It is necessary to distribute and replicate the components in order to achieve scalability and reliability in a distributed environment.

1.4   Distributed Pub/Sub

As mentioned in this chapter, direct notification of subscribers by a publisher is not scalable. Therefore it is vital to develop techniques for distributing the notification process. To this end, a number of pub/sub network designs have been developed.

An *event broker or router* is a component of a pub/sub network that forwards notification messages across multiple hops. An example pub/sub network is presented in Figure 1.2 that shows the layered design. The pub/sub network offers the notification API to subscribers and publishers and utilizes the network API, typically the Sockets API, to disseminate the notification message and take it from the source router to the destination router and subnetwork. The network level routers are responsible for taking the message end-to-end across the Internet. Such overlay designs have favourable characteristics in terms of deployability and flexibility; however, the resulting high level routing may not be efficient in terms of the network level topology.

An event router typically has local clients and neighbouring routers. The algorithms and protocols for local clients and neighbouring routers are different. Both cases require a *routing table* for storing information about message destinations. A pub/sub routing table is an index structure that contains the active subscriptions and typically supports add, remove, and match operations.

The design and configuration of pub/sub networks has become an active area of research and development. We will focus on various strategies for implementing pub/sub networks.

The simplest form of notification in the distributed environment is called *flooding*. With flooding each pub/sub broker simply sends the message to all neighbours except the one that sent the message. Thus the message is introduced at every broker; however, the price of the technique is its inaccuracy. Ideally, we want to prevent the forwarding of a message to a broker that we know does not have subscribers for the message. Moreover, excess and uncontrolled messaging may lead to congestion that in turn may cause notification messages to be dropped.

In order to avoid unnecessary message deliveries, we introduce the notion of *filtering* into the pub/sub network. Filtering involves an interest registration service that accepts filter information from the subscribers. The subscribers can thus specify in more detail what kind of data they desire. The pub/sub network then distributes this filtering information in such a way that minimizes the overhead in notification message delivery. The process of optimizing a pub/sub network is not simple, because the filtering information also introduces overhead into the network. For example, filtering information may need to be updated, and there is propagation delay in setting up and maintaining the routing tables of pub/sub brokers. Later in this book we will consider various techniques in optimizing these networks.

Accuracy is a key requirement for a pub/sub network. The accuracy of event delivery can be expressed with the number of false positives and false negatives.

A *false positive* is a message that is sent to a subscriber that does not match the subscriber's active interests. Similarly, a *false negative* is a message that was not sent to a subscriber, but should have been because it matches the subscriber's active interests.

Various filtering languages and filter matching algorithms have been developed. Filtering involves the specification of filters that are organized into a filtering data structure. A filter selects a subset of notifications based on the filtering language. Thus a filter is a constraint on the notification message and it can be applied in the context of the notification type, structure, header, and content.

Filtering allows the subscribers to specify their interest beforehand and thus reduce the number of uninteresting event messages that they will receive. A filter or a set of filters that describes the desired content is included with the subscription message that is used by brokers to configure routing tables. Many filtering languages have been developed, specified, and proposed. For example, the filtering language used by *Java Message Service (JMS)* is based on *Structured Query Language (SQL)* [6].

Filtering is a central core functionality for realizing event-based systems and accurate content-delivery. Filtering is performed before delivering a notification to a client or neighbouring router to ensure that the notification matches an active subscription from the client or neighbour. Filtering is therefore essential in maintaining accurate event notification delivery.

Filtering increases the efficiency of the pub/sub network by avoiding to forward notifications to brokers that have no active subscriptions for them. Filters and their properties are useful for many different operations, such as matching, optimizing routing, load balancing, and access control. To give some examples, a firewall is a filtering router and an auditing gateway is a router that records traffic that matches a given set of filters.

1.5   Interfaces and Operations

Table 1.1 presents the pub/sub operations used by many event systems [7]. The operations are requested by a client, denoted by X, of the system. There are many ways to define the interests of the subscriber. In our generic API, we denote the general interests by F.

Table 1.1 Infrastructure interface operations

Operation	Description	Semantics
Sub(X,F)	*X* subscribes content defined by *F*	Sub/Adv
Pub(X,n)	*X* publishes notification *n*	Sub/Adv
Notify(X,n)	*X* is notified about notification *n*	Sub/Adv
Unsub(X,F)	*X* unsubscribes content defined by *F*	Sub/Adv
Adv(X,C)	*X* advertises content *C*	Adv
Unadv(X,C)	*X* unadvertises content *C*	Adv
Fetch(X,P)	*X* fetches messages that satisfy the given constraints *P*	Sub/Adv

In expressive content-based routing *F* is typically defined with a Boolean function that selects a subspace of the content space, in which the notifications are defined. Notifications are points in this space. There are also less expressive semantics for subscribing content, such as type-based subscriptions. We will return to these notions shortly.

As presented by the table, the key operations pertain to publishing, subscribing, unsubscribing, and fetching content. It should be noted that the subscribe and unsubscribe operations are idempotent, which means that even if the same operation is executed repeatedly it does not change the state of the system. Publish operation, however, is not idempotent and repetitions will cause many publications to be delivered.

In a large-scale pub/sub system, the API typically supports leases that determine the validity time period for each subscription and advertisement. Leases are useful in removing obsolete state from the pub/sub network, and they are instrumental in ensuring the eventual stability of the network. The unsubscription and unadvertisement are not necessary if leases are supported by the API; however, they may still be useful in terminating a lease before the it expires.

There are two different kinds of operational semantics for a pub/sub system:

- Subscription-driven: Subscriptions are propagated by the pub/sub network and the routing tables are based on filters specified in the subscription messages.
- Advertisement-driven: Publishers first advertise content with advertisement messages that are propagated by the pub/sub network. The pub/sub network then connects subscriptions with matching advertisements to active content delivery across the network.

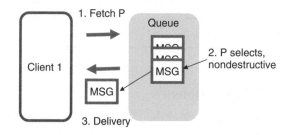

Figure 1.6 Example of the expressive fetch operation.

The table presents the API operations for these two filtering semantics. The advertisement semantics introduces the operations for advertising and unadvertising content. Moreover, the API operations are typically extended with security and quality-of-service properties as well as more expressive notification retrieval strategies. Key extensions pertaining to event retrieval, the fetch operation illustrated by Figure 1.6, include:

- Fetch operation for retrieving a specific number of messages.
- Nondestructive fetch, which leaves the retrieved messages at the server's queue. This is useful when multiple instances of the same software are retrieving messages.
- Fetch operation with query operation that allows specific event messages to be fetched from the queue. This operation is frequently supported by pub/sub standards, for example JMS.
- Fetch the latest event message in the message queue. This is useful when starting an application or recovering from application failure.

In the following section we will investigate the different filtering semantics for targeted information delivery.

1.6 Pub/Sub Semantics for Targeted Delivery

As mentioned above, there needs to be agreement on how notification messages are delivered from publishers to subscribers. There are many possible semantics for selecting notifications that need to be delivered for a given set of subscribers. In this section, we will briefly examine key semantics for targeted notification delivery.

Depending on the expressiveness of the filtering language, a field, header, or the whole content of the notification message may be filterable. In *content-based routing* the whole content of the event message is filterable.

Figure 1.7 illustrates the four key types of message routing semantics. The types are the following: content-based, header-based, topic-based, and type-based. As mentioned above, content-based routing allows the evaluation of filters on the whole event message. Header-based is more limited and only allows to evaluate elements included in the header of the message. Topic-based only allows to evaluate a specific topic field in the message. Topic-based systems are similar to channel-based systems and the topic name can be seen to be the same as the channel name. Typically topic-based systems require that the topic of an event message exactly matches with the requested topic name and thus it is not very expressive. Finally, type-based systems allow the selection of event messages based on their designated type in a type hierarchy. We can take a type hierarchy pertaining to buildings as an example: the root of the hierarchy is the building name, the second level consists of floors, and the third level of the offices. By subscribing to a floor the subscriber receives all events related to that specific floor in the named building.

The different routing semantics are characterized by their selectivity. Type-based systems make the forwarding decision based on a predefined set of message types. In topic-based and channel-based pub/sub, the subscribers are defined by a queue name or a channel name. The notifications are sent to a named queue or channel, from which the subscriber then extracts the messages. An important limitation is that the queue or channel name has to be agreed beforehand. Subject-based systems make the routing decision based

Name	Value
Resource_name	CS Department's home page
Address	www.cs.helsinki.fi
Resource_type	Web page
Content element 1	Data

Content-based routing

Name	Value
Resource_ name	CS Department's home page
Address	www.cs.helsinki.fi
PAYLOAD	

Header-based routing

Name	Value
Topic	CS Department Channel
PAYLOAD	

Topic-based routing

Name	Value
Type	UH\Faculty of Science\CS Department
PAYLOAD	

Type-based routing

Figure 1.7   Examples of message targeting systems.

on a single header field in the notification. Header-based systems use a special header part of the notification in order to forward the message. Content-based systems are the most expressive and use the whole content of the message in making the forwarding decision. Content-based pub/sub is flexible because it does not require that topic or channel names are assigned beforehand.

> Various pub/sub delivery semantics can be implemented with a content-based communication scheme making it very expressive. Header-based routing is more limited, but it has a performance advantage to content-based routing, because only the header of a message is evaluated when making a forwarding decision.

Expressiveness and scalability are important characteristics of an event system [8]. Expressiveness pertains to how well the interests of the subscribers are captured by the pub/sub service. Scalability involves federation, state, and the number of subscribers, publishers, and brokers can be supported as well as the how much notification traffic can the system support.

Other requirements for a pub/sub network include simplicity, manageability, implementability, and support for rapid deployment. Moreover, the system needs to be extensible and interoperable. Other nonfunctional requirements include: timely delivery of notifications (bounded delivery time), support for *Quality of Service (QoS)*, high availability and fault-tolerance.

Event order is an important nonfunctional requirement and many applications require support for either causal order or total order. Causality determines the relationship of two events A and B. In order to be able to determine causality in the distributed system a logical clock mechanism is needed. The two well-known solutions are the Lamport clocks and vector clocks. We will examine these clocks in more detail in Chapter 2.

1.7 Communication Techniques

Event systems are widely used, because asynchronous messaging provides a flexible alternative to RPC [4, 9]. RPC is typically synchronous and one-to-one, whereas pub/sub is asynchronous and many-to-many. Limitations of synchronous RPC calls include:

- Tight coupling of client and server lifetimes. The server must be available to process a request. If a request fails the client receives an exception.
- Synchronous communication. A client must wait until the server finishes processing and returns the results. The client must be connected for the duration of the invocation.
- Point-to-point communication. Invocation is typically targeted at a single object on a particular server.

On the other hand, RPC is a building block for distributed pub/sub systems. Many pub/sub implementations use RPC operations to implement the API operations presented in Table 1.1.

Event delivery between two processes can be realized in many ways depending on the requirements and the operating environment. The two key differing environments are the local and remote communication context. In local event delivery, techniques such as shared resources and local procedure calls or message passing can be used. Remote event delivery is typically implemented with message queuing or RPC.

RPC offers reliability and at-most-once semantics whereas message queuing systems have differing message delivery options. The key reliability semantics are:

- Exactly-once: The highest reliability guarantee in which the message is sent to the remote node exactly once. The message is delivered even if the server crashes or the network fails. This reliability level is not possible to achieve in the typical distributed environment.
- At-least-once: This reliability level guarantees that a message is sent to the remote node, but duplicates are allowed. Duplicates may happen due to network failures and server crashes. The semantics are appropriate for idempotent operations.
- At-most-once: This reliability level guarantees that the message is sent to node once if at all. It does not guarantee that message is delivered. A message can disappear due to network problems or server crashes.

Typically commercially used message queue systems support either at-least-once or at-most-once. The semantics are implemented with sender and receiver side message buffering, sequence numbers, and timers for detecting lost messages and other problems.

Figure 1.8 illustrates the options when implementing pub/sub systems. The key difference between message queuing and RPC is that messaging is asynchronous whereas traditional RPC is synchronous although there are also asynchronous RPC features. Alternative techniques are tuple spaces and distributed shared memory. We will later in Chapter 2 consider Java RMI as one example of an RPC system.

Distributed shared memory can be realized in many ways based on the memory abstraction. A page based abstraction organizes the shared memory into pages of fixed size. The object based abstraction organizes the shared memory as an abstract space for storing shareable objects. A tuple space, on the other hand, is based on the

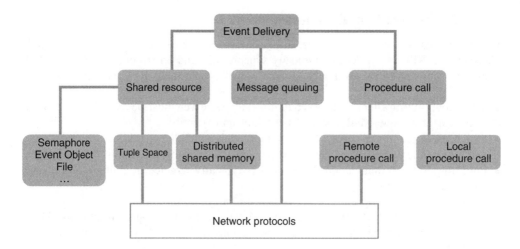

Figure 1.8 Communication techniques for event delivery.

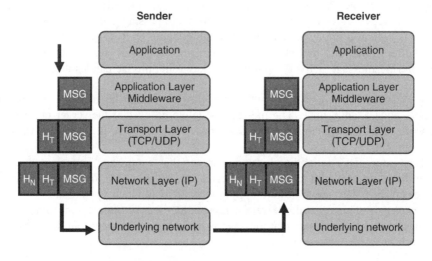

Figure 1.9 Protocol stack with middleware.

tuple abstraction. A coherence protocol is needed to maintain memory coherence in a distributed shared memory implementation. Memory update and invalidation techniques include update-on-write, update-on-read, invalidate-on-write, invalidate-on-read. Typically these systems follow the weak consistency model, in which synchronizations must be atomic and have a consistent order.

Figure 1.9 illustrates the layered nature of the communications environment. Each layer provides functions for the higher layers and abstracts details of the underlying layers. The organization of protocols into a stack structure offers separation of concerns; however, it makes it difficult to optimize system behaviour across layers. As shown by the figure, each layer adds its own header and details to the packets and messages being processed.

In a similar fashion, when receiving a packet, each layer processes its own information and gives the data to a higher layer. Pub/sub systems can be implemented on multiple levels of the stack, starting from the link layer towards the application layer. Most pub/sub systems are implemented on top of TCP/IP and they are offered as middleware services or libraries. A pub/sub system can itself be viewed to be a layered system, in which the higher level functions of distributed routing and forward are based on lower layer message queuing primitives.

1.8   Environments

The pub/sub paradigm can be applied in many different contexts and environments. Early examples include GUIs, control plane signalling in industrial systems, and topic-based document dissemination. The paradigm is fundamental to todays's graphical and network applications. Most programmers apply the paradigm in the context of a single server or device; however, distributed pub/sub is vital for many applications that require the timely and efficient dissemination of data from one or more sources to many subscribers.

The operating environments for pub/sub can be examined from differing viewpoints, for example based on the underlying communications environment and the application type. In the following we summarize key environments for pub/sub technology:

- Local: Event loop, GUI, in-device information delivery.
- Wireless and ad hoc: Event delivery in wireless networks in which nodes can move. Publishers and subscribers are typically run on constrained limited devices, for example mobile phones.
- Sensor: Event delivery in sensor networks from a number of source sensors to sinks that then deliver the events for further processing.
- Embedded and industrial: Event delivery in an embedded or industrial setting, for example within a car or an airplane or a factory.
- Regional: Event delivery within an organization or a region.
- Internet-wide: Event delivery in the wide-area network across organizational boundaries.

In this book we will focus especially on distributed pub/sub systems for the last three categories; however, we do also consider the mobile and wireless domain as well.

Small and wireless devices have limited capabilities compared to desktop systems: their memory, performance, battery life, and connectivity are limited and constrained. The requirements of mobile computing need to be taken into account when designing an event framework that integrates with mobile devices.

From the small device point of view, message queuing is a frequently used communication method because it supports disconnected operation. When a client is disconnected, messages are inserted into a queue, and when a client reconnects the messages are sent. The distinction between popular message-queue-based middleware and notification systems is that message-queue-based approaches are a form of directed communication, where the producers explicitly define the recipients. The recipients may be defined by the queue name or a channel name, and the messages are inserted into a named queue, from which the recipient extracts messages. Notification-based systems extend this model by adding an entity, the event service or event dispatcher, that brokers notifications between producers of information and subscribers of information. This undirected communication supported by the notification model is based on message passing and retains the benefits of message queuing. In undirected communication the publisher does not necessarily know which parties receive the notification.

The pub/sub paradigm and technology can be seen to be a unifying technology that combines the different environments and domains through event delivery. Indeed, pub/sub has been proposed as the new basis for an internetworking architecture; however, there are still many unsolved challenges in applying the paradigm on the global Internet scale. We consider these solutions in Chapter 13.

1.9   History

As mentioned above, pub/sub can applied in many different environments for solving the information dissemination problem. The early applications include the filtering and delivery of Usenet postings as well as being the glue of many GUIs. More recent applications include Internet technologies such as RSS and XMPP as well as the many standards such as JMS, CORBA Notification Service [10], and OMG DDS.

In this section, we will examine the history of pub/sub systems from three viewpoints, namely the research highlights, standardization, and Internet technology. The last category illustrates the importance and applicability of pub/sub for Internet-based applications. Figure 1.10 gives a timeline of the evolution of pub/sub technology for the three categories. In the following, we briefly examine the key developments. We will return to many of these later in the book.

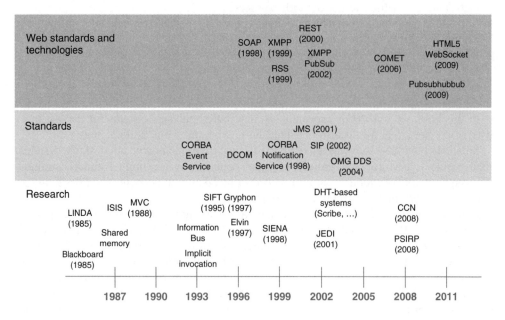

Figure 1.10   Timeline of pub/sub solutions.

1.9.1   Research Systems

The history of pub/sub has its roots in the requirement to process asynchronous events. The notion of the event loop is very old; however, the patterns employed today to realize distributed pub/sub are considerably newer. The first systems were based on the shared memory abstraction. The memory represented rendezvous space for senders and receivers. Processors communicated by posting messages to the shared memory.

The shared memory is very similar to the blackboard pattern frequently used in creating artificial intelligence systems. The blackboard pattern that was proposed in 1985 for solving complex problems with the help of a shared memory abstraction [11, 12]. The LINDA tuple space model is also from this period [5]. LINDA is a coordination and synchronization technique based on the tuple abstraction. LINDA supports communication through a shared memory region called the tuple space. Processes generate tuples and store those in the shared space. Other processes can then monitor the tuple space and read tuples.

Another early example of interprocess communication is the UNIX signal notification system that was implemented in 1986. UNIX processes use signals to notify each other. A process has a unique numeric process identifier and a group of processes have a numeric group identifier. A signal can be directed to a specific process or a group of processes.

The frequently used *Model-View-Control(MVC)* design pattern was developed in the SmallTalk community in 1988. MVC facilitates the communication between the model, view, and control components [13]. The MVC pattern separates concerns over application state (model), user interface (view), and the control aspect. MVC requires that a component is able to subscribe to the state of another component. This subpattern used in the MVC developed into the observer pattern that is widely used. We examine these and other patterns in more detail in Chapter 4.

An early pub/sub service was proposed in the seminal ISIS system in 1987 [14]. This ISIS subsystem was responsible for disseminating news items from publishers to subscribers. The ISIS news service allowed processes to subscribe to system-wide news announcements. The subscriber specified a subject and then received postings under that subject. The ISIS architecture also featured filters that were used on client systems to process incoming messages. The subscribe operation of the news subsystem was implemented with one RPC per posting and the actual posting delivery with one asynchronous multicast operation (with causal or total ordering).

The key contributions of the ISIS system for pub/sub were:

- Reliable atomic multicast communications primitives.
- Causal and total ordering of multicast messages.
- Developing the pub/sub system based on the RPC and multicast primitives.

Another early example of a pub/sub system is the Information Bus proposed in 1993 [15]. This model consists of service objects and data objects. Service objects can have local data objects, and they send and receive them through the datacentric information bus. Each data object is labelled with a subject string. Subjects are hierarchically structured. The Information Bus supports both pub/sub and request/reply APIs. With the pub/sub model, the system decouples the component and subscribers do not need to know the identities of the publishers. The Information Bus model calls this kind of communication subject-based addressing. The system has built-in support

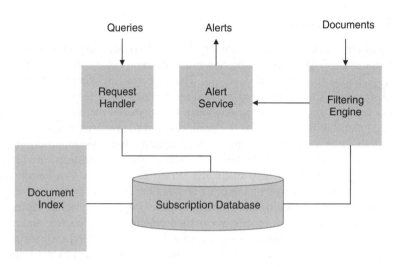

Figure 1.11 Overview of the SIFT system.

for dynamic discovery of participants. This is implemented with two publications, first a query to prospective participants listening a specific subject, and then response publication indicating presence. The system can be extended with adapters that convert information from the data objects to application specific formats.

The ISIS system and Information Bus did not take the content of the messages into account. In the news applications implemented with these systems, the news items were delivered based on the subject that was a configuration parameter. The SIFT information dissemination system is an early example of an alert service [16] that takes the content of the disseminated documents into account. The system proposed the inverted index for matching documents to subscriptions. Figure 1.11 illustrates this system and its key components.

The key idea of the *inverted index* is to allow fast full text searches of the documents. The words need to be extracted during the insertion of the document. Thus the technique places most of the processing cost to the insertion phase. The index structure maps document content, typically the words, into locations in a set of documents. Thus given a query it is easy to determine the matching documents with the inverted index. For example, consider the set of three words {"pub","sub","event"} with "pub" mapping to documents {1,2}, "sub" to documents {1,3,4}, and event to {2,6,7}. Now, a query for "pub" and "event" would result in {1,2} ∩ {2,6,7} = {2}. Thus document number two would be returned.

The system accepts document queries and stores them into a subscription database. Similarly, documents are parsed and an inverted index is stored in the document index. A filtering engine then is responsible for matching documents to the queries with the document index.

The SIFT system did not consider the distributed environment in more detail. IBM's Gryphon project developed a distributed pub/sub system consisting of a network of brokers [17, 18]. The Gryphon system was developed at the Distributed Messaging Systems group at the IBM T.J.Watson Research Center. Gryphon is a Java-based pub/sub message broker intended to distribute data in real time over a large public network. Gryphon uses content-based routing algorithms developed at the research center. The clients of Gryphon use an implementation of the JMS API to send and receive messages. The Gryphon project was started in 1997 to develop the next generation web applications and the first deployments were made in 1999. Gryphon is designed to be scalable, and it was used to deliver information about the Tennis Australian Open to 50000 concurrently connected clients. Gryphon has also been deployed over the Internet for other real-time sports score distribution, for example the Tennis US Open, Ryder Cup, and monitoring and statistics reporting at the Sydney Olympics.

Gryphon supports both topic-based and content-based publish-subscribe, relies on adopted standards such as TCP/IP and HTTP, and supports recovery from server failures and security. In Gryphon, the flow of streams of events is described using an *information flow graph (IFG)*, which specifies the selective delivery of events, the transformation of events, and the creation of derived events as a function of states computed from event histories.

Elvin is another early example of a content-based routing system with expressive semantics [19]. Elvin uses a client-server architecture in notification delivery. Clients establish sessions with Elvin servers and subscribe and publish notifications.

Scalable Internet Event Notification Service (SIENA) is an Internet-scale event notification service developed at the University of Colorado. SIENA balances expressiveness with scalability and explores content-based routing in a wide-area network. The basic pub/sub mechanism is extended with advertisements that are used to optimize the routing of subscriptions [8]. Several network topologies are supported in the architecture, including hierarchical, acyclic peer-to-peer, and general peer-to-peer topologies. Servers only know about their neighbours, which helps in minimizes routing table management overhead. Servers employ a server-server protocol to communicate with their peers and a client-server protocol to communicate with the clients that subscribe to notifications. It is also possible to create hybrid network topologies.

SIENA introduced covering relations between filters to prevent unnecessary signalling. The SIENA system used the notion of covering for three different comparisons:

- matching a notification against a filter;
- covering relation between two subscription filters;
- and overlapping between an advertisement filter and a subscription filter.

Covering relations have been used in many later event systems, such as Scribe [20], Rebeca [21], and Hermes [22, 23]. Scribe and Hermes are examples of *Distributed Hash Table (DHT)*-based pub/sub systems. Scribe is a topic-based system and Hermes supports both topic-based and content-based communication. Scribe and Hermes choose a rendezvous point for each topic or event type in the overlay network topology, and then build and maintain multicast trees rooted at this rendezvous point. We will later return to DHT structures and DHT-based pub/sub systems.

DHTs are a class of decentralized distributed algorithms. They provide a hashtable API and implement the hashtable functionality in a wide-area environment in which nodes can join and leave the network. A DHT maintains (key, value) pairs and allows a client to retrieve a value corresponding to the given key.

The *combined broadcast and content-based (CBCB)* routing scheme extends the SIENA routing protocols by combining higher-level routing using covering relations and lower-level broadcast delivery [24]. The protocol prunes the broadcast distribution paths using higher-level information exchanged by routers.

Java Event-based Distributed Infrastructure (JEDI) [25] is a distributed event system developed at Politecnico di Milano. In JEDI the distributed architecture consists of a set of *dispatching servers (DS)* that are connected in a tree structure. Each DS is located on a node of the tree and all nodes except the root node are connected to one parent DS. Each node has zero or more descendants.

Gryphon, Elvin, SIENA, and JEDI paved way for the next generation of content-based pub/sub systems developed as overlay networks over the Internet. More recent developments have considered also the introduction of pub/sub primitives into the protocol stack design.

SIENA pioneered the notion of content-based networking, in which content demand defines subnetworks and where information is sent. The notion of datacentric networking is similar and has been pioneered by projects such as TRIAD [26] and DONA [27]. These new forms of networking are motivated by the observation that the current Internet architecture has been designed around a host-based model that dates from the 1970s. The aim is to allow the network to adapt to the network usage patterns and improve performance with targeted information delivery and caching.

For example, the *Publish-Subscribe Internet Routing Paradigm (PSIRP)* system [28] and the *Content Centric Networking (CCN)* architecture [29] are based on receiver driven designs. The motivation is that Internet hosts are interested in receiving the proper content rather than who is supplying the content.

1.9.2   Standards

The standards timeline includes systems such as CORBA Event Service, Microsoft's DCOM, CORBA Notification Service, JMS, SIP, and DDS. In this section we briefly examine these developments.

The CORBA Event Service specification defined a communication model that allowed an object to accept registrations and send events to a number of receiver objects. The Event Service supplements the standard CORBA client-server communication model and is part of the CORBAServices that provide system level services for object-based systems. The CORBA Notification Service [10] extends the functionality and interfaces of the older Event Service [30] specification. The Event Service specification defines the event channel object that provides interfaces for interest registration and event notification. One of the most significant additions to the Notification Service is event filtering.

The *Distributed Component Object Model (DCOM)* was the Microsoft alternative to CORBA technology. DCOM facilitates communication between distributed software components. DOM extends the COM model and provides the communication with COM+ application infrastructure. Today DCOM has been replaced with Microsoft .NET Framework.

Standard COM and OLE supported asynchronous communication and the passing of events using callbacks, however, these approaches had their problems. Standard COM publishers and subscribers were tightly coupled. The subscriber knows the mechanism for connecting to the publisher (interfaces exposed by the container). This approach does not work very well beyond a single desktop. The change in the COM+ Event Service was the addition of the event service in the middle of the communication. The event service keeps track of which subscribers want to receive the calls, and mediates the calls.

JMS defines a generic and standard API for the implementation of message-oriented middleware. The JMS API is an integral part of the *Java Enterprise Edition (Java EE)*. JMS is an interface and the specification does not provide any concrete implementation of a messaging engine. The fact that JMS does not define the messaging engine or the message transport gives rise to many possible implementations and ways to configure JMS.

The *Session Initiation Protocol (SIP)* [31] is a text-based, application-layer protocol that can be used to setup, maintain, and terminate calls between two or more end points. SIP is designed to be independent of the underlying transport layer. SIP has been designed for call control tasks and thus the driving application has been telephony and multiparty communications. SIP has been standardized by IETF and adopted widely in the telecommunications industry. SIP was accepted as a 3GPP signalling protocol in November 2000.

The *Data Distribution Service for Real-Time Systems (DDS)* OMG specification defines an API for datacentric pub/sub communication for distributed real-time systems [32]. DDS is a middleware service that provides a global data space that is accessible to all interested applications.

1.9.3 Internet Technology

We briefly consider developments on the Internet technology timeline focusing on Web technologies for building pub/sub systems. One of the earliest examples of a loosely coupled message dissemination system is the Usenet that was created in 1980. Usenet thus precedes many other message dissemination systems. Later developments include W3C's SOAP protocol, the *Really Simple Syndication (RSS)* specifications, the *Extensible Messaging and Presence Protocol (XMPP)*, the *Representational State Transfer (REST)* model, HTML5 from W3C, and the Pubsubhubbub protocol. There are also many other systems that have been proposed and deployed.

RSS is a family of specifications for the definition of Web-based information feeds using XML. RSS is a simple pub/sub system that is based on polling the URL that identifies a feed and then determining if information has changed. RSS builds on existing Web standards, namely HTTP and XML, and it has become ubiquitous. RSS is used to disseminate updates, for example, pertaining to blog entries, news, video and audio resources.

SOAP is a key messaging protocol for Web applications that was designed for XML-based RPC and messaging. SOAP is a one-way message exchange primitive specified and standardized by W3C that is very flexible. SOAP can support various interactions by building on the one-way message exchange primitive and can be used with various message transport protocols, such as HTTP and SMTP.

XMPP [33] (RFC 3920 based on the Jabber protocol) has been designed for instant messaging with support for extensions. Today XMPP can support different message-based communication styles. XMPP extensions include publish/subscribe mechanisms, presence and status updates, alerts, feature negotiation, service discovery, and other features that make it suitable as an asynchronous middleware solution. XMPP is becoming increasingly popular on the Internet with companies such as Google, Twitter,[4] and Facebook[5] using XMPP as a general API.

REST was introduced and defined in 2000 by Roy Fielding in his doctoral dissertation. In this model, clients send requests to servers, servers process requests and return responses. The key idea is that the requests and responses convey representations of resource state. A resource is an entity that can be addressed. The model maps well to the HTTP protocol and has influenced the design of Web application interfaces. For instance, the popular real-time feed service Twitter has a REST API.

Comet is an *AJAX (Asynchronous Javascript and XML)* implementation of a push system over the Web. Comet uses callback functions to handle responses from a server. Comet issues HTTP requests to keep the connection to the server open. A long-lived connection is established that is then used to send and receive event data.

HTML5 is a core new language for defining content for the Web. The specification is still under development at W3C and when completed it will be the fifth revision of the HTML standard originally created in 1990. The key new feature for supporting pub/sub will be the WebSocket interface that allows servers to send asynchronous content to Web browsers. The protocol part of the WebSocket is standardized by the IETF. The Server-sent events is a related specification also part of the HTML5 for providing push notifications from a server to a browser client in the form of DOM events.

PubSubHubbub is an open protocol for distributed pub/sub communication on the Internet. The protocol extends the Atom (and RSS) protocols for data feeds. The main purpose is to provide near-instant notifications of change updates, which would improve on the typical situation where a client periodically polls the feed server at arbitrary interval.

RSS, XMPP, Comet, HTML5, and Pubsubhubbub all introduce pub/sub service features for Web applications. The features are basic in the sense that typically only topic or channel based semantics are supported. The next generation of Web-based pub/sub services is expected to introduce more sophisticated features for supporting Web application development. HTML5 will play a crucial role in supporting the development and deployment of these services.

1.9.4   A Taxonomy

In this section, we present a taxonomy of pub/sub systems that will then be revisited later in the book. We start with a brief survey of taxonomies, and then focus on the taxonomy used in this book.

Typically applications are developed based on modular units of computation, such as classes, modules, and programs. The typical ways of combining the basic units is realized with either explicit or implicit invocation [34, 35]. With explicit invocation, component names are statically bound to the implementations, for example a function invoking another function in some other module via local function call or RPC. This is contrasted by implicit invocation, in which a component publishes an event that then triggers the invocation of the requested functionality. Implicit invocation can abstract the name and location of the component that will perform the requested functionality.

---

[4] www.twitter.com.

[5] www.facebook.com.

Notkin *et al.* were the first to consider the implicit invocation model in detail. They presented design considerations when introducing implicit invocation to traditional programming languages [35]:

- Event declaration that pertains to the vocabulary that is used to define events and to the properties of the vocabulary.
- Event parameters and attributes, which are about the information associated with an event.
- Event bindings, which determine how events are bound to components that process them.
- Event announcement, which determines the invocation model: explicit or implicit.
- Delivery policy defines the rules for event delivery.
- Concurrency pertains to the number of threads and their priorities.

The design consideration for implicit invocation contain the relevant issues from the viewpoint of programming languages; however, they are not sufficient for the development of distributed systems. The distributed environment has been investigated in the distributed pub/sub and event processing community. An early taxonomy of the area is given in [36] and later taxonomies are given in [1] and [37].

We now summarize the history and evolution of publish/subscribe with a taxonomy that demonstrates differences between the ideas and systems considered so far. Figure 1.12 presents the taxonomy that has three core topics, namely research, standards, and Web. In the research category, we have two subcategories, systems and patterns. The systems subcategory contain concrete research proposals, and the pattern subcategory contains abstract architectural and design patterns. Patterns are used to design and implement pub/sub systems. Patterns are also heavily used in the design of standardized solutions, for example the CORBA Event Service and Notification Service are based on the event channel pattern.

The research systems category has two subcategories: centralized/cluster-based solutions, and wide-area systems. The former includes the LINDA, ISIS, and SIFT systems,

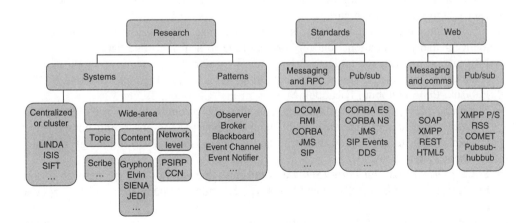

Figure 1.12   A taxonomy of pub/sub solutions.

and the latter include various wide-area systems. The wide-area systems differ based on the operating semantics. We have identified three subcategories, namely topic-based, content-based, and network layer systems. Scribe is an example of a topic-based DHT, SIENA is the classical content-based system, and PSIRP and CCN are examples of recent network layer systems.

The standards category has two subcategories, namely messaging and RPC, and pub/sub. The former contains systems and APIs such as DCOM, RMI, and JMS. The latter contains pub/sub standards, such as CORBA Event Service and Notification Service, JMS, and SIP events.

The Web category has also two subcategories: messaging and communications, and pub/sub. Basic techniques in the first subcategory include SOAP, XMPP, REST, and HTML5. The second subcategory contains XMPP pub/sub, RSS, and other pub/sub related proposals.

The presented taxonomy is coarse grained and can be extended with various functional and nonfunctional details, such as Quality-of-Service support, reliability, fault-tolerance, composite event detection support, state aggregation support, mobility support, etc. We will return to many of these features later in the book.

1.10   Application Areas

In this section, we briefly examine a number of pub/sub application areas. We return to the applications towards the end of the book in Chapter 12 and examine them in more detail. This analysis of the applications will consider how the patterns, protocols, and solutions covered in this book are applied in current applications.

As mentioned, pub/sub solutions can be applied widely in both local and distributed environments. Typical pub/sub applications include the following:

- GUIs, in which pub/sub is typically applied as the glue that connects the various components together. A classical example is the MVC design pattern heavily used in GUIs and its component the observer pattern. These patterns will be presented in more detail later. They allow application components to react to various events, such as a user pressing a touch screen.
- Information push, in which content is pushed to the user. This is a fundamental requirement for many applications that rely on real-time or near real-time data.
- Information filtering and targeted delivery used by alert and presence services (Google Alerts etc.), application stores, RSS brokering services, etc. Examples include XMPP Pub/sub, Pubsubhubbub, Facebook Messenger and Chat, and Twitter.
- Signalling plane, in which pub/sub ensures that asynchronous events are delivered in real-time or near real-time from publishing components to subscribing components. Example applications include industrial and tactical systems. DDS is the key standard for these systems.
- *Service Oriented Architecture (SOA)* and business applications rely on pub/sub in the Enterprise Service Bus (ESB). The ESB is typically implemented with an XML message broker.
- *Complex Event Processing (CEP)* for data analysis. CEP is used extensively in various business applications, for example algorithmic trading and fault detection.

- Cloud computing, in which pub/sub and message queuing is used to connect the different cloud components together.
- Internet of Things, in which pub/sub connects the sensors and actuators together and with Internet resources.
- Online multiplayer games, in which pub/sub is used to synchronize game state across players and servers.

It is evident that pub/sub is employed in differing environments and applications. Thus many different flavours of pub/sub are needed to solve the information dissemination problem for specific environments.

1.11   Structure of the Book

This book examines pub/sub technology and its applications. Our focus is on the design of such systems based on modular building blocks and commonly employed solutions. In order to examine the building blocks and assess their applicability to various scenarios, we first investigate the history of pub/sub systems and how the different solutions have been developed and deployed.

The pub/sub functionality is typically offered in the form a library or middleware service that applications can then utilize. Pub/sub can also be realized on lower layers of the protocol stack as well as in the application itself. Many of the solutions for the the distributed environment are based on overlay networks that are networks that operate on top of the TCP/IP protocol stack. Therefore we address the middleware and overlay systems in detail, but also consider new proposals that introduce pub/sub features in the protocol stack.

Throughout the book the examination includes three differing views to pub/sub, namely standard-based solutions designed to solve specific industrial cases, Web-based solutions developed for the Internet, and research-oriented solutions that consider potential future applications of the technology. These three views are overlapping, and in many cases solutions developed in the academia end up being adopted by specific standards.

Chapter 2 examines the basic technology for supporting distributed pub/sub systems. We consider TCP/IP, naming and addressing, firewalls and NAT devices, and advanced techniques such as multicast, causality, and messaging. TCP/IP as well as store-and-forward messaging solutions are the foundation for pub/sub solutions. We consider two standards based interoperable messaging frameworks, namely Web services and SIP. This chapter provides the essential networking and messaging technology background for later chapters.

Chapter 3 deepens the treatment of networking solutions by illustrating how networks can be created on top of networks in so called overlay solutions. Overlay networks are a robust and scalable solution that does not require the changing of routers or the basic protocol stack. Thus overlay networks are good candidates for supporting various distributed pub/sub systems. We consider Distributed Hash Tables (DHTs) that are a specific overlay solution that have many promising application in information dissemination and content delivery. The DHT-based solutions are examined in subsequent chapters with more details.

Chapter 4 considers the key design principles and patterns for pub/sub systems. We give an overview of the environment, principles, and patterns and then examine pub/sub specific patterns in more detail. Key patterns covered in this chapter are the observer, and event notifier patterns. Indeed, the event notifier pattern is the basis for the distributed pub/sub systems covered in later chapters.

Chapter 5 presents key pub/sub standards and specifications as well as messaging products. We investigate standards such as CORBA Event Service and Notification Service, OMG DDS, SIP Event Framework, JMS and product technologies such as COM+ and.NET, Websphere MQ, and AMQP. The standards exemplify many of the patterns introduced in the previous chapter.

Chapter 6 examines state of the art Web technologies for building pub/sub systems over the Web. These technologies include REST, AJAX, RSS and Atom, SOAP, and XMPP. Web-based protocols are necessary for the creation of real-time and interactive Web pages and applications, and thus an integral part of modern Web application development.

Chapter 7 considers the distributed pub/sub environment in general and presents a number of solutions for meeting various information dissemination requirements. The chapter examines various routing functions including topic, content, and gossip based mechanisms as well as optimization techniques such as filtering, filter covering, and filter aggregation with merging. The chapter together with the following chapters present a toolkit of solutions that can be applied by developers in order to engineer efficient distributed pub/sub solutions.

Chapter 8 investigates content matching techniques and efficient filtering solutions. Content matching is a basic requirement for a pub/sub system and thus it should be efficient and scalable as well as support different filtering constraints. We investigate well-known techniques including counting based algorithms and the poset and forest algorithms.

Chapter 9 examines well-known research prototypes and solutions for pub/sub. The solutions incorporate many of the patterns and solutions introduced in previous chapters, such as the event notifier pattern, filter matching, and filter covering and merging. We consider classical examples such as SIENA, Gryphon, JEDI, Elvin as well as more recent DHT-based solutions. This chapter provides specific examples of systems that utilize solutions presented in previous chapters.

Chapter 10 presents a concrete example of a keyword-based pub/sub system implemented on top of a DHT overlay network. The chapter considers the complexity of the problem and presents an efficient solutions based on rendezvous points on the DHT based network. The chapter illustrates how the already discussed DHT-based solutions can be utilized for keyword based content dissemination.

Chapter 11 consider advanced features of pub/sub systems. We start a number of security solutions for pub/sub. Then we examine topics such as composite subscriptions, filter merging, load balancing, channelization, reconfiguration, mobility support, congestion control, and the evaluation of pub/sub system. Many of the topics pertain to the pub/sub routing topology, its organization and configuration.

Chapter 12 considers applications of pub/sub systems and technology. We consider the role of pub/sub as an enabler of a cloud computing platform, a generic XML-broker for enterprise applications, content advertisement with pub/sub technologies, SOA, CEP, and several Web based applications including Pubsubhubbub, Facebook, and the Apple push service for mobile devices. The patterns and solutions used by the applications are discussed.

Chapter 13 considers new research proposals in adopting the pub/sub paradigm in proposed new protocol architectures that replace TCP/IP with receiver driven protocol suites. The motivation, features, and possibilities of these systems are discussed.

Chapter 14 presents a summary and conclusions of the book. We discuss the role of pub/sub technology as a generic enabler for connecting components across space, time, and synchronization in vast distributed systems.

References

1. Baldoni R, Querzoni L, Tarkoma S and Virgillito A (2009) Distributed event routing in publish/subscribe communication systems. *MiNEMA State-of-the-Art Book*. Springer.
2. Hinze A and Buchmann AP (eds) (2010) *Principles and Applications of Distributed Event-Based Systems*. IGI Global.
3. Tarkoma S (2010) *Overlay Networks – Toward Information Networking*. CRC Press.
4. Eugster PT, Felber PA, Guerraoui R and Kermarrec AM (2003) The many faces of publish/subscribe. *ACM Comput Surv* **35**(2), 114–31.
5. Carriero N and Gelernter D (1989) Linda in context. *Commun ACM* **32**(4), 444–58.
6. Sun (2002) *Java Message Service Specification 1.1*.
7. Pietzuch P, Eyers D, Kounev S and Shand B 2007 Towards a common api for publish/subscribe *Proceedings of the 2007 inaugural international conference on Distributed event-based systems*, pp. 152–157 DEBS '07. ACM, New York, NY, USA.
8. Carzaniga A, Rosenblum DS and Wolf AL (2001) Design and evaluation of a wide-area event notification service. *ACM Transactions on Computer Systems* **19**(3), 332–83.
9. Colouris G, Dollimore J and Kindberg T (1994) *Distributed Systems: Concepts and Design*, 2nd edn. Addison-Wesley, Boston, Massachusetts.
10. Object Computing, Inc. (2001) *CORBA Notification Service Specification v.1.0*. OCI.
11. Fleisch BD (1987) Distributed shared memory in a loosely coupled distributed system. *SIGCOMM Comput Commun Rev* **17**, 317–27.
12. Hayes-Roth B (1985) A blackboard architecture for control. *Artificial Intelligence* **26**(3), 251–321.
13. Krasner GE and Pope ST (1988) A cookbook for using the model-view controller user interface paradigm in smalltalk-80. *J. Object Oriented Program*. **1**, 26–49.
14. Birman KP and Joseph TA (1987) Reliable communication in the presence of failures. *ACM Transactions on Computer Systems* **5**, 47–76.
15. Oki BM, Pflügl M, Siegel A and Skeen D (1993) The information bus – an architecture for extensible distributed systems. Proceedings of the Fourteenth ACM Symposium on Operating System Principles, 5–8 December 1993, Asheville, North Carolina, pp. 58–68.
16. Yan TW and Garcia-Molina H (1999) The SIFT information dissemination system. *ACM Transactions on Computer Systems Database Systems* **24**, 529–65.
17. IBM (2002) *Gryphon: Publish/subscribe over public networks*. (White paper) http://researchweb.watson.ibm.com/distributedmessaging/gryphon.html.
18. Strom RE, Banavar G, Chandra TD, et al. (1998) Gryphon: An information flow based approach to message brokering. *Computing Research Repository (CoRR)*. Available at: http://arxiv.org/corr/home.
19. Sutton P, Arkins R and Segall B (2001) Supporting disconnectedness-transparent information delivery for mobile and invisible computing *CCGRID '01: Proceedings of the 1st International Symposium on Cluster Computing and the Grid*, p. 277. IEEE Computer Society, Washington, DC, USA. 19
20. Castro M, Druschel P, Kermarrec AM and Rowstron A (2002) Scribe: A large-scale and decentralized application-level multicast infrastructure. *IEEE Journal on Selected Areas in Communication (JSAC)*, **20**(8): 1489–99.
21. Mühl G, Ulbrich A, Herrmann K and Weis T (2004) Disseminating information to mobile clients using publish-subscribe. *IEEE Internet Computing* **8**, 46–53.
22. Pietzuch PR (2004) *Hermes: A Scalable Event-Based Middleware*. PhD thesis. Computer Laboratory, Queens' College, University of Cambridge.

23. Pietzuch PR and Bacon J (2002) Hermes: A distributed event-based middleware architecture *ICDCS Workshops*, pp. 611–18.

24. Carzaniga A, Rutherford MJ and Wolf AL (2004) A routing scheme for content-based networking. *Proceedings of IEEE INFOCOM 2004*. IEEE, Hong Kong, China, vol. 2, pp. 918–28.

25. Cugola G, Di Nitto E and Fuggetta A (1998) Exploiting an event-based infrastructure to develop complex distributed systems *Proceedings of the 20th International Conference on Software Engineering*, pp. 261–70. IEEE Computer Society.

26. Gritter M and Cheriton DR (2001) An architecture for content routing support in the Internet. *Proceedings of the 3rd Conference on USENIX Symposium on Internet Technologies and Systems – Volume 3*, p. 4, USITS'01. USENIX Association, Berkeley, CA.

27. Koponen T, Chawla M, Chun BG (2007) A data oriented (and beyond) network architecture. *SIGCOMM Comput. Commun. Rev.* **37**(4), 181–92.

28. Tarkoma S, Ain M and Visala K (2009) The Publish/Subscribe Internet Routing Paradigm (PSIRP): designing the future Internet architecture. *Future Internet Assembly*, pp. 102–111.

29. Jacobson V, Smetters DK, Thornton JD, Plass MF, Briggs NH and Braynard RL (2009) Networking named content. *Proceedings of the 5th International Conference on Emerging Networking Experiments and Technologies*, pp. 1–12. CoNEXT '09. ACM, New York, NY, USA.

30. Object Computing, Inc. (2001) *CORBA Event Service Specification v.1.1*. OCI.

31. Rosenberg J, Schulzrinne H, Camarillo G, et al. (2002) *RFC 3261: SIP: Session Initiation Protocol*. IETF. http://www.ietf.org/rfc/rfc3261.txt.

32. Object Computing, Inc. (2007) *Data Distribution Services, V1.2*. OCI.

33. Saint-André P (2004) *RFC 3920: Extensible Messaging and Presence Protocol (XMPP): Core*. Internet Engineering Task Force.

34. Garlan D, Jha S, Notkin D and Dingel J (1998) Reasoning about implicit invocation. *SIGSOFT Softw. Eng. Notes* **23**, 209–21.

35. Notkin D, Garlan D, Griswold WG and Sullivan KJ (1993) Adding implicit invocation to languages: Three approaches. *Proceedings of the First JSSST International Symposium on Object Technologies for Advanced Software*, pp. 489–510. Springer-Verlag, London.

36. Meier R and Cahill V (2002) Taxonomy of distributed event-based programming systems *Proceedings of the 22nd International Conference on Distributed Computing Systems*, pp. 585–8. ICDCSW '02. IEEE Computer Society, Washington, DC.

37. Blanco R and Alencar P (2010) Event models in distributed event based systems. *Principles and Applications of Distributed Event-Based Systems*, pp. 19–42.

2

Networking and Messaging

The key ingredients of any distributed system are networking and messaging. The solutions of these central fields are the structural blocks reviewed in this chapter. In the typical layered protocol stack application messages are sent and received utilizing the TCP/IP protocol suite. A header is added by each layer and this header encapsulates the higher level data using its own format.

We will first examine here the waist of the protocol stack, meaning IP and TCP/UDP protocols. Our next focus will be on higher level messaging protocols, characteristically used for implementing distributed pub/sub solutions. The primary protocols studied here are REST, Web services and SOAP, and SIP.

2.1 Networking

Internet has been built on the TCP/IP protocol. Structurally Internet may be thought as consisting of five separate layers depicted by Figure 2.1, which are usually abstracted as the application layer, transport layer, network layer, link layer, and physical layer. We meet layered architectures in other places, too, for instance in the case of the seven-layer OSI Reference Model. Application developers use the Sockets API for programming TCP/IP applications. The Sockets API is provided by the transport layer and it offers two key basic data transport primitives, namely reliable data streams, and unreliable datagrams. The former is implemented with TCP and the latter with UDP.

The current Internet was designed closely following the principles given and outlined in RFC 1122 [1]. The main architectural tenets used when designing the Internet were the Robustness Principle and the End-to-End Principle [2]. In the following, we will elaborate the structure and the components of the Internet.

2.1.1 Overview

The general principle behind the Internet is the Robustness Principle. It lays down a rule which all software written for Internet must follow: be conservative in what you do, be liberal in what you accept from others. For instance, it is mandatory that the

Publish/Subscribe Systems: Design and Principles, First Edition. Sasu Tarkoma.
© 2012 John Wiley & Sons, Ltd. Published 2012 by John Wiley & Sons, Ltd.

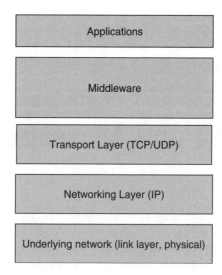

Figure 2.1   A layered protocol stack.

pieces of software written by Internet developers strictly adhere to extant RFCs but are prepared to meet client input which does not necessarily comply with those RFCs. Such nonstandard input must be accepted and parsed by the software just as the standard input. The RFC 1122 mandates that this kind of adaptability must be present in every level of Internet host software.

The question of positioning the maintenance of state and overall intelligence is answered by the End-to-End principle which places these functions at the edges. The core Internet is presumed to remain stateless [3]. Nevertheless, the development of the Internet constraining attributes like firewalls, *Network Address Translation (NAT)* and caches for Web content have made the practical End-to-End principle difficult if not impossible to carry out.

Internet layers are built separate for good reason, aiming to insulate possible failures. Internetworking is realized by the network layer in the TCP/IP model, delivering data from upper layers between end hosts within the IP protocol. Host names are strictly separated from topological addresses by using DNS name resolution in the protocol suite. *DNS (Domain Name System)* converts the (hierarchical) host names to topological IP addresses [4]. Naming and addressing are thus effectively separated and if there is possible failure in the domain name resolution the underlying routing operations are still independent and intact. For the user level an important feature is the ability to define organizational boundaries in DNS irrespective of the network topology.

There are several algorithms responsible for traffic. The *routing algorithm* takes care for setting up and maintaining routing tables. A second building block is the *forwarding algorithm*, which calculates the next hop for a packet when a destination address is given. The packet routing mechanism thus involves the routing and forwarding algorithms. We have actually two main classes of protocols here: the intradomain and the interdomain protocols. Being local tools, intradomain protocols belong to an *Autonomous System (AS)* such as a *Metropolitan Area Network (MAN)* or other regional network. The global tools are interdomain protocols utilized to connect the different AS together. Thus, interdomain protocols facilitate the formation of a global network topology. The standard cases of such protocols are *Open Shortest Path First (OSPF)* used for intradomain operation, and *Border Gateway Protocol (BGP)* for interdomain operation.

Actually Internet offers several different communication models that are commonly used. The unicasting model handles a packet traversing through a set of links from a source to a well-defined destination. Most traffic on the Internet belongs to the unicast variant. Multicasting implements traffic where a packet will selectively traverse through a multiplicity links typically from one source to many destinations. The broadcasting model describes a case where packet is sent through multiple links to all devices on the network. Broadcasting is normally restricted within a designated broadcast domain. Lastly, in anycasting packets are sent to the nearest or otherwise best destination, the suitable chain of links being chosen from a set of candidates.

The currently prevailing IP version 4 protocol only implements the unicasting model globally. In future, the next IP version 6 will include other communication models as well. It is not clear, however, when the IPv6 will actually be universally used; the progress of the deployment has been slower than expected.

Autonomy is a basic concept of the Internet. In the application of hierarchical routing the local AS are connected with each other by peering and transit links. The AS are autonomous also for their routing software. Each AS is free to run its own local routing algorithm, and the interdomain connectivity is achieved with BGP.

The current interdomain practice is based on three tiers: tiers 1, 2, and 3. Tier-1 is an IP network which connects to the entire Internet using settlement free peering. There are a small number of tier-1 networks that typically seek to protect their tier-1 status. A tier-2 network is a network that peers with some networks, but relies on tier-1 for some connectivity for which it pays settlements. A tier-3 network is a network that only purchases transit from other networks.

The applications are separated from the detailed routing and forwarding. This is accomplished by the network layer when it provides both global addressing and end-to-end reachability. On a lower level, the IP protocol itself is able to support a set of link and physical layer protocols, and therefore it is a waist of the protocol stack. On a higher level, features supporting dissimilar operating environments can coexist with the IP. Therefore, the network layer strives to minimize the number of service interfaces and simultaneously to maximize systemic interoperability.

2.1.2 Sockets, Middleware, and Applications

TCP/IP applications use the Berkeley Sockets API to send and receive data. As mentioned above, the API provides a reliable data stream with TCP and unreliable datagram service with UDP. The transport layer uses *port numbers* to associate open connections to active processes. For example, a Web page retrieval is performed with a HTTP request that opens a port for the response from the Web server. The response finds it way to the Web browser and the correct window based on the port number. Similarly, the Web server runs a server socket on the standardized Web port 80 and listens for incoming requests. Thus the transport layer and the Sockets API provide the essential mechanisms for receiving and sending data between distributed processes.

In the layered world of Internet we need a level of software that functions on top of the networking stack but below the application level. This is the middleware that provides services between networking and application levels. Middleware caters for a vast area of viable Internet technology and the majority of overlay technologies belong to it. These middleware blocks make use of the APIs and features of the network and the underlying protocol stack. Furthermore, the application developers can utilize the middleware APIs for their software. The middleware layer is important because it is able to hide and abstract many special features of the lower layers from the applications and thus facilitate the writing and running of distributed software.

2.1.3 Naming and Addressing

Network architectures have great variety of components and some are essential. Of those the naming procedures and the associated namespaces are fundamental for a functional network. The component responsible for handling names and managing the hierarchical domain namespace in the current Internet is the DNS. The protocol for DNS was one of the first Internet ingredients, specified by the IETF in the early 1980s. DNS has a central role in the practical Internet because it provides much of the background flexibility. DNS forms the basic system which creates critical scalability for both network level hierarchical routing and the higher level naming service. Without DNS we could not have the Web or e-mail services.

The management of DNS is distributed over the network. DNS is an overlay that uses both a static distribution tree and a namespace organized hierarchically. The DNS system is implemented as a distributed database system using the client-server model. In DNS sharing, replicating, and partitioning the domain namespace are executed by the nameservers. These also answer client requests. The important characteristics of scalability and resilience are realized by extensive usage of caching and replication. This has a drawback, because due to the system nature, updating DNS records globally may require some time. A second drawback in DNS is the lack of built-in security. This makes the system somewhat vulnerable and opens it to possible attacks.

Any host can can use either a host name (domain name) or a four-byte IP address (with IPv4). The domain name is mapped to an IP address with the help of the DNS system. Thus DNS provides a level of indirection to the network that allows flexible mapping of high level hierarchical names that follow organizational boundaries to network-level addresses. On a higher level, we have the *Uniform Resource Locator (URL)* scheme for referencing Internet resources. URLs include the protocol name, the location name or address, and the resource path and name. URLs are locators, because they include the location of the resource. A URL is a type of *Uniform Resource Identifier (URI)* that can be locators, names, or both. URIs and URLs form the basis for the Web.

IP addresses are directly routable; however, a specific category of addresses only works within private domains. Thus there are public IP addresses and private IP addresses. As the Internet is rapidly becoming larger and more complex, further requirements have emerged, leading to new proposals for the addressing system. Specifically, it has been suggested that the system needs more indirection in the protocol architecture. One notable proposal is the locator-identity split. A split of this kind would allow, for example, mapping cryptographic identifiers [5–7] into IP addresses. This would increase flexibility while reducing the central role of IP as an endpoint identifying system.

2.1.4   Organization

When we are concerned about routing procedures we must note that there are two kinds of networks: static and dynamic. A static network provides an easy platform for routing because each router determines directions for all possible destinations. Instead the situation in dynamic networks is more challenging as the routing tables will be changing continually and a runtime computation is preferable for routing instructions. The location of the data structures and buffers is critical and also the question of how often the state is updated. Usually this is accomplished with a broadcast of routing state to all routers. For instance, link state routing protocols broadcast link state updates for computation of shortest path distances. Because link state updates might cause excessive flooding, the network is divided into separate routing domains and this hierarchy is then used for limiting the update propagation. In the OSPF we find extensive use of areas, here as a network dimensioning instrument. Also in the interdomain context hierarchies naturally occur. Here ASes stem from administrative boundaries.

A node needs forwarding table and this is computed by the routing process. A cost/benefit analysis of a kind is needed because the routing process must estimate the costs of incident links. The node will use these links to communicate with its neighbours. Generally we can define a routing algorithm as a mechanism for specifying the content of information exchange with neighbours and also specifying the exact process used to compute forwarding tables. The routing algorithm should be able to maintain a forwarding configuration allowing nodes to be mutually reachable by forwarding. Here optimality or at least near-optimality of the forwarded packet paths is often desirable [8].

Hierarchy is one of the fundamental tenets of the Internet. Hierarchical clustering is needed to produce scalable routing tables as shown in the seminal work by Kleinrock and Kamoun published in 1975 [9]. In hierarchical routing the central idea is to combine nearby nodes together into clusters, clusters into superclusters, continuing this process bottom-up through the system. The clustering process actually results into a structure where routing tables are freed from unnecessary topological information and therefore the network scales well and notably in the hierarchical routing the table sizes are on the order of $\sqrt{n}$. Because of the advantages, many protocols in both interdomain (BGP, CIDR), and intradomain routing (OSPF) use hierarchical protocols.

2.1.5   Firewalls and NATs

The recent changes in the technological environment of the Internet partly stems from deployment of of devices like firewalls and NAT devices. These are basically technologies used to to control communication between subnetworks. The need to protect systems has led to the ever widening use of firewalls, i.e components selectively blocking incoming connections. Firewalls can be implemented either with hardware or software. The goal is to achieve increased device security and prevent connections that are unauthorized. Another widespread addition to the Internet is NAT. The NAT devices have the ability to carry out conversion between separate address spaces. NATs convert addresses from private to public networks and vice versa. These devices can be seen as a part of the wider movement towards a more secure Internet as they permit the use of private IP address spaces. This partially helps to cater for the growing problem of IP address space exhaustion. There are benefits for some network management problems as well.

NAT is an umbrella term which includes many different devices and also a multitude of network topologies utilizing those devices in diversified environments. Generally all NAT devices give support to local IP addressing domains. These private domains are reachable only through NATs, but not directly. A soft state is created in a NAT device by the client-initiated connections. Responses to requests can then be communicated to the private domain hosts.

2.2 Multicast

While unicast is clearly the most favored mechanism applied within the Internet set of communication models, other models are also gaining in popularity. The best model depends on the real-life needs of communicating entities. A very common type of communication event occurs when a simple, group-based connection is made. The best procedure for this is the multicast.

Multicast effectively decouples the senders and receivers and thus differs radically from unicast. Optimization of the network transmission is easier because data packets can be replicated for the multiple receivers at the last possible moment.

There are two possible levels to implement the multicast functionality. We can insert it into the network level or into the application layer. In the network-level we can use multicast as a complement of unicast in its role of the basic networking primitive. In the application-layer the multicast typically makes use of unicast resources. Here we shall first look at IP level multicast and then make acquaintance with some overlay multicast techniques.

2.2.1 IP (Network Layer) IP-Multicast

IP multicast is a simple, scalable and efficient one-to-many mechanism working in the IP level. In a typical multicast packets are routed from one sender to many receivers. Multicast participants form a group which can be joined and left by sending a packet to the group multicast address using the IGMP (RFCs 1112, 2236, 3376) protocol. *Internet Group Management Protocol (IGMP)* is a protocol that manages IP multicast groups memberships. IGMP is used by IP hosts and adjacent multicast routers to build and maintain multicast groups. According to RFC 3171, addresses 224.0.0.0 to 239.255.255.255 are designated as multicast addresses. IGMP is based on UDP, the common low-level protocol for multicast addressing. It is notable that IP multicast, just as IP itself, is not very reliable. Messages may be lost or delivered in wrong sequence.

The main components of IP multicast are:

- IP multicast group address.
- A multicast distribution tree maintained by routers.
- Receiver driven tree creation.

Multicast typically creates a multicast tree that can be source specific but is often shared by the participating entities. The problem of creating an optimal multicast tree is similar to the Steiner tree problem and we know that creation of a Steiner tree is NP complete

[10]. Therefore heuristic and greedy algorithms are typically employed in the construction of multicast trees, which approximate the optimal solution.

IP multicast is implemented with the help of a group specific distribution tree (or forest), which is maintained and updated when appropriate, that is, when nodes join or leave. There are several multicast protocols in use but perhaps the most widely known is the *Protocol Independent Multicast (PIM)*. PIM creates distribution trees configured for delivery of packets from senders via a multicast group to the receivers. The receivers are members of the group. There are many algorithm specifications usable for multicast activity; of those PIM includes Sparse Mode (SM) (RFC 2362), Dense Mode (DM) (RFC 3973), Bidirectional PIM (RFC 5015), and Source Specific Mode (SSM) (RFC 3569). Let us take a little closer look at these.

If there is need for interdomain multicast and the groups have low numbers of subscribers, PIM-SM is a suitable algorithm. A client wanting to join a PIM-SM multicast group sends an IGMP Join message. The PIM-SM protocol constructs a unidirectional tree from each sender to the multicast group of receivers. For this it uses a *rendezvous point (RP)*. The protocol supports two types of trees: an RP-centered multicast tree which is shared and also source-specific trees with shortest path. Source-specific trees are used for high rate data sources.

The second type of algorithm, PIM-DM is specifically meant for groups with many subscribers. The subscription density is high and therefore routers have the task of processing and forwarding a load of packets published in the multicast groups. PIM-DM algorithm floods packets to the network. After that it prunes the multicast forwarding tree. This pruning is done to lessen the load of the routers by suppressing future messages to routers with no subscribers for the multicast group.

The third type, Bidirectional PIM is a variant of the PIM-SM protocol. The algorithm constructs bidirectional shared trees. For this it uses the rendezvous point to connect multicast sources and the active subscribers.

The fourth type, PIM-SSM algorithm aims to create a source specific delivery tree. The delivery tree is a (S, G) channel, where S is an IP unicast source address and the multicast group address G is the IP destination address. PIM-SM protocol is used to construct the interdomain tree for IP packet forwarding. The tree is rooted at the source S.

IP multicast groupings do not allow expressive membership architectures. They are simply used to partition the IP datagram address-space. A datagram belongs at most to one group. There are also various challenges in deploying IP multicast, for instance in multidomain operation and RP discovery.

Regardless of specific algorithm, as a best-effort unreliable service IP multicast is at disadvantage where reliable transport is needed. Different algorithms for mapping subscribers to multicast groups are presented and evaluated in [11].

Multicast could be be utilized by event systems when delivering notifications to event routers or servers but generally event systems do not use network level IP multicast. However, while multicast (and broadcast) may not be practical in large public networks it is effective in closed networks. In larger environments standards such as TCP/IP and HTTP, which derive benefit from an universal adoption, are probably better alternatives for most communication [12].

2.2.2 Application-Layer Multicast

Because of delays in IPv6 deployment the older IPv4 is still the standard network layer protocol. Unfortunately IPv4 does not provide any native multicast mechanism by default. Therefore the common way to realize multicast is to implement it as an overlay on top of the TCP/IP protocol stack in the application-layer. There are several advantages over the deeper, IP level multicast which requires that the routers maintain states per group and per source for each group. Also in IP multicast routing table entry is mandatory for each unique multicast group address. The multicast addresses are difficult to aggregate and, furthermore, additional solutions for congestion and reliability control must be devised for IP multicast.

In all these problems overlay multicast either avoids the difficulties or ameliorates them. A strong motivation exists, therefore, to develop and deploy viable overlay multicast solutions and many examples of overlay multicast systems are discussed later in this book.

In contrast to native IP-level multicast systems, an application-layer multicast system makes use of prevailing nodal unicast communication tools to effectuate one-to-many communication channel where the replication of data packets is positioned at the end hosts. The multicast overlay protocols are not optimal in the sense of IP multicast because packets may travel multiple times over the same link. The nodes use either UDP or TCP to establish communication and forward messages through those links. The construction algorithm for multicast tree is distributed and built to allow a set of different metrics.

An example of early attempts to create application layer multicast protocols is the Narada protocol [13]. It showed clearly the advantages and feasibility of multicast positioned at the application-layer. Overlay multicast is based on hosts and does not require multicast-capable routers (with greater complexity) of IP multicast protocol. It is far easier to implement over the Internet. While both approaches are tree-based, in IP multicast hosts participate in the tree only as leaves. For reasons given above, IP multicast is less deployed than application level multicast although it could be efficient as a solution where optimal paths and less overhead is required [14].

2.3 Reverse Path Forwarding and Routing

Reverse path forwarding is a frequently employed technique for implementing loop-free communications over a store-and-forward network [15]. The technique is used by multicast algorithms to prevent forwarding loops. When a multicast packet is processed by a router, it will determine the networks that are reachable via the input interface by examining the reverse path of the packet. If a matching entry for the source IP of the packet is found, the packet is forwarded to all other interfaces that are participating in the multicast group. The forwarding of the packet is based on the reverse path of the packet instead of the forward path.

Reverse path routing relies on having information about the path that a packet traverses through the network, and then using the same path for the reverse direction. In general, the technique can be implemented by storing the reverse path information at routers that handle the packet, or by adding the route information in the packet. The latter is problematic from the security point-of-view and hence the technique is typically implemented by storing reverse path state in the routers. Many distributed pub/sub overlay systems are based on the reverse path routing technique.

2.4 Causality and Clocks

In this section we present the basic mechanisms for introducing the notion of time into a distributed system. The basic networking techniques described so far do not support the ordering of messages in the system; however, for many applications ordering of messages is a required feature. Many applications require support for either causal order or total order. Causality determines the relationship of two events A and B. In order to be able to determine causality in the distributed system a logical clock mechanism is needed. The two well-known solutions are the Lamport clocks and vector clocks. Some applications require the more stricter total ordering of events.

2.4.1 Causal Ordering and Lamport Clocks

Lamport's seminal *happened-before* relation orders events based on the potential causal relationship of pairs of events. This ordering can be defined numerically with a Lamport logical clock, which is a monotonically increasing software counter denoted by $C(A)$ for an event A. Given two events A and B, the following relation is true: if X happened-before Y then $C(X) < C(Y)$ [16].

In order to maintain the clocks, the processes need to follow the following simple rules in order to maintain the happened-before relation in the distributed system:

- A process increments its clock before processing an event.
- When sending a message, the process includes its counter value with the message.
- When receiving a message, the process sets its counter to be greater than the maximum value of its own value and the received counter value.

The happened-before relation captures causality and creates a partial order for the events. The relation gives the cause A that has to happen before B that is the effect, but A and B can still be incomparable. Thus the relation only applies in one direction, which is

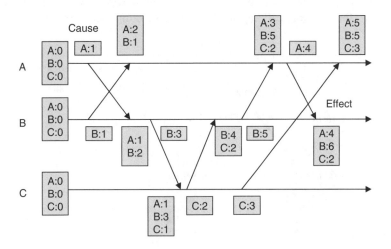

Figure 2.2 Example message trace with Lamport clocks.

called the *clock consistency* condition. The *strong clock consistency* applies for both direction can be implemented with the vector clock technique examined in the next subsection.

Causality is very important for many distributed applications, for example when implementing a collaborative Web application or determining the root cause of a failure. Figure 2.2 illustrates causality with a cause and its effects.

2.4.2 Vector Clocks

As mentioned above, Lamport clocks and the happened-before relation only apply in one direction and given that the relation holds for two events, it cannot distinguish between the two cases that the events are in relation or that they are incomparable. Vector clocks are a more sophisticated technique that can distinguish between these two cases.

This system is based on the notion of a vector clock, which is an array of N logical clocks for a system with N processes. Each process keeps a local copy of the global clock array and follows a set of rules in updating this array. A single clock does not give the strong clock consistency; however, a vector of clocks can do this. The rules are the following:

- The clocks are set to zero initially.
- When a process receives an internal event, it increments its logical clock in the vector.
- When a process sends a message, it increments its logical clock in the vector and then sends the full vector with the message.
- When a process receives a message, it increments its logical clock in the vector and the vector clock. The clock is updated by considering each element in the vector and taking the maximum value between the local vector clock value and the received vector clock value.

2.4.3 Total Ordering

A total order scheme can be implemented in different ways. We briefly examine two well-known techniques for realizing total ordering for a multicast group.

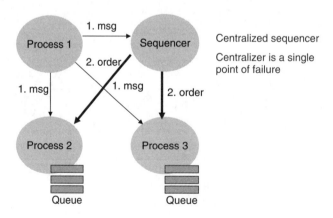

Figure 2.3 Total order with a sequencer.

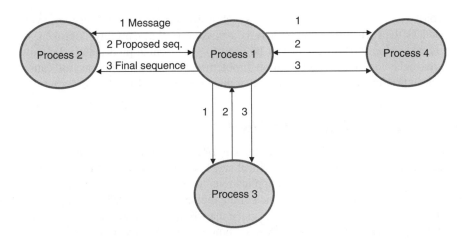

Figure 2.4 Total order in the ISIS system.

In the first technique illustrated by Figure 2.3, a process multicasts a message with a unique identifier to the group including the sequencer. The sequencer receives the message and assigns a unique consecutive sequence number for it. The sequencer then multicasts the order message to the group.

The second technique was proposed in the ISIS system and it requires three communication rounds. Figure 2.4 illustrates this system. First the sender sends the message with its identifier to all the receivers. Then the receivers suggest a sequence number, called priority, and reply to the sender. The sender collects the proposed sequence numbers and decides the final sequence number, and then sends the final sequence number for the message. The receivers then deliver the message with the agreed sequence number.

2.4.4 Discussion

The following list summarizes different ordering requirements for distributed systems:

FIFO-order. Clients have no way of distinguishing the order of event occurrences from the delivery order.

Causal order, partial order. Causally related messages are those messages, which exhibit Lamport's happened-before relation. The publication of one event occurred before the publication of another.

Total order. Messages uniquely ordered using a counter or a timestamp. A globally unique timestamp is associated with each message when it is delivered. Total ordering preservers causal ordering.

Timestamp order. Order is based on timestamps. The granularity is based on the synchronization of the physical clocks.

Unreliable and best-effort notification places less performance penalties and messaging costs for an event system than causal or total ordering of messages. Physical clocks require synchronization algorithm, such as *Network Time Protocol (NTP)* and Cristian's algorithm, that avoids clock drift and minimize the synchronization error [17], however, even with an error less than ten milliseconds there is a possibility that causality is broken. *Global Positioning System (GPS)* devices can be used to synchronize physical clocks, however, wide-area delay causes synchronization problems. The above solutions for causal and total order introduce overhead, but can guarantee the required level of ordering.

2.5 Message Passing and RPC/RMI

A central component of middleware systems is *IPC (Inter-Process Communication)*. IPC is a technique which caters for the critical need of the communication components to be object oriented. Many distributed applications today are object oriented, meaning that software modules used to create a service or an application are distributed objects residing on different computers but connected via a network. The distributed objects use the underlying network and protocol stack to communicate. IPC is a set of techniques which implements the data exchange between one or more processes. Its techniques can local or distributed. The IPC techniques can be divided into four classes:

- message passing,
- synchronization,
- shared memory,
- RPC.

Message passing systems follow the message passing model where copies of data items are sent to a communication endpoint over the network. The endpoint has differing names and definitions in different systems, but the basic principles are the same in all message passing systems. Distributed object systems with remote method invocation include CORBA, RMI, DCOM, SOAP, REST and.NET Remoting. In all these the interaction has typically the same features such as reliability, security, and other transaction related characteristics.

Messaging is a method of communication between software components or applications. A messaging system is a peer-to-peer facility: a messaging client can send messages to, and receive messages from, any other client. Each client connects to a messaging agent that provides facilities for creating, sending, receiving, and reading messages.

Messaging enables distributed communication that is loosely coupled. A component sends a message to a destination, and the recipient can retrieve the message from the destination. However, the sender and the receiver do not have to be available at the same time in order to communicate. In fact, the sender does not need to know anything about the receiver; nor does the receiver need to know anything about the sender. The sender and the receiver need to know only what message format and what destination to use. In this respect, messaging differs from tightly coupled technologies, such as RMI, which require an application to know a remote application's methods.

A vital characteristic of any distributed object system is the serialization of objects, which is the process of creating a representation of an object that would be storable in a storage system or transferable across a network. Similarly, deserialization is the process where the data structure is extracted from a given sequence of bytes. Serialization

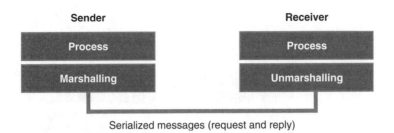

Figure 2.5   Example of marshalling a message.

and deserialization are also known as marshalling and deflating, and unmarshalling and inflating, correspondingly. Figure 2.5 illustrates the marshalling and unmarshalling of messages in a request-reply interaction.

The evolution of SOAP and Web services has made XML serialization very popular. For instance, AJAX Web application uses XML extensively in exchanges of structured data between clients and a server. A lightweight text-based alternative for XML serialization is *JavaScript Object Notation (JSON)*. It utilizes JavaScript syntax, which is widespread and well known. Therefore many programming languages support JSON.

Sometimes it is advantageous if a format is text-based and human-readable. Such a tool is the XML markup, which is practical for persistent and interoperable representation of objects. However, there are limitations in the text-based serialization of XML: increased processing requirements and typically a size overhead. A more compact solution would be desirable in high performance environments, preferably byte stream based. This had led to the development of a number of bit efficient and streamable XML processing systems.

In Figure 2.6 we present an example of a distributed object oriented system, the Java RMI API. Java RMI is an application programming interface for Java. With RMI the object equivalent of remote procedure calls can be performed. RMI consists of the following eight steps:

1. Client Obtains Handle.
2. Stub is called (represents remote object).
3. Marshalling arguments.
4. Data transmission.
5. Skeleton unmarshals.
6. Skeleton calls method.
7. Skeleton marshals result.
8. Stub unmarshals result and passes it to client (type checking).

Two common implementations of the RMI API exist. The original implementation is dependent on class representation mechanisms in Java Virtual Machine (JVM). Therefore it only supports making calls from one JVM to another. The protocol underlying the Java-only implementation is *Java Remote Method Protocol (JRMP)*. Later a CORBA version was developed for those contexts where code must run in a non-Java environment (RMI over IIOP).

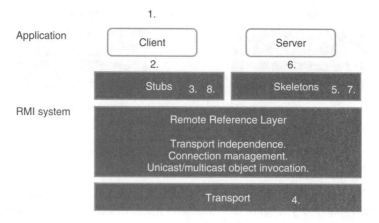

Figure 2.6   Overview of RMI.

2.5.1   Store and Forward

As mentioned above, messaging is a basic ingredient for building loosely coupled distributed systems. Typically, distributed message oriented systems are built based on the *store and forward* pattern, in which intermediate message brokers or relays store incoming message in a message queue for waiting eventual delivery towards the destination. Since the message is stored in a queue, it can be retransmitted until it is successfully delivered to the next hop destination.

Figure 2.7 illustrates this pattern that can cope with unreliable messaging environments. Various messaging semantics can be implemented on top of this basic pattern. Messaging semantics were briefly examined in the previous Chapter 1.

2.5.2   Concurrent Message Processing

Figure 2.8 illustrates a nonblocking system, in which there are two active Sockets and the data flows through the system with two active worker threads. Concurrent message processing is achieved, because each thread reads, processes, and writes content in one direction. This way, it is possible to implement efficient message processing and routing systems that receive messages and then immediately write them to an outgoing Socket.

In order to develop efficient messaging systems, overhead in setting up connections need to be minimized. Therefore it is common practice to establish long lived TCP connections that are used to transport multiple messages. Another approach is to use a UDP-based protocol that does not establish an explicit connection or establishes a very lightweight connection. UDP is used to transport SIP messages in the telecommunications environment.

The system of Figure 2.8 only shows two sources and destinations for messages, but can be extended to support arbitrary number of sources and destinations. This more complex message routing core requires that coordination so that only one thread reads and writes a Socket at a time. The typical implementation solution is to have one thread to read

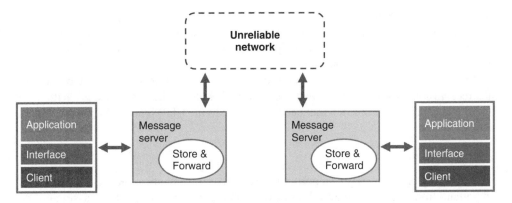

Figure 2.7 Example of store and forward based messaging.

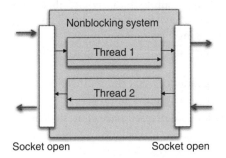

Figure 2.8 Example of non-blocking communications.

the incoming message to a buffer, and then allocate a worker thread to process and send the message that first reads the message from the buffer. Indeed, message buffers and message copying play a crucial role in software-based messaging systems.

Two challenges for message processing systems are:

- Head-of-the-Line Blocking (HOLB),
- Congestion.

The first challenge involves blocking an incoming message due to parallel data transfer and message processing at the destination interface and Socket. In this case, the incoming message is blocked for some time until it can be sent to the destination. One solution is to use small message sizes.

Congestion happens when the message are arriving too fast for the message broker to process them. This results in increasing queue sizes, and ultimately to the dropping of messages due to lack of memory. Congestion can be avoided by tuning the sending rate so that the message broker has less load. Alternatively, congestion can be avoided by adding more brokers to the system.

Congestion happens on multiple levels. On the transport level, TCP takes care of congestion control with the slow start and congestion avoidance algorithms. A TCP sender always uses the minimum of the two possible window sizes, the congestion window and the receiver window size. The latter is responsible for flow control so that the receiver is not overwhelmed. The routers may help senders to reduce the sending rate by marking packets with the explicit congestion notification signal (ECN bit).

On the application layer, TCP's flow control provides protection so that brokers do not receive too many messages that would overflow their message buffers. With UDP additional measures are needed that need to be built to the UDP-based messaging protocol.

2.5.3 Semantics and QoS

A messaging system may support various delivery semantics and QoS properties. Important properties are reliability, priority, ordering and timeliness. For mission-critical operation, it is typically required that no messages are lost or delivered in incorrect order. Furthermore, the arrival of duplicate messages must be prevented in many cases. As discussed above, some applications may require that the messages are delivered in causal or total order to prevent synchronization problems. Causal order is required for synchronizing applications using message passing.

There are four frequently employed messaging semantics:

Best effort. There are no guarantees that the notification will be delivered. Unimportant news items and other noncritical application-level notifications.

At-most-once. A notification is delivered once if at all. The system keeps track of delivered notifications.

At-least-once. The event service must guarantee that the notification is delivered at least once. This applies to all notifications that are idempotent. Retransmit notification if not acknowledged.

Exactly-once. The event service must guarantee that the notification is delivered exactly once. All system notifications and mission-critical uses. Retransmit notification if not acknowledged and filter duplicate notifications.

For TCP/IP, unordered and unreliable communication is analogous to UDP and ordered, reliable communication to TCP. CORBA invocations and asynchronous method invocations (AMI), and Java RMI have at-most-once delivery semantics. CORBA Time-independent invocation (TII) is persistent and supports exactly-once semantics and FIFO ordering. JMS supports two delivery semantics: transient messages have at-most-once delivery and persistent messages support exactly-once delivery [18]. JMS also supports FIFO and priority-based ordering.

2.6 Web Services

The Web services paradigm covers machine readable and accessible content on the Web. In the current model, suitable service components are selected and composed for use in the application level. The core parts of the architecture are the service provider, client or service consumer, and the service registry. In the system complex services are created from basic parts and to facilitate this is the goal of service composition. Essentially the system then supports software reuse. The many composite services of the Web such as personalized portals, electronic commerce service bundles, and mobile services are typical application cases.

2.6.1 Overview

The definition of the W3C Web services is very broad allowing a cloud of differing implementations. However, the term refers most often to clients and servers communicating with XML messages, and typically using the SOAP standard. It is also notable that the system often includes machine-readable description of the service hosted operations. The W3C Web services model utilizes the *Web Services Description Language (WSDL)* for this description, WSDL is comprehensive with a lot of tools and it is able to generate code for the client and server side automatically. New services are continually emerging, such as the RESTful Web services becoming commonplace recently. REST provides tools that are based on lightweight, decoupled, and stateless messages. Typically this architectural paradigm is implemented using HTTP.

W3C Web services architecture consist of four layers here listed (presented in Figure 2.9):

- Transport layer handles the exchange of bytes and messages between network nodes. The HTTP protocol is used in message delivering because it is universal and friendly to firewalls. Also other protocols are utilized, like FTP, SMTP, and JMS.
- XML Messaging layer is responsible for making sure that the sender and receiver will understand the message structure and format. The de facto solution SOAP handles XML messaging in the services architecture of W3C's Web. In SOAP the definition is given to the header and body of the message. It binds in several points to the transport protocols, such as HTTP. There are other alternatives too, like XML-RPC and plain XML (in HTTP POST).
- Description layer provides information about the service interfaces, specifically telling what methods are available. It will also indicate the supported input and output data types. The typical standards based solution used is WSDL.
- Discovery layer provides interfaces across networks for the detection of Web services. The most prominent technology in use is the *Universal Description, Discovery, and Integration (UDDI)*.

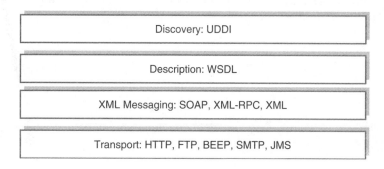

Figure 2.9 Web services stack.

Historically, the Web services have been born out of RPC, but the trend turned strongly towards more decoupled communication. This development is evident in the flexibility of the current SOAP specification and in the emerging REST based services as well. Furthermore, SOAP concepts are very often implemented using existing Web services, therefore the communication is rather based on messages more than on operational side. There is a definite and central need for versatile solutions and this principle has been respected in the design of WSDL version 2.0. It supports binding to all the HTTP request methods.

2.6.2 Asynchronous Processing

Figure 2.10 illustrates the two commonly used communication models, namely the synchronous and asynchronous request models. In the synchronous model, the caller sends a Web service request to an endpoint associated with a business component, and then later receives the response. The request is typically performed with a Web service client (stub) that is responsible for message formatting and checking.

The caller waits for the response. Thus the synchronous request invocation makes the caller block. The model is easy to use in software, because it follows the conventional semantics used in nondistributed software. On the other hand, it requires that the application waits for the response before continuing and thus is cumbersome, because

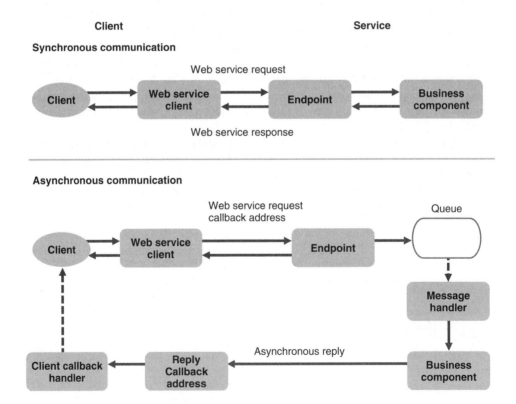

Figure 2.10   Asynchronous request processing.

another thread must be created for concurrent activities. In addition, the thread waiting for response and the other threads need to be synchronized.

The asynchronous request model addresses these points by making the request processing asynchronous with message queues. The caller sends the request and then can perform other activities immediately. The message has an associated reply handler that is responsible for accepting the asynchronous message from the Web service.

The message is handled by each message processing element in the Web service system and ultimately by the responsible business component. A message is then generated for the client and sent to the reply endpoint. The reply handler will receive the message and process it with application specific logic.

The asynchronous model is based on callback handlers that are responsible for processing particular signals. This model is an example of *inversion of control* that contrasts the traditional procedural programming model by having problem and task specific code that is called by the generic core system.

2.6.3   *The Connector Model*

The J2EE Connector architecture has been designed for integrating J2EE products and applications with various *enterprise information systems (EIS)*. The Connector architecture enables a vendor to provide a standard resource adapter for its enterprise information system. This resource adapter can be used in any J2EE-compliant application server. The application server and the resource adapter hide many of the lower level details from the developer, for example security, transactions, and connection management.

The Connector architecture consists of several standardized contracts and interfaces between different entities. The key entities are the backend EIS, the different application components, the application server, and the resource adapter (illustrated in Figure 2.11):

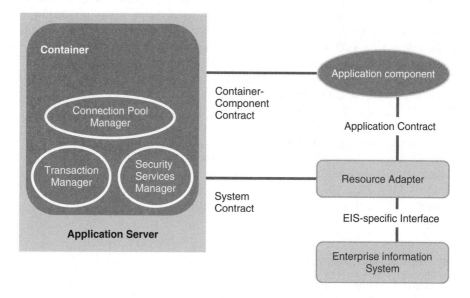

Figure 2.11   J2EE containers, resource managers, and contracts.

- EIS is the backend system that is integrated with the application components.
- Application components are Java or J2EE application components (for example Enterprise JavaBeans).
- Application server provides the runtime environment (container) in which the application components and the resource adapter are executed. It also provides system services such as connection, transaction, and security management.
- Resource adapter implements system level services specific to the EIS and connects application components with the EIS. It is a pluggable component that uses standardized system contracts.

The J2EE Connector architecture defines two types of contracts: system level and application level. The system-level contracts are defined for a J2EE application server and a resource adapter. Application-level contracts are defined for an application component and a resource adapter. The J2EE-compliant application server provides three services: transaction management, security, and connection pooling. The associated contracts in the J2EE Connector Architecture Version 1.0 are:

- Connection management contract that enables an application server to pool connection and allows application components to connect with the system.
- Transaction management contract that gives the transaction manager the ability to manage transactions across multiple resource managers.
- Security contract that enables secure access to the enterprise system.

Additional system contracts were defined by Version 1.5 of the J2EE Connector Architecture. The addition contracts are:

- Life-cycle management that enables an application server to manage the life-cycle of a resource adapter. This contract provides a mechanism for the application server to setup a resource adapter instance, and to notify the resource adapter instance when it should stop.
- Work management enables a resource adapter to perform work by sending work instances to an application server for execution. This allows a resource adapter to delegate thread creation and management to the application server that can efficiently pool resources and can have more control over the run time environment.
- Transaction inflow management enables a resource adapter to propagate an imported transaction to an application server, and perform the necessary transaction related signalling. This ensures that the *Atomicity, Consistency, Isolation and Durability (ACID)* properties of the imported transaction are preserved.
- Message inflow management enables a resource adapter to deliver messages to endpoints residing in the application server independent of the specific messaging style, messaging semantics, and messaging infrastructure. Thus a wide range of message providers (JMS, *Java API for XML Messaging (JAXM)*) can be plugged into any J2EE compatible application server with a resource adapter.

2.6.4   Web Service Platform

A fully fledged Web services platform needs a multitude of components as illustrated in Figure 2.12. Typically the platform must have:

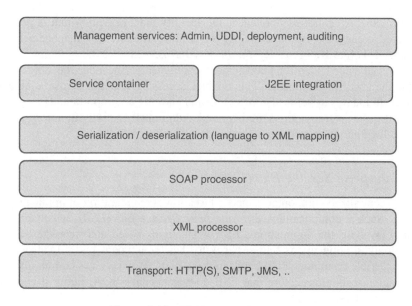

Figure 2.12   XML processing system.

- A set of of transport protocols to support nodal message exchange in the distributed system.
- An XML processing stack is mandatory to provide the capability of sending and receiving XML messages.

The XML processor is intertwined with XML message routing subsystem and also with the security pipeline ensuring the integrity of the security requirements. The security pipeline includes the following components:

- integrity checking;
- message validation;
- content checking to prevent injection attacks;
- authentication of the sender and message parts;
- authorization.

W3C's XML Encryption and Signature specifications are the foundation of basic Web services and XML security. Furthermore, OASIS has standardized the *XML Access Control Markup Language (XACML)*, which provides a usable policy language allowing for better definition of the access control requirements for application resources by the administrators. XACML model consists of *Policy Enforcement Points (PEPs)*, *Policy Decision Points (PDPs)*, *Policy Information Points (PIP)*, and *Security Assertion Markup Language (SAML)* assertions.

In addition to parts related to per-message processing such as XML and security, there is need for a set of other components as well. Figure 2.12 sets apart the management console, which bears the responsibility for support to the design and deployment of security policies. A central component is the Identity management which is provided by the managed credentials stored in LDAP or similar structure. ID management utilizes several protocols like Kerberos and *Public Key Infrastructure (PKI)*.

If a more advanced identity management component is needed, the *Single Sign-On (SSO)* protocol will be supported. This enables accessing services across administrative boundaries. In today's systems also reporting and auditing are important. They are critical especially if business transactions are involved. Additionally activity monitoring, alerting, and secure logging components are central part of modern design.

2.6.5 Enterprise Service Bus (ESB)

A logically centralized component is needed to allow the integration of applications using messaging. Such a component is the *Enterprise Service Bus (ESB)*, which provides an abstraction layer for the applications. ESB is standards based and message based interconnect, dividing applications and services to their constituent parts, It also facilitates decoupled flexible communication between these parts. Web services commonly use ESB implementation technology. Note that ESB does not realize SOA, rather it is a building component for SOA.

There are various features typical of ESBs including the following:

- Support for Invocation in synchronous and asynchronous transport protocols.
- Various routing-related features, such as routing addressability, static/deterministic routing, content-based routing, rules-based routing and policy-based routing.
- Components for mediating protocol transformation and service mapping.
- Message-oriented middle ware with message processing and transformations.
- *Complex Event Processing (CEP)* for event detection, pattern matching and correlation.
- Choreography that where complex business processes are to be realized.
- Orchestration is needed to combine services exposed to users as a single aggregate service.

2.6.6 Service Composition

In literature there are many of treatises on the basic protocols of service composition, but in practice the problem of dynamic service composition is still challenging. This might be seen only logical because of the prevailing loose coupling of Web services and their compositions. The main requirement is that component services can be invoked seamlessly, and for this to be realizable, both component metadata and interface semantics should be developed and published in a heterogeneous environment. WSDL and UDDI are the basic mechanisms for service composition.

WS-BPEL (Web Services Business Process Execution Language) from OASIS is a business process language which is serialized in XML and focuses on abstract processes and interprocess messaging. Tools like this are needed, for example, to make a decision on message sending, on performing a state transition, and coping with failures or time-outs. BPEL is basically an orchestration language because it concentrates on one peer in the system. Another contrasting way to see the system is choreography. It focuses on a global system and takes all entities into consideration. To generate operative orchestration, BPEL supports special programming constructs such as conditional statements, while sequences, flow control, and variable scoping. BPEL is one of the tools heavily depending on Web services, especially on WSDL and XML.

2.7 Session Initiation Protocol (SIP)

New and emerging Internet services are in constant need of new architectures and protocols, adapted, adaptable and optimized for the current and future demands. A very important example is SIP [19]. SIP has been built for tasking the call control and the driving applications are in telephony and multiparty communications. It has been used, for instance in the control of multiparty, multimedia sessions with audio and video.

SIP, standardized by the IETF, is text-based and works in the application layer. It was accepted as a 3GPP signalling protocol in November 2000. It is essentially a tool for setting up, maintaining, and terminating calls between two or more end points. SIP is also independent of the underlying transport layer and therefore conformable to many different transport protocols, for example TCP, UDP, or SCTP. It is an accepted 3GPP signalling protocol since November 2000.

SIP has found wide usage in the telecommunications industry and it is a core part of the IMS architecture for IP based multimedia services. Specifically, it is used as a vehicle to implement features familiar from the traditional telecom systems such as those advanced call processing characteristics in the *Signaling System 7 (SS7)*. SS7 is a centralized protocol with dumb endpoints and but with a complex call processing network. SIP is different because it positions complexity into the endpoints while keeping the network relatively simple.

SIP users are identified by specific application layer addresses, the SIP URIs. A URI of this kind consists of a username and a domain name. These URIs may also contain other parameters encoded in them. Because the SIP URI is mapped to the network address or addresses of a client, the naming system offers useful indirection for implementing various types of mobility: user, terminal, and session [20].

2.7.1 SIP Framework

The SIP protocol framework has several specific entities falling into two classes:

- The *user agents (UAs)* are SIP endpoints. They are managed by users and sessions are established between UAs.
- The SIP servers come in several varieties: registrars, proxies, user agent servers, redirect servers.

A closer look on these different server types reveals the SIP functions and how they differ from some other protocols. Figure 2.13 illustrates the key server types.

A central concept is the registrar. It manages SIP URIs in its domain, handling requests that refer to these addresses. A SIP registrar server handles the acceptance of registration requests from clients. It is the duty of the registrar to associate a client's address to a SIP URI. Registrars also take care of the acceptance of updated locations for the managed set of SIP URIs. This trait allows the mobility of devices from one network location to

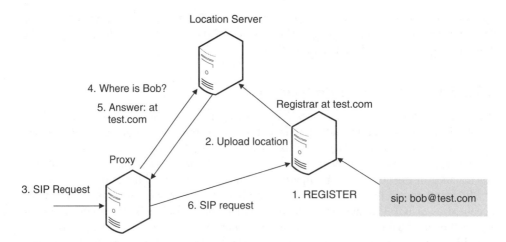

Figure 2.13   SIP entities.

another. The registrar may store the location data in a local database or alternatively, use a special server called the location server for storage of mappings. Note that the interface for location server does not belong to the SIP framework. A registrar server can be combined with the functions of a proxy or redirect server.

A SIP proxy closely resembles an HTTP proxy server and it is responsible for message routing: receiving and processing incoming SIP requests or forwarding them to other servers. Furthermore, the SIP proxies are designated to handle the authentication and authorization of messages. Thus, the SIP routing system is implemented using SIP proxies and they bear the responsibility for relaying a messages to their destinations. Actually the SIP routing system is quite flexible. The message routing is normally based on parameters like the sender identity, the time of the day, subject, session type, etc. There is a special technique called *forking*, which is utilized for replicating a SIP message to multiple destinations. This enables, for instance, a user with two mobile phones to answer calls on either of them.

A SIP redirect server is needed to determine the destinations: where a request is to be forwarded. In a typical case the redirect server is consulted by the proxy to get next hop destination. This is the active function of the redirect server – it does not forward messages.

2.7.2   Method Types

There are six method types for request messages in the SIP specification. There are also additional methods defined in other RFCs that extend the core SIP standard.

The key methods are the following:

- ACK that confirms that a final response has been received in a transaction.
- BYE that is sent by a User Agent Client to indicate to the server that it wishes to terminate the call.

- CANCEL that is used to cancel a pending request.
- INVITE that indicates the user or service is being invited to a session. The body of this message typically includes a description of the session with the *Session Description Protocol (SDP)*.
- OPTIONS that is used to query a server about its capabilities.
- REGISTER that is used by a client to register an address with a SIP server.

2.7.3   Establishing a Session

The session establishment process begins with an INVITE message. This message is sent by a calling user to a callee user and the INVITE message invites the callee user to participate in a session. The message includes information about the session, for example the type of session and its properties. Typically, the *Session Description Protocol (SDP)* is used to describe sessions.

The INVITE message does not necessarily lead to acceptance immediately, rather the caller may receive interim responses before the final session acceptance by the callee user. In the process the caller is normally informed that the user has been alerted, for example, the phone is ringing (in telephony). If the callee user answers the call, the calling client receives an OK response generated by the system. An ACK response message from the calling client then sets the session up, starting the actual data sending and receiving. The dissolution for a session begins when one of the users hangs up, generating a BYE message that is sent to the other client. This is then confirmed by the other client who sends an acknowledgment, ending the call.

2.7.4   Extensions

The original SIP protocol (RFC 2543) has received a number of extensions and enhancements, including some new messaging methods. The extensions include several methods for event notification, instant messaging and call control, listed below with examples based on the Windows Messenger application:

- SUBSCRIBE method enables a user to subscribe to chosen events. The user will be informed when a matching events occur.
- NOTIFY method informs the user of the occurrence of a subscribed event. For example, the Windows Messenger utilizes the SUBSCRIBE method in requests concerning contacts and groups and also allowing or blocking lists from the server. NOTIFY also provides a way toget the presence of contacts in a group.
- MESSAGE is utilized for instant messaging in SIP. If a user wans to send an instant message to another user the caller sends a request using the MESSAGE method. The MESSAGE request also carries the actual text in the SIP packet body.
- INFO method transfers information during a session, for instance data on user activity. For example, in Windows Messenger the INFO method is invoked to indicate that a user is typing on the keyboard.

- SERVICE method carries a SOAP message as the payload. Windows Messenger uses the SERVICE method for adding new contacts and groups onto the server. One can also use SERVICE method to search for contacts in the SIP domain.
- NEGOTIATE method finds use in negotiating the choice of several types of parameters, for instance security mechanisms and algorithms.
- REFER method enables the sender to advise on third party contacts; the receiver is instructed to take a contact using the details provided in the REFER request.

We will examine the event extension in Chapter 5 as well as many other state of the art messaging and pub/sub protocols and systems.

2.8 Summary

In this chapter, we examined the layered network architecture, TCP/IP, notion of time in a distributed system, and state of the art messaging solutions. These mechanisms are building blocks for distributed pub/sub systems that we will examine later in the book. Next, we will investigate overlay networks that are the essential substrate for wide-area pub/sub systems.

References

1. Deering S and Hinden R (1998) RFC 2460: Internet Protocol, Version 6 (IPv6) Specification.
2. Clark DD (1988) The design philosophy of the DARPA internet protocols. *SIGCOMM*, pp. 106–114 ACM, Stanford, CA.
3. Saltzer JH, Reed DP and Clark DD (1984) End-to-end arguments in system design. *ACM TOCS* **2**(4), 277–88.
4. Mockapetris P and Dunlap KJ (1988) Development of the domain name system. *SIGCOMM Comput. Commun. Rev.* **18**(4), 123–33.
5. Aura T (2005) Cryptographically generated addresses (CGA). *RFC* 3972, IETF.
6. Komu M, Tarkoma S, Kangasharju J and Gurtov A (2005) Applying a cryptographic namespace to applications. *Proceedings of the First ACM Workshop on Dynamic Interconnection of Networks*, DIN, pp. 23–7.
7. Nikander P, Ylitalo J and Wall J (2003) Integrating security, mobility, and multi-homing in a HIP way. *Proceedings of Network and Distributed Systems Security Symposium (NDSS03)*, The Internet Society, San Diego, CA.
8. Levchenko K, Voelker GM, Paturi R and Savage S (2008) Xl: an efficient network routing algorithm SIGCOMM '08: *Proceedings of the ACM SIGCOMM 2008 Conference on Data Communication*, pp. 15–26. ACM, New York, NY.
9. Kleinrock L and Kamoun F (1975) Hierarchical routing for large networks. *Computer Networks* **1**, 155–74.
10. Winter P (1987) Steiner problem in networks: a survey. *Netw.* **17**(2), 129–67.
11. Opyrchal L, Astley M, Auerbach J, Banavar G, Strom R and Sturman D (2000) Exploiting IP multicast in content-based publish-subscribe systems. *Middleware '00: IFIP/ACM International Conference on Distributed Systems Platforms*, pp. 185–207. Springer-Verlag New York, Inc., Secaucus, NJ.
12. IBM (2002) Gryphon: Publish/subscribe over public networks (White paper). http://researchweb.watson.ibm.com/distributedmessaging/gryphon.html.
13. Fox G and Pallickara S 2002 The Narada Event Brokering System: Overview and extensions PDPTA '02: Proceedings of the International Conference on Parallel and Distributed Processing Techniques and Applications, pp. 353–359. CSREA Press.
14. Jin X, Tu W and Chan SHG (2009) Challenges and advances in using IP multicast for overlay data delivery. *IEEE Communications*, **47**(6): 157–63.
15. Dalal YK and Metcalfe RM (1978) Reverse path forwarding of broadcast packets. *Commun. ACM* **21**: 1040–8.

16. Lamport L (1978) Time, clocks, and the ordering of events. *Communications of the ACM* **21**(7), 558–65.

17. Colouris G, Dollimore J and Kindberg T (1994) *Distributed Systems: Concepts and Design*, 2nd edn. Addison-Wesley, Boston, Massachusetts.

18. Sun 2002 Java Message Service Specification 1.1.

19. Rosenberg J, Schulzrinne H, Camarillo G, *et al.* (2002) RFC 3261: SIP: Session Initiation Protocol IETF. http://www.ietf.org/rfc/rfc3261.txt.

20. Schulzrinne H and Wedlund E (2000) Application-layer mobility using SIP. SIGMOBILE *Mob. Comput. Commun. Rev.* **4**(3), 47–57.

3

Overlay Networks and Distributed Hash Tables

This chapter deepens the treatment of networking solutions by illustrating how networks can be created on top of networks in so called overlay solutions. Overlay networks are networks that have been designed and deployed on top of existing underlying networks [1]. Many recent network algorithms and designs have been developed as overlay networks, because they do not require changes to routers and network equipment. Thus overlay networks offer flexibility in the creation of new distributed systems and services that extend the current features of the Internet and TCP/IP.

Overlay networks are good candidates for supporting various distributed pub/sub systems. There are two main categories of overlay solutions, namely unstructured and structured. Unstructured overlays can process content on any node and the search function typically relies on flooding messages. Structured, on the other hand, make assumptions regarding content placement and thus can support more efficient lookups.

In this chapter, we consider structured *Distributed Hash Tables (DHTs)* in particular that have many promising application in information dissemination and content delivery [2]. DHT-based solutions are examined in subsequent chapters with more details.

3.1 Overview

An overlay network is a network that is built on top of an existing network. The overlay relies on an *underlay* network for basic networking functions. Most overlay networks are built in the application layer on top of the TCP/IP protocol stack. Overlay technologies can be used to extend the features of the network layer, for example to offer new routing and forwarding functions. Nodes in an overlay network are connected by using logical connections that may consists of many network level hops.

Peer-to-peer (P2P) networks are a category of overlay networks. A P2P network consists of nodes that cooperate in order to provide services to each other. A pure P2P network consists of equal peers that are both clients and servers. The P2P model is different from the *client-server* model where clients access services provided by servers.

Publish/Subscribe Systems: Design and Principles, First Edition. Sasu Tarkoma.
© 2012 John Wiley & Sons, Ltd. Published 2012 by John Wiley & Sons, Ltd.

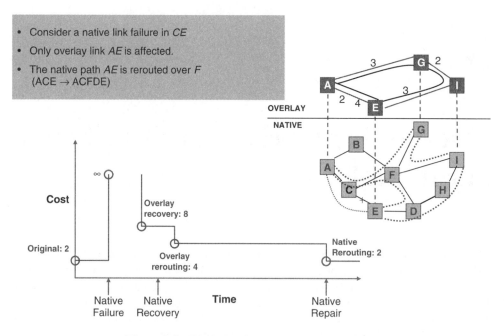

Figure 3.1 Example of resilient overlay routing.

Thus, an overlay network consists of a set of distributed nodes on the Internet. Following the end-to-end model of the Internet, these nodes are located at the edge of the Internet. The overlay nodes are expected to support the network in the following ways:

1. Providing infrastructure for the overlay network and supporting the execution of distributed applications.
2. Participating in the overlay network and perform routing and forwarding of messages and packets.
3. Deployed over the Internet in such a way that they are reachable by third parties.

Figure 3.1 illustrates how an overlay network can support resilient routing and go around network level reachability problems [3]. In the figure, the overlay network has been deployed over the underlay, namely the Internet. The overlay links can span several network level hops. The timeline illustrates a network layer router fault (native fault on link C-E), which breaks communication in the overlay network (link A-E). The overlay system is able to rapidly recover from the native fault and resume communications (rerouting through F) with some overhead cost denoted by the y-axis in the timeline figure. The underlay is eventually able to correct the native fault problem, and the overlay can then return to the more optimal configuration (native repair). This example demonstrates the benefits of overlay routing in increasing fault tolerance.

Typically, overlay networks are based on a distributed algorithm for creating scalable routing tables. A common technique is to build an overlay using DHTs that are scalable lookup structures. A DHT partitions a hashtable over multiple nodes on a network,

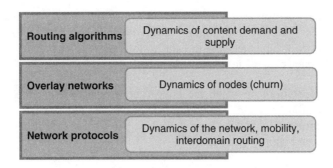

Figure 3.2   Dynamic properties on multiple layers.

and provides a hashtable API for adding and retrieving elements. Modern DHT algorithms can provide logarithmic lookup cost in terms of the number of hops needed to find content with logarithmic routing table sizes. Recent algorithms can take also network proximity into account and handle *churn*. Churn denotes situations in which nodes join and leave the network rapidly. A DHT algorithm is typically expected to tolerate churn.

DHTs typically support lookup using semantic free labels. With semantic free labels, any data blob can be transformed into a label by using a hash function. The SHA-1 hash algorithm is one frequently used hash function. Labels can also have security properties. A frequently used technique adds a public key signature to a flat label. This is called *self-certification*. A self-certified label consists of hash of data signed by a public key and the public key. A node receiving a self-certified label can then verify that some data actually hashes to the label, and that the public key correctly verifies the signature.

A generic DHT API is simple and features three operations: *put*, *get*, and *delete*. The put operation places data into the structure with the given key. Get retrieves the data given the key. Delete removes the data associated with the given key.

Figure 3.2 presents three layers for distributed pub/sub systems. The lowest layer consists of the network layer protocols, typically the TCP/IP protocol suite. This layer needs to cope with dynamics of the network, mobility issues, and interdomain routing. The overlay network layer builds on this layer and introduces new addressing schemes and offers flexibility to higher levels. Our previous example illustrated the potential benefits of overlay networks; however, the overlay network must address concerns with node dynamics (churn) and security. On top of the overlay network, various pub/sub routing algorithms and systems can be developed. One example is the Scribe pub/sub system implemented on top of the Pastry DHT. The solutions need to consider the dynamics of content demand and supply. Therefore issues pertaining to dynamic operation need to be investigated at each layer.

3.2   Usage

Overlay networks can be used in different operating environments. The two typical environments are the clusters and the Internet (wide-area). Clusters can typically be assumed to be reliable, secure, and administered. In addition, they are not prone to network partitions or unpredictable operating conditions. Wide-area environments, on the other hand, lack these qualities. They are unreliable, unpredictable, insecure, prone to network partitions, and do not have centralized administration. DHTs balance hop count with the size of the routing tables, network diameter, and the ability to cope with changes.

Typical requirements for wide-area overlays include:

- Decentralized operation, in which the nodes cooperate and form the system without central coordination.
- Scalability to millions of nodes.
- Fault tolerance and support for churn.

Typical applications of overlay networks include:

- Content search and file transfer.
- Distributed directories with efficient lookups.
- Content routing over the Internet.
- Publish/subscribe.
- Distributed storage systems.
- Substrates for multiplayer games.

The two important categories for overlay networks are [1]:

- Unstructured networks that do assume a particular way of distributing content to nodes and building links between nodes. Example systems include the classical P2P systems such as Gnutella and Freenet as well as gossip algorithms discussed later in this chapter.
- Structured networks that make more assumptions regarding the links between nodes and how content is placed on the nodes. Example systems are the DHT algorithms that will be presented next.

3.3   Consistent Hashing

The seminal consistent hashing technique was proposed in 1997 as a solution for distributing requests to a dynamic set of web servers [4]. The technique maps incoming requests to servers in a consistent manner that allows servers to be added and removed at runtime. The scheme is more flexible than, for example, linear hashing that allows incremental scalability, but allows servers only to be added in a specific order. Consistent hashing is the basis of the DHT algorithms discussed next.

Consistent hashing allows the addition and removal of servers with approximately K/n elements being relocated with each change, where K is the number of keys and n is the number of servers. The idea is to map buckets and items to the unit interval, and then place a data item to the closest possible bucket. In order to cope with network partitions, the scheme assumes that each client of the system has a limited view to the global state. Each view contains a subset of the servers. In order to ensure information availability, the buckets need to be replicated across servers.

A bucket is replicated $\kappa \log(C)$ times, where C is the number of distinct buckets, and κ is a constant. When a new bucket is added, only those items are moved which are closest to one of its points.

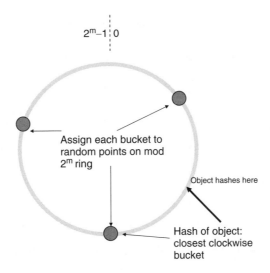

Figure 3.3 Example of consistent hashing.

The three important properties in consistent hashing are the following:

- Smoothness, which pertains to the load distribution across the servers. The expected number of items that are relocated when adding or removing a server should be minimized.
- Spread, which is the number of servers responsible for an item over all views. This number of servers should be kept small.
- Load, which defines the number of items assigned to a particular bucket over all views.

Figure 3.3 illustrates the circular address space of consistent hashing. Each of the buckets has a random position in the address space. Each bucket is responsible for storing objects that hash to the address space preceding the bucket. The scheme localizes transfers between buckets when a bucket is added or removed.

3.4 Geometries

Research on parallel interconnection networks has resulted in a collection of topologies over static graphs. The main application area of the these networks has been the development of efficient hardware systems. More recently, the interconnection geometries have been applied in the creation of overlay networks and DHTs. The static graph structures cannot be directly used, but extended to support dynamic and random geometries [5].

Figure 3.4 presents a summary of the frequently used geometries. The frequently used overlay topologies are: *trees*, *tori* (k-ary n-cubes), *butterflies* (k-ary n-flies), *rings*, and the *XOR geometry*. These structures form the basis of different DHT algorithms. The differences between some of the geometries can be very small and subtle. For instance, the static DHT topology emulated by the DHT algorithms of Pastry and Tapestry are Plaxton trees; however, the dynamic algorithms resemble hypercubes.

	Tree	Hypercube	Ring	Butterfly	XOR
Neighbour selection	Yes	No	Yes	No	Yes
Route selection	No	Yes	Yes	No	Some
Sequential neighbours	No	No	Yes	No	No
Guaranteed independent paths	No	No	No	No	No
Example system	Plaxton mesh	CAN, Tapestry, Pasty	Chord	Viceroy	Kademlia

Figure 3.4   Summary of typical DHT geometries.

The two important characteristics of the geometries are the network degree and network diameter. High network degree implies that joining, departing, and failing may affect more nodes. The geometries can be grouped based on the network degree into two types: constant-degree geometries and nonconstant degree geometries. In the former, the average degree is constant irrespective of the size of the network. The latter type involves average node degrees that grow typically logarithmically with the network size.

3.5   DHTs

DHTs are based on consistent hashing and they aim to support information lookup in decentralized wide-area systems. The early canonical DHT is the Plaxton's algorithm and the cluster-based consistent hashing technique. After this the first four DHTs, namely CAN, Chord, Pastry, and Tapestry, were introduced in 2001.

DHT-based systems typically guarantee that any data object can be located using $O(\log N)$ overlay hops on average, where N is the number of nodes in the overlay network. The underlying network path between two peers can be significantly different from the path used by the overlay network.

In the Chord DHT, the participating nodes have unique identifiers and form a one-dimensional ring. Each node maintains a pointer to its successor and predecessor node (determined by their identifiers). As an optimization they maintain additional pointers (called *fingers*) to other nodes in the network. A message is forwarded greedily towards a closer node in the finger table with the highest identifier value less or equal than the identifier of the destination node.

Pastry and Tapestry are designed after the Plaxton's algorithm and thus they are tree-based solutions. In a similar fashion to Chord, a message is routed to a node who is closer to the destination (by one more digit). Pastry uses prefix routing whereas Tapestry is based on suffix routing.

CAN differs from the other mentioned DHTs in that it is based on a d-dimensional Cartesian coordinate space and each node has an associated zone. Each node knows its neighbours in the logical topology. Messages sent to a coordinate are delivered to the node that is responsible for the zone that contains the coordinate. Each node forwards the message to the neighbour that is the closest to the destination.

3.5.1   DHT APIs

The typical API of a DHT is quite simple. The aim of the API is to abstract distribution details from the developer. Thus the DHT APIs look like conventional data structure APIs.

DHTs can be seen to have a common element called the *key-based routing service (KBR)* [6]. This service offers efficient routing to identifiers defined based on a large identifier space. The KBR abstraction can be layered to create more complex key-based operations. KBR abstractions include lookups, anycast, multicast, and object location and routing [6]. The KBR API is offered by structured DHTs and it implies that each overlay node maintains a routing table and can forward messages towards their destinations.

An DHT KBR is expected to provide the following operations:

- join(q): current node contacts node q to join the overlay network.
- leave(): current node leaves the overlay network.
- lookup(key): current node searches for the node responsible for the given key.

3.5.2   Chord

Chord is a distributed and decentralized lookup service [7]. Each Chord node has a unique m-bit node identifier (ID), which is obtained by hashing (using, for example, SHA-1) the node's IP address and a virtual node index. IDs occupy a circular identifier space. Keys are also mapped into this ID space by hashing them to m-bit key IDs. Following the consistent hashing model, the node responsible for a given ID is the node with the smallest ID that is greater or equal to the object ID. In order to maintain coherence in the mapping of IDs during node join operation, some IDs are assigned to the new node. When a node leaves the Chord system, all keys assigned to the leaving node are transferred to the new responsible node.

Consistent hashing is efficient with constant time lookups if there is global knowledge of all other nodes. Chord provides a scalable version of consistent hashing based on routing. Chord routes in $O(log(n))$ messages and requires information about $O(log(n))$ nodes. The routing table (called as finger table) maintains at most m elements. Figure 3.5 illustrates the Chord ring and the finger table. The i:th entry in the table at node n identifies the first node s that succeeds n by at least $2^{(i-1)}$ on the identifier circle. A finger table entry includes both the Chord ID and an IP address of the node [7]. The Chord algorithm consults the finger table in order to forward a message closer to its destination. As mentioned above, a message is forwarded in a greedy manner towards a closer node in the finger table with the highest identifier value less or equal than the identifier of

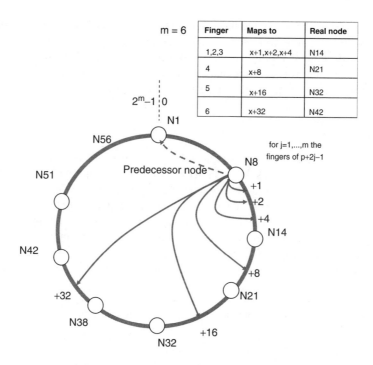

Figure 3.5   Example of routing with Chord.

the destination node. The overlay update processes are responsible for keeping the finger tables up-to-date when nodes join and leave the network.

3.5.2.1   Internet Indirection Infrastructure

The Internet Indirection Infrastructure (i3) [8] is an overlay network based on Chord [9] that aims to provide a more flexible communication model than the current IP addressing [8]. In i3 each packet is sent to an identifier. Packets are routed using the identifier to a single server in the distributed system. The server, a i3 node, maintains triggers which are installed by receivers that are associated with identifiers. When a matching trigger is found the packet is forwarded to the associated receiver. An i3 identifier may be bound to a host, object, or a session unlike the IP address, which is always bound to a specific host.

The i3 system can support a number of interactions, including unicast, multicast, anycast, and service composition. The overlay provides a level of indirection that can be used for supporting mobile and multihoming hosts.

In i3 unicast a host R inserts a trigger (id, R) in the i3 infrastructure to receive all packets with identifier id. Figure 3.6 illustrates the i3 multicast primitive. An application can build a multicast tree using a hierarchy of triggers. The packet is replicated at triggers in order to realize the multicast.

Figure 3.7 illustrates the i3 anycast primitive. Applications can specify a prefix for each trigger identifier. Packets are then matched to the identifiers according to the longest matching prefix rule.

Multicast trigger primitive

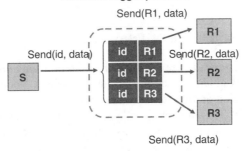

Figure 3.6   Example of i3 multicast.

Anycast trigger primitive

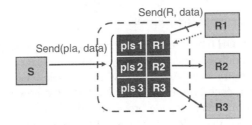

Figure 3.7   Example of i3 anycast.

3.5.3   Pastry

Pastry is a classical example of a scalable DHT structure [10]. Pastry is similar to other well-known DHT systems, such as Chord, CAN, and Tapestry. Pastry assigns nodes and objects random identifiers from a large identifier space. Node identifiers (NodeIDs) and data keys are 128 bit strings. In essence, they are sequences of digits in base 2^b, where b is a configuration parameter.

Pastry is based on consistent hashing and the Plaxton's algorithm. The key aim of the system is to provide an object location and routing scheme. Pastry is a prefix based routing system, and it takes the network proximity into account in the routing. The message routing process is straightforward: A message is always sent to a numerically closer node. The expected average hop count is $\log(N)$ hops for reaching a given destination.

Following the Plaxton's algorithm, Pastry routes a message to the node with the NodeID that is numerically closest to the given key. The destination node is said to be the key's root.

The key Pastry API functions are:

- *NodeID = pastryInit(Credentials)*. This function allows the current node to join the Pastry network. The function returns a NodeID for the current node.
- *route(msg,key)*. This function routes a message to the node with NodeID numerically closest to the given key.

- *send(msg,IP-addr)*. This function sends a message to the node with the given IP address.

Applications use the system with the Pastry API:

- *deliver(msg,key)*. This upcall is invoked by the system when a message destined to the current node is received. In this case, the current node's NodeID is numerically the closest to the destination address specified in the message.
- *forward(msg,key,nId)*. This function is called by the system before a message is sent to the node that has the identifier nId. This function can be used by the application to inspect the message, modify the contents, and prevent the system from sending the message.
- *newLeafs(leafSet)*. This function is called by the system when there is a change in the set of neighbouring nodes.

3.5.3.1 Joining and Leaving the Network

A node joins the Pastry DHT by determining a new node identifier (NodeID) by hashing the node's current IP address. The node then sends a message to this NodeID via the topologically nearest Pastry node A0. The new node copies the neighbour set from the initial node A0. Then at each hop towards the NodeID, the new node receives a routing table row from an existing node. A node processing the request will also include the new node in its routing table if the address is numerically smaller than an existing address for a given prefix. The node that sends the last routing table row also sends the leaf set to the new node. Some entries need to be then checked for consistency. Finally, the new node sends its routing table to each neighbour.

The join consists of the following steps:

- Create NodeID and obtain neighbour set from the topologically nearest node.
- Route message to NodeID.
- Each Pastry node processing the join message will send a row of the routing table to the new node. The Pastry nodes will update their long-distance routing table if necessary.
- Receive the final row and a candidate leaf set.
- Check table entries for consistency. Send routing table to each neighbour.

A Pastry node may fail or leave the system without prior notification. The Pastry DHT can manage such situations. A failed node needs to be replaced in the routing tables of other nodes. This can be done by replacing the failing node with the largest corresponding index in the failing node's routing table. Thus replacements can be found by consulting the routing table.

3.5.3.2 Routing

A Pastry node maintains three crucial data structures: the routing table, the neighbour set, and the leaf set. The first stores the long distance links to other nodes. A node's routing table has $l = \lceil \log_{2^b} N \rceil$ rows and 2^b columns. The entries in row r of the routing table refer to nodes whose NodeIDs share the first r digits with the current node's NodeID.

The $(r + 1)$th NodeID digit of a node in column c of row r equals c. The column in row r corresponding to the value of the $(r + 1)$th digit of the local node's NodeID is empty.

The second structure contains nearby nodes based on the scalar proximity distance between nodes.

The third structure contains numerically close nodes ($l/2$ smaller and $l/2$ larger keys) and can used to find nearby nodes in the numerical addressing space. The basic long-distance routing table is thus supported by the leaf set, which can be seen to introduce a ringlike structure in addition to the hypercube-like routing table.

A node sends an incoming message to a node whose NodeID shares with target address a prefix that is at least one digit longer than the prefix that the key shares with the current node's identifier. If such a node cannot be found, the message is then sent to a node whose NodeID shares the prefix and has a numerically closer identifier than the current node. The expected number of routing hops is less than $\log_{2^b} N$.

The Pastry overlay takes network proximity into account by measuring the network level distances between nodes. Given that there are many possible choices for a given routing table entry, the node with a low network delay is selected. This results in an overlay topology that takes the lower level network topology into account. Thus Pastry has a low delay penalty.

Figure 3.8 presents a Pastry routing table with a node that has the address $65a1x$, $b = 4$, and $l = 4$ (with base 16). In the routing table elements 'x' represents an arbitrary suffix.

Figure 3.9 illustrates the prefix-based routing [11]. Node $65a1fc$. routes a message to destination $d46a1c$. The message is routed to the nearest node in the identifier circle that is responsible for the address space of the destination.

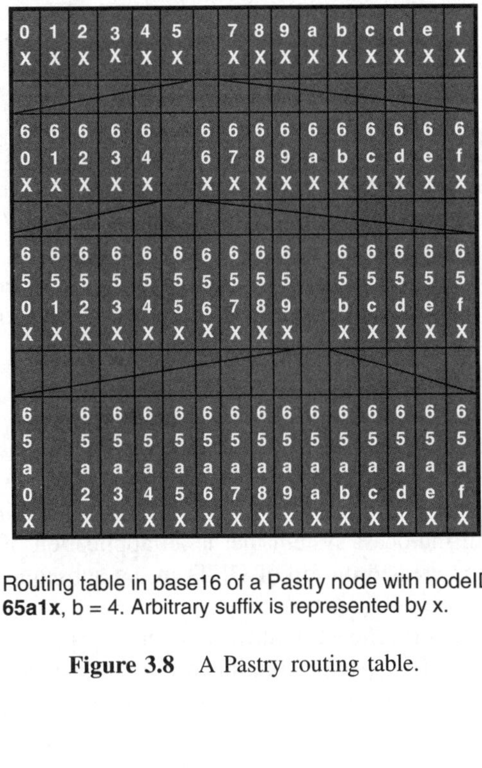

0	1	2	3	4	5	6	7	8	9	a	b	c	d	e	f
0X	1X	2X	3X	4X	5X		7X	8X	9X	aX	bX	cX	dX	eX	fX
60X	61X	62X	63X	64X		66X	67X	68X	69X	6aX	6bX	6cX	6dX	6eX	6fX
650X	651X	652X	653X	654X	655X	656X	657X	658X	659X		65bX	65cX	65dX	65eX	65fX
65a0X		65a2X	65a3X	65a4X	65a5X	65a6X	65a7X	65a8X	65a9X	65aaX	65abX	65acX	65adX	65aeX	65afX

Routing table in base16 of a Pastry node with nodeID
65a1x, b = 4. Arbitrary suffix is represented by x.

Figure 3.8 A Pastry routing table.

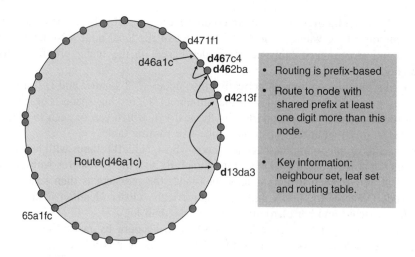

Figure 3.9   Prefix routing with Pastry.

Pastry routing algorithm for destination K at node N.

- If K is in the leaf set, then route packet directly to the node.
- If K is not in the leaf set, then determine the common prefix (N, K).
- Search for the entry E in the routing table with prefix $(E, K) >$ prefix (N, K) and forward the message to E.
- If this is not possible, then search for node E in the routing table with longest prefix (E, K) based on the merged set of the routing table, leaf set, and the neighbour set. Forward to E.

3.5.3.3 Characteristics

On average, Pastry requires $\log(N)$ hops in order to reach a given destination. With concurrent node failures, eventual delivery is guaranteed unless $l/2$ or more nodes with adjacent NodeIDs fail simultaneously.

A Pastry routing table has $(2^b - 1) * \lceil \log_{2^b} N \rceil + l$ entries. After a node failure or the arrival of a new node, the routing tables can be updated with $O(\log_{2^b} N)$ messages.

3.5.3.4 Applications

Figure 3.10 gives several examples of the applications of the Pastry DHT. Scribe is a topic-based pub/sub system implemented on top of Pastry. Another example is the PAST network file storage service that utilizes Pastry for storing and caching data [12]. Next, we consider the SplitStream multicast system that is an application of Pastry. We will later examine pub/sub systems created on top of DHTs in the subsequent chapters, especially in Chapter 9.

SplitStream aims to support efficient multicast when nodes participating in the peer-to-peer network are dynamic. The specific solution is to stripe the content across a forest of

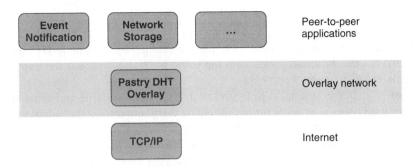

Figure 3.10   Pastry applications.

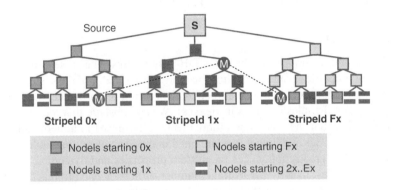

Figure 3.11   Overview of SplitStream.

interior-node-disjoint multicast trees. These trees are then responsible of distributing the forwarding load among participating peers [13].

A set of trees is said to be interior-node-disjoint if each node is an interior node in at most one tree, and a leaf node in the other trees.

SplitStream splits the content into k stripes. The system uses a separate multicast tree to distribute a given stripe. Peers can then join the stripe specific multicast trees in order to receive the content. Each peer has an upper bound on the number of stripes that they can forward to other peers. Given that the original content has bandwidth requirement B, each stripe has a bandwidth requirement of B/k. The peers can control their inbound bandwidth in increments of B/k.

Figure 3.11 illustrates SplitStream's construction of the forest. The source generates the stripes based on the content and then multicasts each stripe using its corresponding multicast tree. The stripe identifier o starts with a different digit. The node identifiers of interior nodes share a prefix with the stripe identifier. Thus they must be leaves in the interior-node-disjoint multicast forest.

The main challenge is to build and maintain this forest of multicast trees such that an interior node in one tree is a leaf node in all the remaining trees and the bandwidth constraints specified by the nodes are met. SplitStream builds on the properties of Pastry and Scribe overlay routing to build interior-node-disjoint trees. Scribe trees have a disjoint set of interior nodes when the identifier for the trees all differ in the most significant digit. The value of b for Pastry needs to chosen so that it results in a suitable value for k.

In SplitStream, the expected amount of state maintained by each node $O(log|N|)$ and the expected number of messages to build the forest is $O(|N|log|N|)$ if the trees are well balanced or $O(|N|^2)$ in the worst case.

Overcast is a similar system to SplitStream that provides scalable and reliable single-source multicast using a protocol for building data distribution trees that adapt to network conditions [14]. Overcast organizes dedicated servers into a source-rooted multicast tree using bandwidth estimation measurements to optimize bandwidth usage across the tree. The main differences between Overcast and SplitStream are that Overcast uses dedicated servers while SplitStream utilizes clients. Moreover, SplitStream assumes that the network bandwidth available between peers is limited by their connections to their *Internet Service Provider (ISP)* rather than the network backbone.

3.5.4 Discussion

Figure 3.12 presents a summary of three key DHT algorithms. The presented table summarizes the salient properties of the algorithms, namely the geometry of the system, the routing and matching principles, the system parameters, the routing performance, the routing state, and how the system handles churn (joins/leaves). We can observe that the three compared systems are similar, but there are some fundamental differences in how the address space is divided between nodes and how it is maintained in the form of routing tables. The underlying geometry thus greatly influences the system properties and dynamics.

The DHT properties, namely the average number of hops needed to reach a destination and the routing table size, are related. Figure 3.13 examines the asymptotic tradeoff curve between the routing table size and the network diameter [15]. The figure does not consider tolerance to churn and how well the overlay topology maps to the underlying network topology. These two are also important parameters in characterizing overlay networks.

	CAN	Chord	Pastry
Geometry	Multidimensional space (d-dimensional torus)	Circular space	Plaxton-style mesh
Routing	Maps (key,value) pairs to coordinate space	Matching key and node identifier	Matching key and prefix in node identifer
Key parameters	Number of peers N, number of dimensions d	Number of peers N	Number of peers N, base of peer identifier B
Routing performance	$O(dN^{1/d})$	$O(log\ N)$	$O(log_B\ N)$
Routing state	$2d$	$log\ N$	$2Blog_B\ N$
Churn (joins and leaves)	$2d$	$(log\ N)^2$	$log_B\ N$

Figure 3.12   Summary of well-known DHT algorithms.

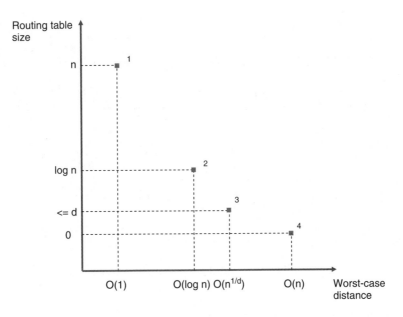

Figure 3.13   Tradeoffs with routing table size and network diameter.

Many well-known DHT algorithms, such as Chord, Pastry, Tapestry fall into the middle portion of the graph with the typical characteristics of logarithmic diameter and routing table size. The CAN algorithm can offer constant routing performance with larger routing table sizes. Analysis of the tradeoffs suggests that the routing table size of $\Omega(\log n)$ is a threshold point that separates two state-efficiency regions. This threshold point is in the middle of the asymptotic curve. If the routing table size is asymptotically smaller or equal, the demand for congestion-free operation prevents the overlay from achieving the smaller asymptotic diameter. When the routing table size is larger, the demand for congestion-free operation becomes less limiting for the network.

3.6   Gossip Systems

Gossip is a technique for delivering messages in the distributed environment in a probabilistic way to some set of destination nodes [16, 17]. Communicating nodes exchange information by selecting the next hop destinations based on a probability distribution, for example the uniform distribution.

3.6.1   Overview

A node in a gossip system selects a subset of peers using the distribution and then sends the input message to them. Nodes receiving a gossip message repeat this process. Ultimately after sufficient rounds of communication, all the target nodes have received the message. Gossip therefore floods the messages and the flooding process is parameterized by the probability distribution that is used to select the next hop destinations.

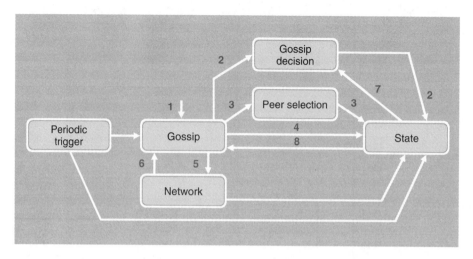

Figure 3.14   Example of gossip-based dissemination.

Figure 3.14 illustrates how gossip works [18]:

1. A node receives an incoming message or a periodic trigger expires, and the gossip process of the message starts.
2. A gossip decision needs to be made. A node gossips a message with some given probability.
3. Given that a positive gossiping decision was made, the node needs to select a subset of nodes in the network to which the message is sent. The peer selection process consults the current state of the node in order to determine the receivers.
4. The receivers are then contacted and the message is sent to the receivers.
5. The system state is updated.

The three key metrics for a gossip system are:

• the latency of the communication;
• the probability of reaching all the proper nodes in the network.

The reachability probability and latency can be adjusted by tuning the replication factor (the *fanout*) of the message forwarding. The replication factor determines the size of the subset of nodes to contact for each round of gossiping. The random selection of destination nodes to contact is determined by using local information available at the node. If local information is used to choose the nodes the gossip protocol is called *informed*.

The reliability of the system stems from the observation that the peer sets of nodes in the network are independent of each other. In addition, the redundancy and randomization go around possible failures in the network and avoid expensive reconfiguration of the network.

Gossip-based unstructured overlays can be used as a building block for various network and service management applications, especially when support for dynamic operation and eventual consistency is required [19]. Gossip has been used for monitoring and configuration in the AstroLabe system [20], and for achieving eventual consistency in Amazon's Dynamo [21] that is a core system for the company's services. Gossip lends itself well to monitoring in large-scale networks, where each node monitors a small random subset of other nodes thus distributing the monitoring cost. Gossip is also suitable for managing routing tables in a large-scale P2P network.

Gossip-based protocols can be divided into three main categories:

- Dissemination protocols. This category uses gossip to spread data across the network by using probabilistic flooding. A technique called *rumour-mongering* chooses a node at random to which the rumour is spread. The technique uses gossip for some predetermined time. The time parameter needs to be selected so that it is high enough to ensure that the data is delivered across the network to all expected receivers. Rumour-mongering is a basic building block for reliable multicast protocols.
- *Anti-entropy* protocols are used to replicate data through pair-wise interactions. These protocols are based on an opportunistic strategy that compares replicas between nodes and then reconciles observed differences. Information is gossiped until it is made obsolete by newer information. This approach is useful for applications that require eventual consistency.
- *Data aggregation* protocols compute an aggregate value by sampling information at the nodes with the aim of computing a system wide aggregate value. The aggregation function needs to be computable by fixed-size pairwise information exchanges. Typically a logarithmic number of communication rounds are needed for computing the aggregate. Well-known applications of the technique include counting, sorting, and summing of values in the network [22].

3.6.2   View Shuffling

View shuffling is a commonly used technique in building gossip-based systems. As mentioned above, a node needs to have knowledge of a subset of other nodes that can be used for gossiping. This is called a view and all the nodes' views should be independent. The aim of the view shuffling process is to periodically update a node's view and keep the views independent. To this end, nodes randomly exchange their views through pairwise interactions. The simple basic shuffling algorithm assumes that there are no failures and that the neighbourhood information, the view, of a node is available [23, 24]. Each node has a set of continuously changing neighbours, and at certain times, typically periodically, each node contacts a random neighbour to exchange some of their neighbours. The nodes maintain a neighbour table that includes the network addresses of a subset of other nodes in the network.

The shuffling operation exchanges the views of two nodes X and Y, and it reverses the relation between the nodes. This is illustrated by the following six steps (from the viewpoint of the node X):

1. Select a random subset of l neighbors from the local node's (X) neighbour table. Select a random peer, Y, within the table. The parameter l is the *shuffle length*.
2. Replace Y's address with X's address (to remove Y from the table).

3. Send the updated subset of length l to Y.

4. Receive from Y a subset of no more than l of Y's neighbors. Y randomly selects this subset of its own neighbours and updates its table. As a result Y is no longer a neighbour of X.

5. Remove entries pointing to X, and entries that are already in X's neighbour table.

6. Update X's neighbour table to include the remaining entries. First populate empty slots, and then replace entries that were originally sent to Y.

The connectivity resulting from the shuffle algorithm is guaranteed assuming that the environment is fail-free. An overlay network created by using the shuffling algorithm cannot be divided into two disjoint networks as a result of the shuffling operation [19].

3.6.3   Gossip for Pub/Sub

Gossip is suitable overlay dissemination substrate for building scalable pub/sub systems. A number of gossip algorithms for pub/sub systems have been proposed in [25–31]. Many of the proposed systems combine gossip with hierarchical structure in order to achieve better guarantees and mapping to the underlying network topology. We briefly consider two examples of gossiping pub/sub systems and later revisit this technique in Chapters 7 and 9.

Eugster and Guerraoui proposed the *probabilistic multicast (pmcast)* system that is an example of informed gossip for pub/sub [32]. The system optimizes the gossiping process by avoiding subscribers who are not interested in the content. This requires that the nodes are organized in a hierarchy of groups that are built based on the physical proximity of nodes. Event messages are delivered using the hierarchy by gossiping depth-wise, starting at the root of the hierarchy. The hierarchy follows the network topology in order to reduce the number of network boundaries that are crossed. Each node maintains a view that includes the subscriptions of its neighbours in a group. Selected delegate nodes are responsible for aggregating subscriptions within a group.

Costa and Picco in [30] present a hybrid system that combines deterministic and probabilistic operation. Subscriptions are disseminated only in the proximity of a subscriber. If subscription information is not available, events are sent in a probabilistic fashion over the overlay network.

The SpiderCast system is based on an overlay network in which nodes manage both randomized links and semantic links [27]. The semantic links represent shared interests between the nodes. Event delivery is implemented by utilizing semantic links if they exists and then as a fallback measure using gossiping with random links.

Baldoni *et al.* propose the Tera system, which is a two layered topic-based dissemination system [26]. The system aims to uniformly distribute information about nodes' interests. Event delivery consists of a two phase process. In the first phase a random walk is used over the low-level overlay that connects the nodes with the goal of locating a node that is subscribing to the topic. The second phase starts when such a subscriber has been found and it is based on an upper layer overlay that connects all the subscribers of the topic. The event is then disseminated in this second phase.

Voulgaris *et al.* present the multilayer *SUB-2-SUB* architecture [31]. This system has on three layers:

- The lower layer uses a gossip protocol to exchange subscription information.
- The middle layer maintains semantic relations between the subscribers and clusters them based on their interests.
- The upper layer is a logical ring structure that connects all nodes.

The SUB-2-SUB system relies on the overlapping intervals of range subscriptions and uses their structure to cluster subscriptions and create the middle layer. A publisher needs to find at least one subscriber in the middle layer in order to start the event delivery process. Once the subscriber has been located, the subscribers collaborate in delivering the event by using shortcut links. The shortcut links are built by gossiping, and also by using the ring topology that is needed to find all the potential subscribers.

3.7 Summary

In this chapter, we examined overlay solutions that are frequently used as a basis for pub/sub systems. We focused on structured DHTs and also examined unstructured gossip systems. Most solutions discussed later in the book are based on overlay networks. We will examine pub/sub systems that build directly on top of the network layer, and systems that build on top of DHT solutions. Pub/sub systems are frequently built on top of DHTs, because they abstract network level details and provide reliability, churn support, and fault tolerance for the higher levels.

References

1. Tarkoma S (2010) *Overlay Networks – Toward Information Networking*. CRC Press.
2. Lua EK, Crowcroft J, Pias M, Sharma R and Lim S (2005) A survey and comparison of peer-to-peer overlay network schemes. *IEEE Communications Surveys and Tutorials* **7**: 72–93.
3. Andersen D, Balakrishnan H, Kaashoek F and Morris R (2001) Resilient overlay networks. *SOSP '01: Proceedings of the Eighteenth ACM Symposium on Operating Systems Principles*, pp. 131–45. ACM, New York, NY.
4. Karger D, Lehman E, Leighton T, Panigrahy R, Levine M and Lewin D (1997) Consistent hashing and random trees: distributed caching protocols for relieving hot spots on the World Wide Web. *STOC '97: Proceedings of the Twenty-Ninth Annual ACM Symposium on Theory of Computing*, pp. 654–63. ACM, New York, NY.
5. Manku GS (2003) Routing networks for distributed hash tables. *PODC '03: Proceedings of the Twenty-Second Annual Symposium on Principles of Distributed Computing*, pp. 133–42. ACM, New York, NY.
6. Dabek F, Zhao B, Druschel P, Kubiatowicz J and Stoica I 2003 Towards a Common API for Structured Peer-to-Peer Overlays Proceedings of the 2nd International Workshop on Peer-to-Peer Systems (IPTPS03), Berkeley, CA.
7. Stoica I, Morris R, Karger D, Kaashoek MF and Balakrishnan H (2001) *Chord: A Scalable Peer-to-Peer Lookup Service for Internet Applications*. ACM SIGCOMM.
8. Stoica I, Adkins D, Zhuang S, Shenker S and Surana S (2002) Internet indirection infrastructure. *Proceedings of the 2002 Conference on Applications, Technologies, Architectures, and Protocols for Computer Communications*, pp. 73–86. ACM Press.
9. Stoica I, Morris R, Karger D, Kaashoek F and Balakrishnan H (2001) Chord: a scalable peer-to-peer lookup service for internet applications. *Computer Communication Review* **31**(4), 149–60.
10. Rowstron A and Druschel P (2001) Pastry: Scalable, decentralized object location and routing for large-scale peer-to-peer systems. *IFIP/ACM International Conference on Distributed Systems Platforms (Middleware)*, pp. 329–50.

11. Castro M, Druschel P, Kermarrec AM and Rowstron A (2002) One ring to rule them all: service discovery and binding in structured peer-to-peer overlay networks. *EW10: Proceedings of the 10th Workshop on ACM SIGOPS European Workshop*, pp. 140–5. ACM, New York, NY.

12. Rowstron A and Druschel P (2001) Storage management and caching in PAST, a large-scale, persistent peer-to-peer storage utility. *18th ACM Symposium on Operating Systems Principles (SOSP'01)*, pp. 188–201.

13. Castro M, Druschel P, Kermarrec AM, Nandi A, Rowstron A and Singh A (2003) SplitStream: High-bandwidth multicast in a cooperative environment. *19th ACM Symposium on Operating Systems Principles (SOSP'03)*, pp. 298–313.

14. Jannotti J, Gifford DK, Johnson KL, Kaashoek MF and O'Toole, Jr. JW (2000) Overcast: reliable multicasting with on overlay network. *OSDI'00: Proceedings of the 4th Conference on Symposium on Operating System Design & Implementation*, pp. 14–14. USENIX Association, Berkeley, CA.

15. Xu J, Kumar A and Yu X (2003) On the fundamental tradeoffs between routing table size and network diameter in peer-to-peer networks. *IEEE Journal on Selected Areas in Communications* **22**: 151–63.

16. Boyd S, Ghosh A, Prabhakar B and Shah D (2006 Randomized gossip algorithms. *IEEE/ACM Trans. Netw.* **14**(SI), 2508–30.

17. Ganesh AJ, Kermarrec AM and Massoulié L (2003) Peer-to-peer membership management for gossip-based protocols. *IEEE Trans. Comput.* **52**(2), 139–49.

18. Lin S, Taïani F and Blair GS (2008) Facilitating Gossip programming with the GossipKit framework. *DAIS*, pp. 238–52.

19. Voulgaris S, Gavidia D and van Steen M (2005) CYCLON: Inexpensive membership management for unstructured P2P overlays. *J. Network Syst Manage* **13**(2): 197–217.

20. Van Renesse R, Birman KP and Vogels W (2003) Astrolabe: A robust and scalable technology for distributed system monitoring, management, and data mining. *ACM Trans. Comput. Syst.* **21**(2): 164–206.

21. DeCandia G, Hastorun D, Jampani M, et al. (2007) Dynamo: Amazon's highly available key-value store. *SIGOPS Oper. Syst. Rev.* **41**(6): 205–20.

22. Jelasity M, Montresor A and Babaoglu O (2005) Gossip-based aggregation in large dynamic networks. *ACM Trans. Comput. Syst.* **23**(3): 219–52.

23. Bakhshi R, Gavidia D, Fokkink W and Steen M (2009) An analytical model of information dissemination for a gossip-based protocol. *ICDCN '09: Proceedings of the 10th International Conference on Distributed Computing and Networking*, pp. 230–42. Springer-Verlag, Berlin, Heidelberg.

24. Stavrou A, Rubenstein D and Sahu S (2004) A lightweight, robust P2P system to handle flash crowds. *IEEE Journal on Selected Areas in Communications* **22**(1): 6–17.

25. Baehni S, Eugster PT and Guerraoui R (2004) Data-aware multicast. *Proceedings of the 2004 International Conference on Dependable Systems and Networks (DSN 2004)*, pp. 233–42.

26. Baldoni R, Beraldi R, Quema V, Querzoni L and Tucci-Piergiovanni S (2007) TERA: topic-based event routing for peer-to-peer architectures. *DEBS '07: Proceedings of the 2007 Inaugural International Conference on Distributed Event-Based Systems*, pp. 2–13. ACM, New York, NY.

27. Chockler G, Melamed R, Tock Y and Vitenberg R (2007) SpiderCast: a scalable interest-aware overlay for topic-based pub/sub communication. DEBS '07: *Proceedings of the 2007 Inaugural International Conference on Distributed Event-Based Systems*, pp. 14–25. ACM, New York, NY.

28. Costa P, Migliavacca M, Picco G and Cugola G (2003) Introducing reliability in content-based publish-subscribe through epidemic algorithms. *Proceedings of the 2nd International Workshop on Distributed Event-Based Systems (DEBS'03)*.

29. Eugster PT, Guerraoui R, Handurukande SB, Kouznetsov P and Kermarrec AM (2001) Lightweight probabilistic broadcast. DSN '01: *Proceedings of the 2001 International Conference on Dependable Systems and Networks (formerly: FTCS)*, pp. 443–52. IEEE Computer Society, Washington, DC.

30. Picco GP and Costa G (2005) Semi-probabilistic publish/subscribe. *Proceedings of 25th IEEE International Conference on Distributed Computing Systems (ICDCS 2005)*, pp. 575–85.

31. Voulgaris S, Rivire E, Kermarrec AM and Steen MV (2006) Sub-2-Sub: Self-organizing content-based publish subscribe for dynamic large scale collaborative networks. *IPTPS06: the Fifth International Workshop on Peer-to-Peer Systems*.

32. Eugster PT and Guerraoui R (2002) Probabilistic multicast. *Proceedings of the 2002 International Conference on Dependable Systems and Networks*, pp. 313–24. DSN '02. IEEE Computer Society, Washington, DC, USA.

4

Principles and Patterns

Distributed pub/sub systems are based on a set of design principles and patterns that have been found to work well in specific environments and settings. Architectural and design patterns provide solutions for well-defined problems, and digress the various dimensions of the problem [1].

In this chapter, we examine key principles and patterns for efficient and scalable pub/sub systems. We first examine the general model and high-level principles guiding the development of pub/sub systems. Then, we examine key architectural and design patterns including the central event notifier pattern.

4.1 Introduction

First we should define the key vocabulary to be able to understand the concepts used in the later exposition:

- A thematic principle is a central design tenet which presupposes a certain state or attribute of the system. Starting from the principles we can format rules and norms to support the design. One cannot subdivide principles and thus they are minimal or atomistic parts of the design effort. All rules and norms may be traced back to a principle, but not vice versa, nor are the principles reducible to each other. As such they are the axioms of design.
- An architecture of a system is built by starting from the principles and using the architectural patterns for guidance. An architecture typically consists of a set of components and a set of rules and constraints governing the relations between the components.
- Patterns are generally design features of software engineering with a successful implementation history. Patterns are not confined in any single context of design but they must offer a solution for a well-defined problem. Furthermore, they extend across the various dimensions of the problem.

Publish/Subscribe Systems: Design and Principles, First Edition. Sasu Tarkoma.
© 2012 John Wiley & Sons, Ltd. Published 2012 by John Wiley & Sons, Ltd.

Patterns can be classified according to their level of abstraction:

- Architectural patterns handle the task of creating architectural designs; for example the broker pattern in the CORBA architecture.
- Design patterns work in the medium level and handle language independent object-oriented design strategies.
- Idioms work at the programming-language level where they support generating of acceptable solutions.

We can illuminate the relations between these concepts like this: the principles guide the architectural development, complemented by the architectural patterns which support the development of architectural specifications. The design patterns support the realization of principles in protocol and system design stage.

Patterns can be defined using the various features of a design problem, for example the underlying motivation, specific problem, systemic structure, consequences, implementations, and the eventual users. If we consider all patterns applicable in a particular domain we can form a collection of these. This kind of superset is called a pattern language.

4.2   General Pub/Sub Model

In the general model of pub/sub event notification, subscribers define their interests using topics, channel names, or filters and these constraints are applied on published event messages in order to select a subset of messages for a specific subscriber. This general model builds on several principles that we will examine in this section.

4.2.1   Principles and Characteristics

The event service is a logically centralized component that provides the interfaces for interest expression and notification. The producers or suppliers are responsible for monitoring particular types of phenomena and notifying the event service [2]. Thus this *logically centralized service* interface is a crucial design principle for pub/sub systems. The interface can be implemented within a single component, for example the source of events, or as a generic component managing events for many publishers and subscribers. We will examine these options, namely direct and infrastructure notification, later in this chapter.

Typically, it is desirable to that event service decouples the communicating components and offers anonymity for the communicating entities. Thus decoupling of the communicating endpoints is our second key principle.

Depending on the expressiveness of the event system, a specific field, header or the whole content of the event may be filterable. Interest in an event is realized by invoking the *interest registration* interface provided by the event service. This is our third key principle.

The two important characteristics for an event service are expressiveness and scalability [3]. Expressiveness deals with how well the interests of the subscribers are captured by the notification service, and scalability deals with federation, resources and issues such as how many users can be supported and how many servers are required. Our fourth principle, the *filtering mechanism*, determines the expressiveness of an event system and with the distribution mechanism they determine the scalability of the system.

In a distributed environment, filtering must be implemented in an efficient manner over the routers and brokers. A frequently used principle is *upstream filtering and downstream replication*. This principle states that filtering should be performed as close to the content sources as possible in order to minimize unnecessary delivery of events. In addition, events and messages should be replicated as close to the subscribers as possible thus minimizing the number of messages needed to deliver the content to the subscribers.

In order to cope with different application requirements as well as operating environments, the pub/sub system organization needs to be modular and extensible. This *modularity principle* requires that the pub/sub system allows either compile time or runtime configuration of the different components, such as the routing algorithm, the event format, serialization engine, and so on. In the ideal case, these are components that can replaced at runtime thus allowing the system to adapt to various requirements and operating conditions.

In addition to expressiveness, scalability, and modularity, the event framework needs to have other desirable characteristics as well:

- simplicity in order to be implementable and manageable;
- support for rapid deployment over existing infrastructure;
- exhibit fairness towards publishers so that the available bandwidth in a fair manner divided between the publishers;
- interoperable between to work together with other systems. Interoperability between application defined event types and different event systems is typically desirable.

Other nonfunctional requirements are: timely delivery of notifications (bounded delivery time), support for Quality of Service (QoS), security, high availability and fault-tolerance. Event order is an important nonfunctional requirement and many applications require support for either causal order or total order.

We briefly summarize the key principles:

- P1: Pub/sub is provided by a logically centralized service that offers the necessary API methods and hides the distribution and configuration of internal service components.
- P2: The event service decouples the communicating entities and hides their identities. Hence the service supports anonymity.
- P3: Interest in an event is realized by invoking the interest registration interface provided by the event service.
- P4: The interest registration service allows the specification of data filters. This API hides the internal organization of filtering.
- P5: Distributed filtering follows the upstream filtering and downstream replication model, which aims to minimize unnecessary event forwarding by filtering events close to the publishers.
- P6: The organization of the pub/sub system needs to be modular and extensible in order to cope with various requirements and environments. Thus the pub/sub service can be realized as a basic core with a component-based extension system.

4.2.2   Message Service

A messaging service that supports both asynchronous and synchronous communication is the basic building block of any system that aims to disseminate information. Messaging in general was discussed in Chapter 2. The messaging service needs to provide reliable, efficient, and interoperable one-way and two-way communication and cope with possible disconnections. For example, transport level connections may be lost, because of transmission errors, timeouts, and changes in network-level addressing.

The basic design pattern for a messaging service is the Acceptor-Connector [1], which decouples connection management of cooperating peers from the processing and services provided by the peers. The pattern supports the use of various protocols, which share the same connection establishment and initialization mechanism. Also other design patterns are needed to cope with QoS issues and mobility support.

4.2.3   General Patterns

There are two general patterns for event notification:

- direct notification, and
- infrastructure-based distributed notification.

Generally, direct notification requires that the subscribers have a mechanism to locate objects, and the infrastructure models provide this location mechanism as an internal service. The resolution mechanism is typically a lookup service, such as the Jini [4] solution. Some architectures, however, contain properties from both, for example the CORBA event channel [5], which connects the consumers and suppliers, but each channel object needs to be discovered separately. A separate specification defines how channels are connected to form communication topologies [6]. There is always some property that needs to be located and identified, be it the event service access node, the source object, the source channel, or the event type. Content-based systems route events based on their content and thus an explicit channel or topic is not required.

In infrastructure-based notification an entity called the event service is responsible for accepting registrations and delivering notifications. The general distributed application-level event service consists of a number of servers running the event service server-software. Dotted lines indicate network-level connections. The event service is an overlay network, running on top of the network and transport layers. Each server supports a number of clients. Generally, the model is typically two-fold; the clients employ a client-server protocol to connect to the event servers, publish events and receive notifications. The servers use a server-server protocol to synchronize and distribute information.

4.2.4   Event Notification Patterns

The three well-known patterns for event notification are:

- the observer or publish-register-notify pattern [7];
- the event-channel pattern; and
- the notifier pattern [8].

The publish-register-notify pattern and the observer pattern are examples of direct notification, in which the event source publishes its interface. The interface is defined using an *Interface Definition Language (IDL)*, which is not necessarily the IDL used in CORBA.

This interface includes the events it is capable of notifying. A client invokes the object synchronously and can register for events by indicating parameters or wildcards. The object accepts registrations and notifies the clients that match the registration template. The notification is performed, when the event firing conditions and access restrictions are satisfied. The paradigm supports direct source-to-client event notification. Publish-register-notify is used in the *Cambridge Event Architecture (CEA)* [7, 9], and the observer pattern is used, for example, in the Java event models.

The observer pattern is a very basic pattern, often combined with the broker arrangement, mediator pattern, and proxy usage. It can be used to build complex distributed and infrastructure based pub/sub systems.

The challenge with the observer design pattern is that it does not solve how the objects of interest (the event publishers) are located and the number of remote references wishing for a notification may grow very large in wide-area systems. The pattern does not scale to large numbers of subscribers per object; however it allows the use of a mediator that improves flexibility of the system. The notifier pattern that we examine later in this chapter can be realized by using the observer pattern and mediators or proxies [10].

Infrastructure-based approaches usually solve the problem of locating subscribers by propagating a subscription message within the infrastructure. The infrastructure also handles nonfunctional requirements, such as QoS and disconnected operation, and hides the entities from each other thus supporting transparent group communication.

A common and important feature in the event-channel and notifier patterns is the ability to decouple subscribers and publishers. This is accomplished by introducing an event mediating component, the broker. Both patterns also support some nonfunctional requirements, like QoS and disconnected operation. In most respects the event-channel is similar as the notifier pattern; the difference in functionality results from the ability of the notifier pattern to abstract both the location and distribution of event brokers. In the event channel solution the client has to get the reference of the channel as the first stage.

In the event notifier systems the event service (event dispatcher) is responsible for brokering subscriptions and the event notifications between publishers (producers of events/information) and the subscribers. This notification-type communication model is undirected and utilizes message passing while retaining all benefits of message queues. Typically, in an undirected communication model the publisher does not have to know the identities of the receiving parties. The principle also applies to message-oriented middleware such as JMS (supporting publish/subscribe type of communication).

The ability to decouple publishers and subscribers (producers and consumers) is a fundamental feature. Furthermore in these systems additional tools are often used, such as filtering and pattern detection. Those tools strongly reduce the amount of transmitted information and also improve the accuracy of event notifications.

4.3   Architectural Patterns

Software engineering often relies on well established solutions when encountering architectural problems. These are the architectural patterns. Such a pattern must describe the elements of the problem and the surrounding constraints. Moreover, the pattern will show how elements are used together for reaching a solution to the problem. An architectural pattern is thus the structural schematics for a (sub)system solving the problem. We must remember, though, that a pattern is not an architecture in itself; still the architectural patterns generally will be more complicated and large-scale than mere design patterns.

In the following list we give a short overview of most important architectural patterns. The first three pertain to the interaction model between the communicating components:

- *Client-Server*. In distributed computing we normally encounter the dichotomy where the clients utilize resources and servers provide the services. The client-server pattern is the most universally used of the architectural patterns.
- *Peer-to-peer*. In this pattern which has become very popular, each node (peer) in the network has both client and server roles. In most cases a peer has symmetric functionality but the roles can also be unevenly distributed as is the case in those peer-to-peer systems with so called super peers. The super peers provide supplementary services to other peers.
- *Multitier*. A multitier architecture is basically a client-server system. An application in a multitier system will be executed by several distinct software agents. An example of this is the case where middleware is utilized to service data requests between a user and a database. A three-tier system is perhaps the most widespread class of the multi-tier architectures.

The organization of components and tasks is important and there are several pertinent patterns:

- *Multilayer*. A software architecture may use several layers to allocate the responsibilities among the applications. Layering is very typical for communication architectures. Many distributed pub/sub solutions examined in this book build on top of the basic TCP/IP protocol suite and a middleware overlay network layer.
- *Pipeline (or pipes and filters)*. When the processing elements (processes, threads, coroutines, etc.), have been linked together so that the output of an element is the input of the next, we speak of a pipelined structure. A normal feature in such a setup, where the stream often consists of records, bytes or bits, is usually some amount of buffering provided between consecutive nodes.
- *Broker*, is a component that handles communication, typically requests and responses, and maintains the separation (decoupling) of clients and servers.

Then we have patterns pertaining to the data flow within the environment:

- *Blackboard* system. In this pattern we use a common knowledge base, called the blackboard. A set of specialist knowledge sources updates this blackboard iteratively. The process always starts from a problem specification and ends with a solution to the problem. A partial solution to the problem is provided by every knowledge source and this partial solution updates the blackboard when the blackboard state matches the internal constraints of the knowledge source. This kind of mechanism is particularly useful when the problems are ill-defined, large and complex.

- *Model-View-Control (MVC)* is an exception because it functions both as an architectural pattern and a design pattern, depending on the specific usage. MVC facilitates rapid and easy alterations to the user interface irrespective of the business model in usage. Both the application data and the rules (business rules) used for its manipulation are included in the model where the elements of the user interface generate the view. The MVC controller ensures that the user's control signals are handled properly according to the information stored in the model and given in the view.

- *Inversion of control (IoC)* is a principle-level concept defining behaviour in some software architectures where the normal systemic control flow, such as is seen in traditional software libraries, will be inverted. Typically a programmed flow of control is expressed with a series of instructions or procedure calls. Normally this is accomplished by specifying a sequence for decisions and procedures during the lifetime of a process. In an IoC framework the user instead sets the responses as linked to specific events or data requests. The calling order is controlled by external components which will also manage the additional operations carried out for the execution the desired process.

4.4   Design Patterns

Here we introduce some useful and prevalent patterns sectioned into three categories:

1. structural patterns,
2. behavioural patterns,
3. concurrency patterns.

The first class includes system patterns such as the adapter, decorator, facade, and proxy. They generally provide mechanisms that modify and extend the basic system structure. Behavioural patterns are used to augment the component behaviour of a system. Patterns such as chain of responsibility, mediator, observer, and strategy belong to this category. Concurrency patterns are typically used to manage concurrency in systems, for example singleton and guarded. In the following, we briefly examine these patterns and then consider pub/sub specific patterns in more detail.

4.4.1   Structural Patterns

Adapter converts a class interface into another expected by the clients, enabling cooperation between classes with otherwise incompatible interfaces. Generally this is accomplished by implementing an adapter class that converts the requests between interfaces. It is possible to use either a class adapter or an object adapter implementation.

The decorator pattern dynamically attaches supplementary responsibilities to an object providing a resilient alternative to using subclasses where functionality is to be extended.

Facade defines a common interface for a subsystem group of interfaces, providing a higher-level interface that facilitates the use of the subsystem. A Facade component thus offers a simple higher-level interface for using the subsystem. This generally reduces the communication and dependencies between subsystems. The Facade receives the requests from the clients and translates them into a set of requests, resending these to appropriate objects within the subsystem. The Facade greatly simplifies the usability of the subsystem functionality while not completely hiding the lower-level functionality from the programmer.

The proxy pattern means that a surrogate is created in the system for another object for access control and additional services.

4.4.2 Behavioural Patterns

The chain of responsibility pattern generally enlarges the set of allowed request handler components from one to multiple objects, thereby enabling the decoupling of sender and receiver from one another. Typically the receiving objects are linked into a chain and the request is passed along the chain until a handling object is found.

Mediator is an object that carries the rules how a set of components will interact. Mediator extricates the objects from the need of identifying or referring each other explicitly and thus enables loose coupling or full decoupling. For example, events pass from the objects to the mediator which processes the event and creates notification to a further set of components. When the system changes, only the mediator has to be modified.

Strategy pattern is basically a family of algorithms where the pattern enables the encapsulation and interchangeability of each algorithm, making them independent of the clients that use it.

4.4.3 Concurrency Patterns

Singleton pattern ensures the uniqueness of class instantiation, simultaneously specifying the access point of this instance.

Guarded suspension is a design pattern of concurrent programming. It is used to manage operations where two features are required:

- acquiring a lock, and
- satisfying a precondition

before the execution of the operation.

4.5 Design Patterns for Pub/Sub

Next, we focus on pattern that are especially useful in the design of distributed pub/sub systems. We start with the broker pattern that is useful in decoupling components, and then proceed to the basic observer pattern that is the canonical example of direct notification. After this we present the MVC pattern that is extensively based on the observer pattern. Finally, we consider three interesting patterns, namely rendezvous, handoff for mobility support and then client-initiated connection for supporting push communications. We will later examine example systems following these patterns. The central event notifier pattern is investigated in the next section. It has a fundamental role in the design of distributed pub/sub systems.

4.5.1 Broker

Broker is a system component for effective decoupling of clients and servers (e.g. information subscribers and publishers). The brokered system functions like this:

- Servers register themselves with the broker using a special interface. They offer their services to clients using other interfaces provided by the broker.
- Clients use the broker interfaces to send requests for the server functions.

The broker (event service) is tasked with responsibilities including finding the location of the appropriate server and forwarding the requests. The broker also transmits results and exceptions to the client.

The broker pattern makes it possible for an application to access distributed services easily by sending messages to a specific object. Applications can thus avoid focusing on interprocess communication which would be low-level traffic. Furthermore the broker architecture has the clear advantage of flexibility, because dynamic changing of objects is possible, likewise addition, deletion, and relocation of objects.

One of the main benefits of the broker pattern is the reduced complexity of developing distributed applications. The broker pattern essentially will hide distribution problems from the developer making them transparent. This is achieved because the broker is based on a model where distributed services are strictly hidden and confined within objects. For example, the CORBA architecture uses the broker pattern (Figure 4.1).

4.5.2 Observer

This pattern defines a one-to-many dependency, component-wise, where all dependent objects are notified at the occurrence of a state change in the object under observation.

A commonly encountered problem, in systems with many communicating components, is the need of establishing a consistency between the states of these components. This may be guaranteed if we set additional links between the components for notifications of state changes. Nevertheless, the resulting system would have tight coupling between the components, and this would reduce the maintainability and reusability in the system.

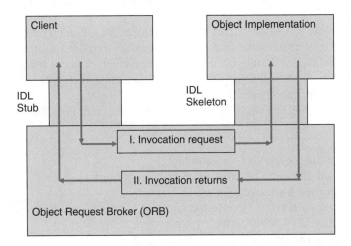

Figure 4.1 Overview of the CORBA broker.

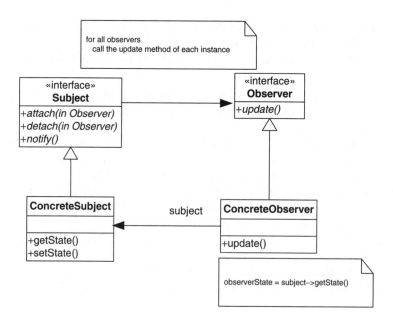

Figure 4.2 The observer pattern.

Thus the outcome is clearly undesirable, quite unlike the observer pattern which supports the component decoupling.

In Figure 4.2 we have given the overview of the observer pattern. The pattern aims to decouple the components and to this effect it uses two different kinds of objects, namely the Observer and the Subject. The subject operates in the system by setting up an interface where an unlimited number of observers can subscribe (and unsubscribe) from the subject. This interface is also used for notifying the subscribers (currently subscribing components) about any state change happening in the subject. The logic therefore makes the subject an active part of system, promoting and favouring itself as the target of observations.

The willing observers on their part must implement the Observer interface in order to receive notifications about state changes. The interface provides a method used by the subject to update available state information. When changes occur the interface is used to send a message to all registered Observer components, which will request the current state from the Subject component.

In this kind of mechanism there are both benefits and disadvantages. The main benefits are:

- The subject and the observers are reusable.
- The subject and observers are abstracted from each other.
- The interfaces used for interaction are abstracted.
- The subject has no need to identify the concrete classes of the observers.
- Group communication is supported by the system.
- Observers are unknown to the subject beforehand.

There are also some liabilities, listed below:

- A single state change in the Subject can start an unexpected sequence of updates, because the observers have no primary knowledge of the other observers nor of their dependencies.
- The system does not necessarily postulate that the information about the changed state of the Subject is sent with the notification to the observers. Only the fact that a change has occurred is always delivered. Then, in some cases, the observers have to send a request to the subject for obtaining the changed information.
- Because the observer pattern allows direct registering of the subscribers with a producer, the system essentially couples system components together.
- The pattern does not describe how the producers are located.

The observer pattern is difficult to scale when there are large numbers of subscribers for one object. It is possible, however, to use a mediator to improve the flexibility of the pattern. Common uses for the observer pattern include the Java and Jini event models [4]. There is a most similar pattern called the publish-register-notify, used in the Cambridge Event Architecture [7, 9].

4.5.3   Model-View-Control (MVC)

MVC is a pattern often used in applications where interaction is important, because it facilitates the changing of the user interface. To this end the pattern casts the application into three parts:

- the controllers which handle user input;
- the model which provides the core functionality;
- the views which display the information to the user.

The MVC pattern ensures the creation of the user interface by the view and also the consistency of the model with the controller. Figure 4.3 presents an overview of the pattern.

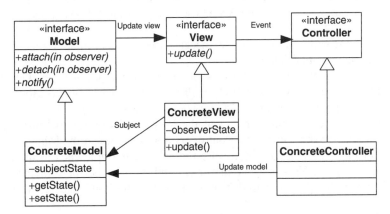

Figure 4.3   The Model-View-Control pattern.

Web applications frequently use architectural patterns and also MVC belongs to this group of patterns. Then the view is the HTML content and the code creating the HTML content has the role of the controller. All data sources governing the generation of the content, such as databases, XML resources and the business rules are representations of the MVC model. The pattern is used because it provides the essential decoupling of models and views. This augments to reduced complexity and improves the flexibility.

Whenever the User Interface (UI) needs to be flexible or the UI changes often, the MVC pattern is convenient for executable applications. For instance this is the case if

- a system has multiple interfaces implemented; or
- the implemented functionality forces corresponding changes in the UI.

In application interfaces which are tightly coupled with the core functionality, all changes in the UI are often difficult. Furthermore, the mere process of UI changing may cause programming errors. Additionally, a modified UI may have some effects in other system components.

All key functions, such as data structures, of the application are embedded in the Model component. Interfaces for invoking application-specific services and accessing the component data are also embedded in the Model.

A set of components of the View are needed to retrieve the Model data for displaying it to the user. Every View has its own distinct Controller component which processes user input issuing appropriate service requests to the Model. In certain applications the Controller receives from a View the functionality to manipulate the display directly.

One of the mechanisms specified in the MVC is the change-propagation, where the Views and Controllers registering with the Model will receive notifications about structural changes. Model changes then induce notifications for the registered Views and Controllers.

It is also possible to implement auxiliary Views and Controllers without modifications in the Model. Because these Views and Controllers can be implemented as pluggable components they can be dynamically added and removed. Furthermore, because the change-propagation mechanism is used, the Model is always synchronized with the changeable Views and Controllers. This augments the portability of MVC pattern.

There are also drawbacks when MVC is used, such as the following:

- The applications will be more complex because there is a set of additional components in use with their mutual interactions.
- The Model may generate a number of superfluous change notifications. Generally, the views are interested only in a subset of all possible changes of the Model. Filtering techniques can be used, though, to minimize the redundancy.
- The Model involves a tight coupling of the Views and Controllers, possibly causing the need to change the Views and Controllers after changes in the Model.
- Because displaying the data might cause the need of multiple calls to the Model, the data access from the View may be inefficient and create under performance.
- To achieve portability for the applications, changes in the Views and Controllers might be needed.

4.5.4 *Rendezvous Point*

We describe a frequently used pattern for structuring a distributed pub/sub system and optimize its behaviour. A rendezvous point is a designated node in the distributed system responsible for certain event type or subspace of the content space in which events are published. Rendezvous has been applied in many proposed overlay and pub/sub systems [11–13] to reduce communication costs.

There are many ways to determine a rendezvous point for a given event. Typically the point is obtained by hashing the event type or content into a flat label. The node that has the numerically closest identifier then becomes the rendezvous point for the content. It is straightforward to connect the publishers and subscribers of content through the rendezvous point. The point may or may not participate in the actual forwarding of the content. In some cases, the point is only used to coordinate the propagation of routing state.

The rendezvous point may become a bottleneck if most of the traffic goes through it. The point may also become a target for attacks and thus replication of the point is needed to ensure system availability. If the rendezvous points are chosen according to the uniform distribution and the content popularity is not uniform there may be a load mismatch between different points in the network.

Scribe is an example of an application layer multicast system that is based on rendezvous points. These points are the roots of a multicast three in Scribe. Content flows on the reverse path from the root towards subscribers based on the overlay routing tables. Hermes is an example of a content-based pub/sub system that uses rendezvous points in a similar fashion to Scribe [14, 15].

Next we present a useful pattern based on rendezvous for managing mobile clients. We will return to these patterns later in the book.

4.5.5 *Handoff with Rendezvous*

A handoff is related to state transfer between different brokers. Many types of handoffs (or handovers) have been specified in the mobile computing context where they are important because they enable essentially seamless connectivity in wireless and mobile communications systems.

Here we give an overview of an abstract protocol enabling state transfers. Mobile 2G and 3G standards use patterns like this to facilitate roaming between access points and access networks; furthermore handoff is also defined in higher level mobility protocols. Examples of these are Mobile IP, SIP, and still other systems.

The state transfer or handoff pattern has the following main objects:

- A client,
- Two access points (AP),
- A rendezvous (indirection) point.

In this pattern the client roams (relocates) from one access server to another. Thus there emerges need to to transfer the state from the old access server to a new server. A state transfer necessitates signalling between the access servers.

We could effect the state transfer without any indirection point. Unfortunately, this does not fully guarantee the reachability of the client. To avoid this problem, an indirection point may be used. This keeps track of the current location of mobile nodes. There are both fixed and nonfixed indirection points and there also may be several points for indirection.

In the handoff pattern, the client must establish connectivity with the old access point before the state transfer is started and also attach itself to a new AP. The new AP executes a location update for the client enabling all nodes that need to communicate with the mobile node to lookup the new AP address. Furthermore the new AP is able to contact the old AP if it wants to transfer states related to connections made by the mobile node. Afterwards the connectivity with the old AP is not needed and it can be discontinued.

There might be ongoing conversations and in order to maintain these the state transfer pattern defines update messages from the mobile client to correspondent nodes. The messages contain the active and valid addresses information for the client.

It is possible to use different mechanisms to implement the state transfer pattern. In a typical scenario state transfer pattern may be started by a mobile client which detects a new AP. In another case the state transfer pattern is generated by the network when a mobile client has moved to a location assigned to the new AP.

There are several main benefits accruing from the implementation of the handoff pattern:

- It will facilitate mobility between different regions, when they are assigned to different access servers.
- It enables to load balance clients from one access server to another.
- The state will be transferred from one access server to another.
- The rendezvous point is always updated.

The handoff protocol is favoured by several systems, for example, Mobile IP (the networking layer) and SIP (the application layer). Many of the pub/sub mobility solutions that will be presented in Chapter 11 are based on this pattern as well.

4.5.6 Client-Initiated Connection

The client-initiated connection is a very simple pattern, but it is very significant for engineering systems that support push communications. In many cases the clients of a server are behind private addressing domains, for example, behind NAT devices. Thus the server cannot contact the clients unless they have contacted the server first. One way to overcome this limitation is to have the clients establish long-lived connections to the server, and have the server use these connection in pushing data to the clients. This is the core of the pattern, in which such a long-lived connection is established and maintained. The server needs to have resources for maintaining the connections. The state requirement and management is a liability of this pattern.

The client-initiated connection pattern is applied extensively in many Web-based pub/sub and push systems, such as AJAX Web applications, Facebook Messenger, push e-mail, etc. It is also applied in the mobile context on smartphones.

4.5.7   Other Patterns

There are many pub/sub specific patterns and solutions not covered in this chapter. We examine specific solutions later in the book in Chapters 7 and 11. Here we briefly mention patterns and solutions that will be examined later. Some of these are fundamental for distributed pub/sub solutions, such as reverse path routing, separation of routing and forwarding, reconfiguration support, and scoping.

- Reverse path routing is a frequently employed pattern for implementing multicast tree formation. Pub/sub solutions implement the pattern by recording the source broker of a subscription message and then sending matching notifications in the reverse direction. Therefore the subscription messages build the content delivery tree or graph. This system is applicable to advertisement messages as well. The liability of the pattern is the state requirement at the brokers and that this solution does not work well if the underlying topology changes frequently. There are solutions, such as DHT overlays, that can be used to mitigate the impact of network topology changes. Well known systems include SIENA and Hermes.
- Separation of routing and forwarding is a frequently employed pattern that separates the pub/sub network maintenance from the actual forwarding of events and messages. This separation of concerns allows the implementation of efficient routing and forwarding algorithms.
- Reconfiguration of the middleware is required to adapt the pub/sub system to meet various application requirements and to work in different environments, such as the wired Internet or the wireless domain. The reconfiguration requires that there is a component that monitors the network and the operating context, and then dynamically tunes and configures the system. Well known systems include REDS and GREEN.
- Reconfiguration of the topology is needed to optimize content delivery in the pub/sub network. Typically this means that inefficient links are replaced by more efficient links. There are many ways to do this configuration, for example by utilizing pub/sub API operations, or custom messages that update broker routing tables. Well-known systems include REDS, GREEN, and Rebeca.
- The *subscriber-requested publisher-offered* pattern supports QoS negotiation between publishers and subscribers. This patterns starts with the publisher offering a certain QoS contract that is then matched against the QoS specification requested by the subscriber. The event service is responsible for performing the matching. This pattern is used by the DDS system.
- Leases are a frequently used technique for supporting eventual consistency in the system. When a subscription lease expires, it is removed from the routing tables and the distributed system. Thus explicit unsubscriptions are not needed. We discuss leases in the context of system stability and soft state in Chapter 7.
- Scoping is a useful pattern and strategy for introducing structure into a pub/sub network. Scopes support the creation of visibility domains in the system. Thus scopes enable the creation and maintenance of organizational boundaries and structures. Scoping can be done, for example, based on department, group, unit, and company. The pub/sub system then enforces that only brokers and clients belonging to a publication's scope can see and process it. This pattern is used in the Rebeca pub/sub system.

- Quench introduced in the Elvin system is a pattern that allows publishers to query the pub/sub system whether or not specific events have subscribers. Thus the publisher may refrain from publishing if there is no interest in the events it plans to publish. The publisher can also subscribe a notification when subscribers for particular content become available.

4.6   Event Notifier Pattern

The following example motivates the need and usage of event notifier in a networked system. A typical network consists of distributed components: computers, routers and software. Although the system is distributed we must watch and manage the separate components centrally. In a large and complex system these managed components often cause problems which are infrequent – they have a stochastic time distribution but they cannot be predicted. Obviously we want to know when a disturbance or failure happens but just as evidently it is inefficient to poll all component continuously. Accordingly the system should notify us when anything critical happens but only then.

We could have a fixed system of notification with an event line from a component to the central monitor but this is very rigid – it cannot be modified easily and it is essentially unscalable. Furthermore this kind of system would be very error prone. Thus we want a scalable system which is adaptable to the system component set – or in fact unaware of it.

4.6.1   Overview

The observer pattern is a good starting point for building pub/sub systems. Changes in the observer components may happen without referencing the rest of the system. The centralization that always decreases the freedom of various components does not bind observer system unlike the mediator system. However, the computational burden of target components is increased as they have to maintain the observer list and use it for notifying events – in fact sending the same event message many times.

Nevertheless, a troubling problem remains: the interested parties must have information about the managed objects in order to choose a component and register themselves. In a large and growing network the number and identification of target components are typically not a very well defined. Thus this system still requires too much a priori knowledge. We should have a mechanism without any inter-object knowledge, a priori or otherwise.

An efficient system would combine the benefits of both the mediator and observer approaches and avoid their shortcomings. This is the event notifier system [16]. The approach is based on the mediator which functions as a central event server receiving event information from the publishers, which send the notification only to the event server and do not maintain an observer list. The observer list resides in the mediator (event server) node and this node has the responsibility to notify the interested parties (the subscribers). Thus, the registration of interest remains free and dynamical but does not burden the target nodes, saving a large amount of computation and network usage.

This system removes all necessity of interobject knowledge, either of publishers about subscribers or vice versa. The subscribers only will know about the events but they do not need know anything about the publishers of these events. This enables a subscriber to get all events of a certain type (say a malfunction in a hardware system) thus establishing a useful level of abstraction between the sources of events and the information users.

Thus the central features of the event notifier system are the following:

- A node can produce an event without any need to know who are the subscribers.
- A node can subscribe to an event (or event type) without any knowledge of possible publishers of this kind of event.
- The event set can be dynamically enlarged to include new events or event types.
- There can exist a multitude of possible publishers of an event type.
- The event stream can be filtered according to arbitrary criteria.
- The subscribers are able to define the classification of interesting events as broad or as narrow they want.
- The subscribers can draw a list of the events critical to them and subscribe to all.
- The notification system is wholly dynamical and the sets of publishers and subscribers are independent.

Thus subscribers and publishers are independent of each other and furthermore, they do not have any knowledge of each other per se. All interaction works using a set of allowed event types, the semantics an event, and the specific event data in relation to an event type. Event notifier systems do not require that subscribers would know in advance the possible publishers of the specified event type; there can be multiple publishers (and of course also subscribers) for a given kind of event. Subscribers and publishers are fundamentally transient concepts as new subscribers and publishers may appear/disappear freely. This has no impact on other components of the system. The relocation of services in a distributed and mobile environment is thus supported.

4.6.2   Structure

The event notifier system has following types of objects [16]:

- Event: A specific data producing occurrence, and also the descriptions of this event
- Publisher: An object which emits or produces events.
- Subscriber interface: An interface for registration of objects that need to know about events.
- Subscriber: An object which registers itself at event service using the Subscriber interface.
- Event Service: An object which mediates events between subscriber and publisher.
- Filter: Code which handles the rejection/acceptance of events.

The participating objects interact typically with each other this way:

- The subscriber starts by invoking the subscription (registration) on the event service. The subscriber specifies the relevant event type(s) and filter for the event information. It gives also a reference to the receiver of notice (mainly itself).
- When a publisher has an event, it notifies the event service and releases an event object.

- The event service uses the subscription data to find the set of interested subscribers. It will also apply the provided filters to get the accept/reject information for each subscriber.
- If a specific filter returns TRUE for the event/subscriber pair, the event service passes the information to the subscriber.

The event service handles all operations: publication and subscription, maintaining the full set of information needed about the subscribers' interest in a specific event type. Thus, the publishers and subscribers are not aware of each other. The subscription interface is defined in the system so that anyone may publish events or subscribe to events using the interface. No auxiliary communication logic is needed. To access the event service a well-known point of access must be known so that subscribers and publishers can use the same event service. A common access point can be provided for instance by registering the event server with a name service.

Filtering is an important part of event processing allowing the subscriber to fine tune the types of events of interest. Actually the specification of subscribed events proceeds in two stages:

1. Event type specification, that is, the subscriber selects a known class of events.
2. Setting up a filter to further narrow the choice of events. This requires giving values or ranges to a set of event attributes. These values are used to restrict certain events (for instance prohibiting events from a group of sources).

Figure 4.4 presents an UML diagram of the event notifier pattern. The diagram allows subscribers to register filters that are evaluated by the EventService class.

Although subscriptions in event notifier system are normally based on the type of events, not the source, a subscriber might also be interested only for events produced by a subset of publishers (for example quality news sources) and this can be taken into account by defining an attribute identifying the acceptable source(s) or their qualities. Then the event server will discard events from uninteresting sources.

The types of events are handled as an inherited hierarchy objects and this facilitates a further characteristic: the subscriber interested in a particular event type is notified when any of the event hierarchical subtypes occur. Using the subtypes enables the targeting of subscription interest as broadly or narrowly as necessary. The time dimension is also feasible because the subscription process is fully dynamic and any profile can be changed at any time.

New event types may be added to the system dynamically without any disturbance. If a subscriber has already registered for the supertype of the new event, he will receive the new subtype events without any further subscription operations and without any need to rebuild/attune the used application even partially. No change in the event server codes is required.

There are weak points certainly. The single event server might be a point of failure or form a bottleneck for the whole event notifier system. These problems of centralization can be mitigated by distributing the event server function in the system. Typically an event notifier will use the push semantics. In this model the publisher pushes events to the event service, and similarly the event service will push events to the subscriber.

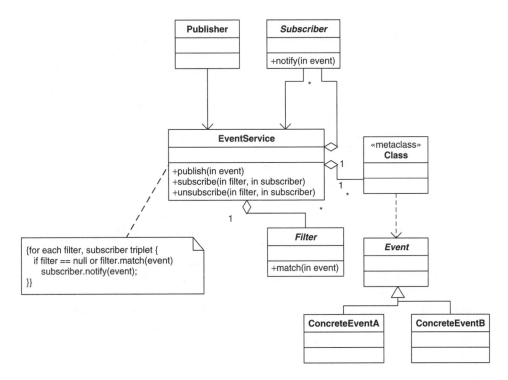

Figure 4.4   The notifier pattern.

It is still possible to use either partial push model where the subscriber pulls information from the event service. Also the event service may pull the information from a passive publisher.

As to the communication safety we must remember that event notifier keeps the components separate and decoupled from each other. The interfaces used are generic to the event types and therefore easily changeable by nature. In the rigid distributed systems changes typically entail rewriting/recompilation which traverses through the system; event notifier minimizes the alterations and therefore supports scalability. The cost is a certain amount of lost safety (attached to the per-node type-checking) but the flexibility gains normally compensate for this and there are ways to create more static safety at event checking.

4.6.3   Distributed Event Notifier

In a large distributed system the event notifier receives additional benefits from the use of local proxies for a remote event service. A local server can function as a proxy maintaining a list of local subscriptions. The proxy will mediate the local subscriptions to the remote event service. It is not necessary, however, for the local proxy to maintain a list of subscribers; this can be delegated, as all other operations, to the remote event service. On the other hand it is possible to store all local information in a local event server proxy and only the identities of proxies in the remote event service; the proxies then appear as subscribers.

An event proxy is essentially an insulator between publishers and complexities of a distributed network. Many benefits ensue from this role [16]. An event proxy handles congestion resulting from inaccessibility of the primary event server; in case of network failures it can store the event stream for later delivery. Also the proxy functions as a preserver of publishing order making sure that events are delivered to subscribers in the right sequence. A local proxy also provides additional level of information hiding because there is no need for publishers or subscribers to know the location of primary event server. Furthermore, subscribers can reside inside subnets which are not accessible from the primary server. Proxies may also function as nodes for additional security and data storage quite transparently to subscribers or publishers. Network traffic is effectively reduced when the proxy subscribes events for multiple local customers.

Furthermore, in practice an event is often important only in the local environment while some events may be of universal interest. These functions – local publishing and universal publishing – can be separated for such events that form a clearcut dichotomy. In the local case the proxy will function as a standalone event server without reference to the primary server. It is possible to define an attribute of strict locality in the event type hierarchy. When a tagged local event is subscribed the proxy server handles its publication alone without contacting the remote event service.

Figure 4.5 presents a distributed version of the event notifier pattern. The DistributedEventService is realized through EventProxy classes that mediate events between distributed proxies. The proxies extend the subscription and notification interface by allowing proxies to subscribe events and forward notifications to other proxies.

4.6.4 Design Considerations

There are several important design issues to note when one builds an event notifier system. The following list includes important issues:

- Event models,
- Type-checking,
- Advertising the event set,
- Performance and fault-tolerance.

We discuss here briefly these topics giving attention to the implementation techniques.

4.6.4.1 Event Models

Events in a networked system occur in the form of event objects. There are two main issues here:

- Event messages are generic but they should convey some form of standard information.
- All event messages should be hierarchical subclasses of an Event class, regardless of the vast differences in the original nature of real world events.

The common event superclass can therefore force certain useful attributes to the whole set of events. For instance the timestamp of the original event is such a characteristic

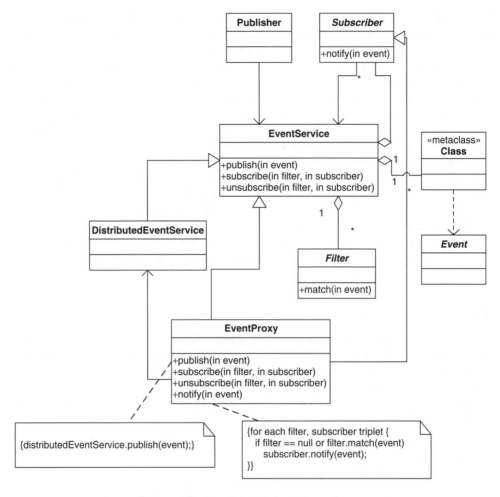

Figure 4.5   The distributed notifier pattern.

without which the system would be quite inefficient. Such an event hierarchy also enables type-based subscription: subscribers can specify a nonleaf node in the event tree to be informed of all events of that type or below.

When using an object-oriented language as customary to code system objects and applications, the inbuilt inheritance can be used to advantage. The inheritance mechanism allows the subscriber to use the full tree format to formulate his information needs and this is a very strong tool. The event type tree node at which the subscriber uses to register itself generates the subtree for subscriber's interest. Also the attributes which are part of the filter can be packaged within the event object. A certain amount of error checking (names check) also may be associated with the inheritance mechanism.

The hierarchy model is not necessarily unique, because the events are manifold and there are lots of possible ways to arrange them into hierarchies. Basically one should choose a model which is best applied to the specific world of events encountered in the real-world mechanism. In a technical event system, for example, the subscriber might be interested in fault events (hardware or software) and/or in performance events or both. Different attributes are then important forming different hierarchies.

We must note that a characteristic of an event can often be modelled either on the event type level or at the attribute (filtering) level. These levels differ on the subscription abilities: one can subscribe based on the type/subtype, but the filter is used only after the event type is accepted for a certain subscriber. Still both mechanisms, namely type and filter, overlap and one can get the same event stream at subscriber level using either one.

As mentioned earlier sometimes it is useful to filter the events using the source identification (for instance for quality/reliability reasons). In a simplistic example a triggered fire alarm is an event but it is important to know which fire alarm was triggered. The action often depends on the identification of the source. Furthermore, knowing the source usually reveals us immediately a set of event attributes related to the source which otherwise would have to be coded into the filter. Also the an event source handle given as an event attribute allows a subscriber to acquire more information directly from the event source. Of course this means that the decoupling between publishers and subscribers is not complete.

4.6.4.2 Type-Checking

We have to do a tradeoff between efficiency and security when we use the generic event features in the event notifier. This concerns especially the implementation of the event type information. Say, the subscriber receives an event of certain type. First it needs to validate the type to prevent errors (especially if it is subscribing many event types). Then it will need to downcast the type (which might be a supertype of the event type actually wanted by the subscriber) into the target event type.

4.6.4.3 Advertising the Event Set

Often it is useful or even mandatory to know which event types are available in the system. This can be done by querying the Event server. For example, in the subscription phase the subscriber normally does not have the most up-to-date information of possible events or even might not have complete view of how to optimize its awareness of all critical information conveyed by the event system. To fully optimize its interest profile, the subscriber then has to know the available event types. Event notifier can handle this with a modification in the publishing mechanism, which allows the publishers to tell the event server – using a specific interface – which events they want to be advertised for the subscribers. Furthermore the event server will provide an interface for the subscribers to query of the set of advertised events. This knowledge can then be used to subscribe to the suitable events.

4.6.4.4 Performance and Fault-Tolerance

A large event notifier system relying on a single event server can fail or become a service bottleneck. We should somehow be able to increase the fault-tolerance of the system. One way is multiplicity. Use of proxies as above for local traffic or a true system of multiple event servers are certainly helpful. In the case of multiple servers it is rational to allocate the work to servers not based on locality but based on event types: one server brokering for news, one for financial services, one for system failures etc. This has the advantage that failures do not propagate across systemic event type boundaries because the servers only know and handle their own set of types.

We could also use a weakened version of the notifier network where the subscription information (list) is distributed among the subscribers. It is true that above we discarded this approach for inefficiency but in fact it is an operational solution if we take into account the advertising of event types by the publishers. When the centralized event service knows about the association between a publisher and an advertised set of event types the server can admit subscriptions, passing the subscriber information to all publishers which have advertised the wanted event. The event server maintains this association information. The publishers will use the list of subscribers to their events and inform them directly.

This type of distributed notifier service which does not channel the event stream through a single node, diminishes the impact which failures would have on the system. There are no evident bottlenecks in the system because the publication rate (from many publishers to many subscribers) is vastly greater than the centralize subscription rate (from many subscribers to a unique server). Also the publication list does not change as frequently as the subscription list and thus we can easily use multiple backup event servers making the system more failure proof.

We will investigate different techniques for distributed event notification in Chapter 7.

4.6.4.5 Summary

In brief the event notifier system can be summarized as follows:

- Knowledge of publishers and subscribers is localized in the event service.
- Subscribers handle events independently.
- Message flexibility is based on different event types and their attributes.
- Subscribers register their interest dynamically at the event server.
- Subscribers can specify filters for fine-grained message selection.
- Publishers and subscribers are decoupled by the event server. The event server can provide anonymity to the communicating components.
- There can be many publishers and many subscribers so the basic setup is many-to-many.
- The event service can consists of multiple event servers.

4.7 Enterprise Integration Patterns

Enterprise integration patterns document commonly used solutions for integrating various enterprise systems [17]. These patterns deal mainly with asynchronous messaging that has proven to be an excellent basis for enterprise integration due to its loosely coupled nature. Enterprise integration patterns are used in message-oriented middleware solutions and Web services. The aim of the patterns is to help the design and implementation of various enterprise systems as well as address well known pitfalls and challenges. Asynchronous operation is different from the traditional synchronous way of creating applications and it introduces new considerations that need to be addressed.

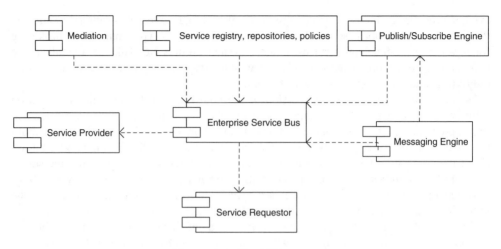

Figure 4.6   The ESB pattern.

The key messaging based architectures include:

- Web services specifications. Key specifications include SOAP, WS-Addressing for routing and forwarding messages in a distributed environment, and other WS-* specifications.
- Service-oriented architectures that allow the creation of loosely-coupled systems composed of individual components and services.
- ESB (Enterprise Service Bus) illustrated in Figure 4.6 provides basic messaging, routing, transformation. ESB is one core component of SOA.

The well-known Enterprise Integration Patterns book organized patterns into the following categories:

- *Integration Styles* pertains to different ways applications can be integrated. In this context, these refer to more historical examples of integration technologies.
- *Channel Patterns* pertain to messaging systems and thus they are a fundamental component of loosely coupled enterprise integration and messaging systems.
- *Message Construction Patterns* are about the intent, syntax, and content of messages in a messaging system. The key pattern is the Message pattern.
- *Routing Patterns* examine routing and forwarding mechanisms. Routing is the process that takes a message closer to its ultimate destination. The key pattern is the Message Router pattern.
- *Transformation Patterns* modify the information content of a message. Transformations are needed for integration, because in many cases the message format and content need to adapted from one system to another. In this category, the Message Translator is the key pattern.
- *Endpoint Patterns* define the behavior of messaging system clients. They are about the different ways that a message recipient can be specified.

- *System Management Patterns* are about the interfaces and feature needed to manage a message-based system. The different routing and transformation subsystems need to be configured and managed.

4.8 Summary

In this chapter, we have examined the key principles and patterns for designing pub/sub systems. The principles included a logically centralized service for decoupling the components, an interest registration service for accepting subscriber interests, a filtering mechanism for selective information dissemination, and a modular event service core. They important characteristics included expressiveness and scalability that contrast each other. Additional important characteristics are simplicity and interoperability.

There are two general patterns for event notification, namely direct notification and infrastructure-based distributed notification. We examined the key patterns under these two categories. The observer pattern sets a one-to-many dependency between objects where dependent objects are automatically notified/updated whenever the observed object changes state. For example, the MVC pattern employs the observer pattern to notify the Views and Controllers when changes happen in the Model. The event notifier pattern combines the observer and mediator patterns into a logically centralized service that fully decoupled the subscribers and publishers and that is suitable for distributed environments.

The principles and patterns presented in this chapter are the foundation for the systems examined in subsequent chapters. We elaborate specific optimization patterns and strategies in Chapter 7 when presenting solutions for distributed pub/sub.

References

1. Schmidt D, Stal M, Rohnert H and Buschmann F (2000) *Pattern-Oriented Software Architecture. Vol. 2: Patterns for Concurrent and Networked Objects*. John Wiley & Sons, Inc. New York.
2. Eugster PT, Felber PA, Guerraoui R and Kermarrec AM (2003) The many faces of publish/subscribe. *ACM Comput. Surv*. **35**(2), 114–31.
3. Carzaniga A, Rosenblum DS and Wolf AL (1999) Interfaces and algorithms for a wide-area event notification service. Technical Report CU-CS-888-99, Department of Computer Science, University of Colorado. Revised May 2000.
4. Roberts S and Byous J (2001) *Distributed Events in Jini Technology*. Sun Microsystems. Available at: http://java.sun.com/developer/technicalArticles/jini/JiniEvents/.
5. Object Computing, Inc. (2001) *CORBA Event Service Specification v.1.1*. OCI.
6. Object Computing, Inc. (2001) *Management of Event Domains Specification*. http://www.omg.org/cgi-bin/doc?formal/2001-06-03.
7. Bacon J *et al*. (2000) Generic support for distributed applications. *IEEE Computer* **33**(3), 68–76.
8. Gupta S, Hartkopf J and Ramaswamy S (1998) Event notifier, a pattern for event notification. *Java Report* **3**(7), 19–36.
9. Hayton R, Bacon J, Bates J and Moody K (1996) Using events to build large scale distributed applications. *Proceedings of the 7th ACM SIGOPS European Workshop on Systems Support for Worldwide Applications*, pp. 9–16.
10. Yu H, Estrin D and Govindan R (1999) A hierarchical proxy architecture for internet-scale event services. *Proceedings of 8th International Workshop on Enabling Technologies: Infrastructure for Collaborative Enterprises (WETICE '99)*, pp. 78–83, Palo Alto, CA.

11. Rowstron AIT and Druschel P (2001) Pastry: scalable, decentralized object location, and routing for large-scale peer-to-peer systems. *Middleware 2001: Proceedings of the IFIP/ACMInternational Conference on Distributed Systems Platforms Heidelberg*, pp. 329–50. Springer-Verlag, London.

12. Stoica I, Morris R, Karger D, Kaashoek F and Balakrishnan H (2001) Chord: A scalable peer-to-peer lookup service for internet applications. *Computer Communication Review* **31**(4), 149–60.

13. Zhao BY, Kubiatowicz JD and Joseph AD (2002) Tapestry: a fault-tolerant wide-area application infrastructure. *SIGCOMM Comput. Commun. Rev.* **32**(1), 81.

14. Pietzuch PR (2004) *Hermes: A Scalable Event-Based Middleware*. PhD thesis Computer Laboratory, Queens' College, University of Cambridge.

15. Pietzuch PR and Bacon J (2002) Hermes: A distributed event-based middleware architecture. *ICDCS Workshops*, pp. 611–18.

16. Riehle D (1996) The event notification pattern - integrating implicit invocation with objectorientation. *TAPOS* **2**(1), 43–52.

17. Hohpe G and Woolf B (2003) *Enterprise Integration Patterns: Designing, Building, and Deploying Messaging Solutions*. Addison-Wesley Longman Publishing Co., Inc., Boston, MA.

5

Standards and Products

In this chapter, we examine well known message oriented middleware and pub/sub standards and products. We start with the CORBA event service and notification service, *Data Distribution Service (DDS)*, and the SIP event framework and then proceed through Java related technologies, such as the *Java Message Service (JMS)* to message-oriented middleware products. Finally, we summarize the key similarities and differences of the examined solutions.

5.1 CORBA Event Service

The CORBA Event Service specification defines a communication model that allows an object to accept registrations and send events to a number of receiver objects [1]. The Event Service supplements the standard CORBA client-server communication model and is part of the CORBAServices that provide system level services for object-based systems. In the client-server model the client makes a synchronous IDL operation on a specified object at the server. The event communication is unidirectional (using CORBA one-way operations). The Event Service extends the basic call model by providing support for a communication model where client applications can send messages to arbitrary objects in other applications. The Event Service addresses the limitations of synchronous and asynchronous invocation in CORBA.

The specification defines the concept of events in CORBA: an event is created by the event supplier and is transferred to all relevant event consumers. The set of suppliers is decoupled from the set of consumers, and the supplier has no knowledge of the number or identity of the consumers. The consumers have no knowledge of which supplier generated the event. The Event Service defines a new element, the event channel, which asynchronously transfers events between suppliers and consumers. Suppliers and consumers connect to the event channel using the interfaces supported by the channel. An event is a successful completion of a sequence of operation calls made on consumers, suppliers, and the event channel.

Publish/Subscribe Systems: Design and Principles, First Edition. Sasu Tarkoma.
© 2012 John Wiley & Sons, Ltd. Published 2012 by John Wiley & Sons, Ltd.

The event channel performs the following functions:

- It allows consumers to register interest in events and stores the registration information.
- It accepts events generated by suppliers.
- It forwards events from suppliers to registered consumers.

The Event Service is defined to operate above the ORB architecture: the suppliers, the consumers, and the event channel may be implemented as ORB applications and events are defined using standard IDL invocations.

The CORBA Event Service supports different implementations of the Event Channel, and this allows a wide range of approaches for implementing Quality of Service and delivery issues. Moreover, the event consumer and supplier interfaces support disconnection. The CORBA Event Service addresses some of the problems of the standard CORBA synchronous method invocations by decoupling the interfaces and providing a mediator for asynchronous communication between consumers and suppliers. The supplier does not have to wait for the event to be delivered to the consumer. Moreover, the event channel hides the number and identity of the consumers from suppliers using the proxy objects (transparent group communication). The supplier sends events to its proxy consumer, and the consumer receives events from its proxy supplier. However, the specification does not address several important issues, such as Quality of Service support. Applications may have requirements for event notification in terms of reliability, ordering, priority, and timeliness. Furthermore, the specification does not provide a system for event filtering. Event filtering needs to be implemented using a proprietary system within the event channel by adding a mechanism for selective event delivery. Event channels can be composed, because they use the same consumer/supplier interfaces. An event channel can push an event to another event channel. Typed event channels can be used to filter events based on event type.

In addition, the specification does not address compound events, but suggests that complex events may be handled by creating a notification tree and checking event predicates at each node of the tree. The drawback of the tree is that the number of hops needed to deliver an event increases. This motivates the use of a centralized filtering service. The use of proprietary event service implementations restricts the interoperability of applications. Applications that use one proprietary event service implementation may not interoperate with another application that is based on a different event service implementation.

5.2 CORBA Notification Service and Channel Management

The CORBA Notification Service [2] extends the functionality and interfaces of the older Event Service [1] specification. The Event Service specification defines the event channel object that provides interfaces for interest registration and event notification. One of the most significant additions to the Notification Service is event filtering. Filters allow consumers to receive particular events that match certain constraint expressions. Filtering reduces the number of events sent to the consumers and improves the scalability of the event handling system.

Figure 5.1 presents the components of the CORBA Notification Service, which derive from the Event Service discussed in the previous section. The event channel has been

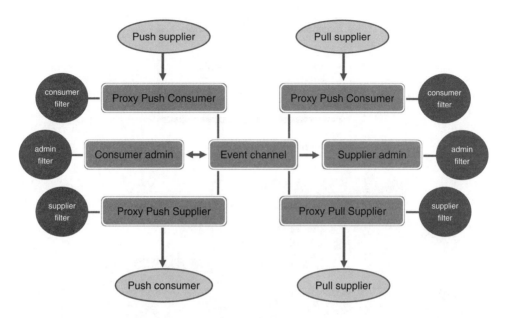

Figure 5.1   CORBA Notification Service model.

extended to support a number of admin objects. The Notification Service allows the definition of filters at the proxies. Moreover, each admin object is seen as the manager of the set of proxies it has created. Admin objects may be associated with QoS properties and filter objects. The QoS properties and filter objects of the admin object are transferred to each proxy it creates, however, the QoS properties may be changed on a per-proxy basis.

An event is created by the event supplier and is transferred to all relevant event consumers. The set of suppliers is decoupled from the set of consumers, and the supplier has no knowledge of the number or identity of the consumers. The consumers have no knowledge of which supplier generated the event. The Event Service and Notification Service are based on the event channel pattern. The event channel is responsible for asynchronously transferring events between suppliers and consumers. Suppliers and consumers connect to the event channel using the interfaces supported by the channel. An event is a successful completion of a sequence of operation calls made on consumers, suppliers, and the event channel.

Internally, both Event Service and Notification Service utilize the proxy pattern heavily. For each external supplier they create an internal proxy consumer, and for each external consumer they create an internal proxy supplier. In the Notification Service the proxies can have their own QoS and filtering rules derived from a common administration object that was used to create the proxies.

Filters are CORBA objects that support the addition, modification, and removal of constraints. Figure 5.2 gives an example of a filter. Constraints are used to match event message values and refer to variables that are part of the event notification message. Constraints are either event types or written in a constraint language. Variable names can refer to all parts of the current notification. The current notification is expressed with the dollar sign '$'.

A sample notification constraint:

$.type_name == StockAlert
$.market_name == 'NASDAQ'
$.ticker == 'Company'
$.price > '100' or $.price < 70

Figure 5.2   Example filter.

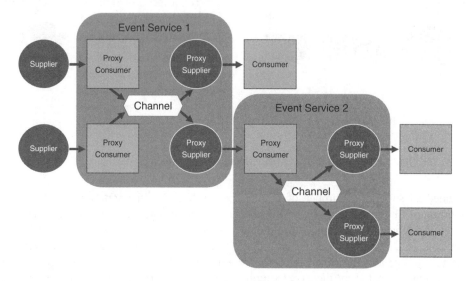

Figure 5.3   Example of channel federation.

The centralized nature of the Event Channel as a CORBA object limits its scalability. All the registered consumers and suppliers are managed by the channel, which may limit the number of active entities and also the maximum number of notifications that the event channel is capable of processing in a given timeframe. Therefore it becomes important to create, manage, and specify federations of event channels (Figure 5.3). Each event channel has a master queue and a number of consumer queues. Each queue has some maximum capacity, which may be enforced using QoS policies supported by the specification. One way to relieve the bottleneck of the centralized event channel is to distribute these queues as CORBA objects; however, this kind of solution is still centralized. Since NS supports the federation of channels by connecting the supplier and consumer proxies, the system supports scalability.

Channel federation can be used to:

- Improve performance by distributing consumers on several event channels. Since an event channel is a CORBA object, it may become a bottleneck if the number of consumers (or producers) becomes large. Event channels may also be used to enhance local delivery by assigning to each event channel only local subscribers. In this case there is only one network invocation and a number of local invocations.

- Improve reliability by having multiple event channels for the same information. If one event channel fails, it does not necessarily prevent consumers from receiving the notifications.
- Improve flexibility by grouping consumers and producers into logical units (event channels).

CORBA Event Service and Notification Service do not specify an event discovery service or a mechanism to federate event channels. Moreover, the procedure for connecting event channels is complex. The OMG Telecommunications Domain Task Force addresses these issues in the CORBA Management of Event Domains Specification [3], which specifies an architecture and interfaces for managing event domains. An event domain is a set of one or more event channels grouped together for management, and for improved scalability.

The specification defines two generic domain interfaces for managing generic typed and untyped channels. Moreover, a specialized domain for both channels and logs is defined by the OMG Telecom Log Service specification.

The specification addresses:

- connection management of clients to the domain;
- topology management;
- sharing the subscription and advertisement information in an event domain, even when connections between event channels change at runtime;
- event forwarding within a channel topology; and
- connections between event channels.

The specification supports the creation of channel topologies of arbitrary complexity, allowing cycles and diamond shapes in the graph of interconnected channels. However, if events may reach a point in the graph by more than one route duplicate events need to be detected and removed. Moreover, if no timeouts are specified, events in a cycle will propagate infinitely. Therefore, the specification defines mechanisms that are used to detect cycles or diamonds in the network topology. Graph topology enforcement is done at channel connection time, and the domain management refuses illegal connections.

5.3   OMG Data Distribution Service (DDS)

The *Data Distribution Service for Real-Time Systems (DDS)* OMG specification defines an API for datacentric pub/sub communication for distributed real-time systems [4]. DDS is a middleware service that provides a global data space that is accessible to all interested applications. The specification describes the service using UML. The latest version is 1.2 and it was released in January 2007.

DDS v1.2 API standard includes two key APIs:

- A lower *DCPS (Datacentric Publish-Subscribe)* level that specifies APIs and mechanisms for efficient pub/sub.
- An optional higher *DLRL (Data Local Reconstruction Layer)* level intended for integrating DDS with applications.

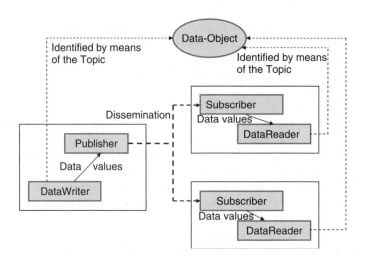

Figure 5.4   Overview of DDS.

5.3.1   Overview

Figure 5.4 illustrates the key elements of the DDS model, namely the publisher, subscriber, and data-object. DDS uses the combination of a Topic object and a key to uniquely identify instances of data-objects. In this model, the subscriptions are decoupled from the publications. DDS creates a name space that allows participants to locate and share objects. In case a set of instances are under the same topic, these different instances must be distinguishable.

The key DDS entities are:

- DomainParticipantFactory. This is a singleton factory that is the main entry point to DDS.
- DomainParticipant is an entry point for the communication in a specific DDS domain. It acts as a factory for the creation of DDS Publishers, Subscribers, Topics, MultiTopics and ContentFilteredTopics.
- TopicDescription is an abstract base class for Topic, ContentFilteredTopic and Multi-Topic.
- Topic is a specialization of TopicDescription that is the most basic description of the data to be published and subscribed.
- ContentFilteredTopic is a specialized TopicDescription like the Topic that additionally allows content-based subscriptions.
- MultiTopic is a specialization of TopicDescription like the Topic that additionally allows subscriptions to combine/filter/rearrange data coming from several topics.
- Publisher is the object responsible for the actual dissemination of publications.
- DataWriter allows the application to set the value of the data to be published under a given Topic.
- Subscriber is the object responsible for the actual reception of the data resulting from its subscriptions.

- DataReader allows the application to declare the data it wishes to receive (by making a subscription using a Topic, ContentFilteredTopic or MultiTopic) and to access the data received by a Subscriber.

DDS uses a key to distinguish between these instances. The key consists of the values of some data fields. These fields need to be indicated to the middleware. Different data values with the same key value represent successive values for the same instance. Different data values with different key values represent different instances. If no key is provided, the data set associated with the Topic is restricted to a single instance.

A ContentFilteredTopic may be created for content-based subscriptions. In addition, the MultiTopic can be used to subscribe to multiple topics and combine/filter the received data. The filter language syntax is a subset of the SQL syntax.

DDS supports different consistency models, for example weak consistency and eventual consistency. DDS can be configured to support an eventual consistency model that guarantees that all matched reader caches will be eventually identical to the respective writer caches. The chosen durability, reliability, lifespan, and ordering parameters affect the choice of the consistency model.

5.3.2 QoS Policies

The QoS usage follows the subscriber-requested publisher-offered pattern. In this pattern, the subscribers request desired QoS properties and these are matched against those offered by the producers.

DDS QoS policies include:

- Deadline defines the maximum interarrival times for data samples.
- Latency budget specifies the maximum delay for a write.
- Transport priority defines the priority for the lower level data protocols.
- Reliability determines the reliability level of the data transport, which can be reliable or best effort.
- Ownership that determines whether or not multiple data writers can operate on the same data instance, and what are the policies if this is allowed.
- Durability, which can be used to specify how published data is stored by the system. The data can be volatile, tied to the writer, transient to be available after the data writer, and persistent even in the case of system restart.
- Lifespan that determines the lifetime of data. The default validity period is infinite.
- History that determines how recent data DDS delivers to subscribers. For example, DDS can deliver the most recent data sample, or a specific sample in the past.

5.3.3 Real-Time Communications

Real-time communications is an important part of the QoS support. QoS policies pertaining to real-time include deadline for a message, latency budget, and transport priority. The DEADLINE QoS policy allows to define the maximum interarrival time between data samples. With this policy, the DataWriter indicates that the application commits a write at least once every deadline period. The DataReaders are notified when the QoS contract

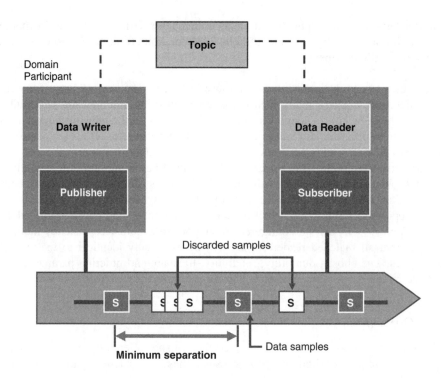

Figure 5.5   Real-time communication with DSS.

is violated. The LATENCY_BUDGET_QoS policy determines the maximum acceptable delay from the time the data is written until the data is inserted into the subscriber's application cache.

Figure 5.5 illustrates real-time communication with DDS. The publishers and subscribers communicate through the topic with the DataWriter and DataReader objects. Data is published and delivered across the network to the subscriber. The QoS specification determines the requirements for the real-time communication. There is a minimum delay separation between published samples. Samples that do not meet the separation requirement are discarded. Thus late data elements are not delivered to the subscriber.

The TRANSPORT_PRIORITY QoS policy is related to these and can be used to instruct the infrastructure to prioritize the communications with the underlying transport.

5.3.4   Applications

DDS applications are typically written with the following phases:

1. Definition of the topics and the domains.
2. Determination of the QoS policies, such as transport priority, deadlines, and durability.
3. Identifying topic readers and writers.

4. Defining QoS requirements for the readers and writers, such as history, latency budget, transport priority, and deadlines.
5. Implement the system with the chosen programming language and DDS implementation.

DDS is suitable for datacentric applications that require timely signal, data, and event propagation. Signals represent continuously changing data, for example from a sensor. In this case, publishers may set the reliability property to best-effort and the history QoS property to retain the last signal (KEEP_LAST). Data delivery, such as exchanging the state of a set of objects, can be realized by using reliable communication and requiring that the last data elements are stored by the system. Events are streams of values and publishers typically use reliable delivery and require that the system keeps a history of all messages (KEEP_ALL). Thus DDS QoS parameters can be tuned for the specific use case to meet application requirements.

DDS is extensively applied in industrial and defense applications, such as high-speed stock trading, telecommunications, manufacturing, power generation, medical devices, mobile asset tracking, and air traffic management. For example, the standard is recommended by US Navy for Open Architecture and EuroControl for air traffic control center operational interoperability.

5.4   SIP Event Framework

The SIP event framework (RFC 3265) defines an extensible event notification system. The SIP event framework allows SIP user agent's to create, modify, and remove subscriptions. The event framework can be extended by using modular *Event Packages*, which are additional specifications that define additional syntax and semantics for the framework. The key applications of the SIP Event Framework include automatic callback services.

The SUBSCRIBE and NOTIFY methods in the SIP extensions are utilized by the SIP event package. They aim to enable a client to subscribe to a set of desired events and receive notification messages upon the occurrence of the expected events (RFC 3680). A subscription can be removed or modified by using the SUBSCRIBE method and by indicating the desired action using header parameters. A final NOTIFY message will be sent regarding the resource's state in the case of unsubscription. SIP events are identified by using three fields: the Request URI, Event Type, and optionally message body. The Request URI of a SUBSCRIBE request must contain the sufficient information to route the request to the destination.

The SIP event package is mainly applied in callback services, message waiting indications and presence-of-buddy lists. All notifications must be subscribed beforehand, and the SIP event package does not allow unsolicited messaging. However, the call-control traffic and other SIP signalling are not affected by the characteristics of the event package. Moreover, SIP protocol leaves unspecified the exact method to distribute and store the subscription state in the system.

An example of the SIP Event Package implementation has been shown in Figure 5.6 in a voice mail application. The notifier server in voicemail accepts subscriptions. In this case, Bob sends a SIP SUBSCRIBE message to the server to get updates of voicemail

Figure 5.6   Voicemail updates with the SIP Event Package.

status. Bob specifies the Event header field to indicate this and sets its value to message-summary. The server sends a 200 OK response to tell that the SUBSCRIBE transaction has been received correctly and also processed. Then the information content is delivered with a SIP NOTIFY transaction. Here the notification tells Bob that he has three new messages and seven old ones. There are urgent messages, one of the new message set and three of the old messages. When the voice mail server receives new messages, it updates the message counters respectively and then notifies Bob with a SIP NOTIFY transaction, allowing Bob to keep up with his voicemail status.

5.5   Java Delegation Event Model

The *Java Delegation Event Model* was introduced in the Java 1.1 *Abstract Windowing Toolkit (AWT)* and serves as the standard event processing method in Java. The model is also used in the Java Beans architecture. In essence, the model is centralized and a listener can register with an event source to receive events. An event source is typically a GUI element and fires events of certain types, which are propagated to the listeners. Event delivery is synchronous, so the event source actually executes code in the listener's event handler. No guarantees are made on the delivery order of the events. The event source and event listener are not anonymous, however, the model provides an abstraction called an adapter, which acts as a mediator between these two actors. The adapter decouples the source from the listener and supports the definition of additional behavior in event processing. The adapter may implement filters, queuing, and QoS controlling.

5.6   Java Distributed Event Model

The *Distributed Event Model* of Java is based on Java RMI that enables the invocation of methods in remote objects. This model is used in Sun's Jini architecture [5]. The architecture of the Distributed Event Model is similar to the architecture of the Delegation Model with some differences. The model is based on the Remote Event Listener, which is an event consumer that registers to receive certain types of events in other objects.

The specification provides an example of an interest registration interface, but does not specify such.

The Remote Event is the event object that is returned from an event source (generator) to a remote listener. Remote events contain information about the occurred event, a reference to the event generator, a handback object that was supplied by the listener, and a unique sequence number to distinguish the event globally. The model supports temporal event registrations with the notion of a lease (Distributed Leasing Specification). The event generators inform the listeners by calling the listeners' notify method.

The specification supports Distributed Event Adaptors that may be used to implement various QoS policies and filtering. The handback object is the only attribute of the Remote Event that may grow to unbounded size. It is a serialized object that the caller provides to the event source; the programmer may set the field to null. Since the handback object carries both state and behavior it can be used in many ways, for example to implement an event filter at a more powerful host than the event source.

5.7   Java Message Service (JMS)

JMS [6] defines a generic and standard API for the implementation of message-oriented middleware. The JMS API is an integral part of the Java Enterprise Edition (Java EE). JMS is an interface and the specification does not provide any concrete implementation of a messaging engine. The fact that JMS does not define the messaging engine or the message transport gives rise to many possible implementations and ways to configure JMS.

JMS API allows applications to create, send, receive, and read messages. Designed by Sun and several partner companies, the JMS API defines a common set of interfaces and associated semantics that allow programs written in the Java programming language to communicate with other messaging implementations.

The JMS API minimizes the set of concepts a programmer must learn to use messaging products but provides enough features to support sophisticated messaging applications. It also strives to maximize the portability of JMS applications across JMS providers in the same messaging domain.

The JMS API enables communication that is not only loosely coupled but also

- Asynchronous. A JMS provider can deliver messages to a client as they arrive; a client does not have to request messages in order to receive them.
- Reliable. The JMS API can ensure that a message is delivered once and only once. Lower levels of reliability are available for applications that can afford to miss messages or to receive duplicate messages.

A JMS application is composed of the following parts.

- A JMS provider is a messaging system that implements the JMS interfaces and provides administrative and control features.
- JMS clients are the programs or components, written in the Java programming language, that produce and consume messages.

- Messages are the objects that communicate information between JMS clients. Administered objects are preconfigured JMS objects created by an administrator for the use of clients.

5.7.1   Two Communication Models

Before the JMS API existed, most messaging products supported either the point-to-point or the pub/sub approach to messaging. Figure 5.7 illustrates the two communication models. The JMS Specification provides a separate domain for each approach and defines compliance for each domain. A standalone JMS provider may implement one or both domains. A J2EE provider must implement both domains.

In fact, most current implementations of the JMS API provide support for both the point-to-point and the pub/sub domains, and some JMS clients combine the use of both domains in a single application. In this way, the JMS API has extended the power and flexibility of messaging products. In the point-to-point model only one receiver is selected to receive a message, and in the publisher/subscriber model many can receive the same message. The JMS API can ensure that a message is delivered only once. At lower levels of reliability an application may miss messages or receive duplicate messages.

A *point-to-point (PTP)* product or application is built around the concept of message queues, senders, and receivers. Each message is addressed to a specific queue, and receiving clients extract messages from the queue(s) established to hold their messages. Queues retain all messages sent to them until the messages are consumed or until the messages expire.

Instead, in a pub/sub product or application, clients address messages to a topic. Publishers and subscribers are generally anonymous and may dynamically publish or subscribe to the content hierarchy. The system takes care of distributing the messages arriving from a topic's multiple publishers to its multiple subscribers. Topics retain messages only as long as it takes to distribute them to current subscribers.

Pub/sub messaging has the following characteristics.

- Each message may have multiple consumers.
- Publishers and subscribers have a timing dependency. A client that subscribes to a topic can consume only messages published after the client has created a subscription, and the subscriber must continue to be active in order for it to consume messages.

A standalone JMS provider (implementation) has to support either point-to-point or the pub/sub approach, or both. Normally, JMS queues and topics are maintained and created by the administration rather than application programs. Therefore the destinations are seen as long lasting. The JMS API also allows the creation of temporary destinations that last only for the duration of the connection.

The JMS API allows clients to create durable subscriptions. Durable subscriptions introduce the buffering capability of the point-to-point model to the pub/sub model. Durable subscriptions can accept messages sent to clients that are not active at the time. A durable subscription can have only one active subscriber at a time. Messages are delivered to clients either synchronously or asynchronously.

Synchronous messages are delivered using the receive method, which blocks until a message arrives or a timeout occurs. In order to receive asynchronous messages, the client creates a message listener, which is similar to an event listener. When a message arrives the JMS provider calls the listener's onMessage method to deliver the message.

JMS clients use *Java Naming and Diretory Interface (JNDI)* to look up configured JMS objects. JMS administrators configure these components using facilities specific to a provider. There are two types of administered objects in JMS: ConnectionFactories, which are used by clients to connect with a provider, and Destinations, which are used by clients to specify the destination of messages. JMS messages consist of a header with a set of header fields, properties that are optional header fields (application-specific, standard properties, provider-specific properties), and a body that can be of several types.

5.7.2 Message Types and Selection

JMS supports five different messages types:

- Map,
- Object,
- Stream,
- Text,
- Bytes.

MapMessage is a set of name/value pairs, where names are strings and values are primitive Java types. ObjectMessage is a message containing a serializable Java object. StreamMessage is a stream of sequential Java primitive values. TextMessage represents an instance using the java.lang.String class and can be used to send and receive XML messages. BytesMessage is a stream of bytes.

Message selection is supported by filtering the message header against the given criteria using an SQL grammar. Figure 5.8 presents the frequently used message selection

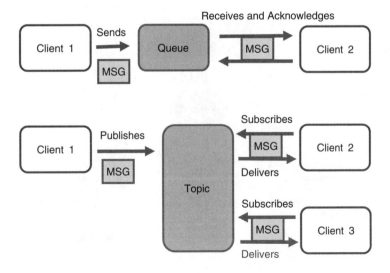

Figure 5.7 Overview of queue and topic based communications.

Operand	Data type
=	All
<, ≤, ≥, >, ≠	Arithmetic
BETWEEN	Arithmetic
AND, OR, NOT	Boolean
IN, LIKE	String

Figure 5.8 Frequently used JMS SQL-92 message selection operands.

operands. A JMS message selector allows clients to define the messages they are interested in. Headers and properties need to match the client specification in order to be delivered to that client. Message selectors cannot reference values embedded in the message body.

An example is JMSType=`stock' AND company=`abc' AND stockvalue > 100.

JMS allows clients to inspect messages in a queue. This is achieved with the Queue-Browser object that allows the browsing of the queue and to inspect message headers. Message sent to a queue are stored there until the message consumer of the queue consumes them.

5.7.3   JMS Process

Figure 5.9 illustrates the key objects of the JMS architecture. The ConnectionFactory is used to create a Connection object. The Connection object is then used to create a Session,

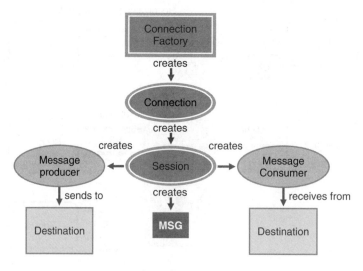

Figure 5.9 Key JMS objects.

which then is used to send and consume messages based on Destinations. The Session object provides methods for creating both MessageProducers and MessageConsumers, which send and receive messages based on explicitly defined Destinations. A Destination maps to a queue that stores the messages before their consumption by MessageConsumers.

Typically a JMS client creates a Connection, one or more Sessions, and a number of MessageConsumers and MessageProducers. Connections are created in the stopped mode. After a connection is started (start() method) messages start arriving to the consumers associated with that connection.

A MessageProducer can send messages while a Connection is stopped. A Session is a single-threaded context for consuming and producing messages. Sessions act as factories for creating MessageProducers, MessageConsumers, and temporary destinations. JMS defines that messages sent by a session to a destination must be received in the order in which they were sent.

JMS messaging proceeds in the following fashion:

1. Client obtains a Connection from a Connection Factory.
2. Client uses the Connection to create a Session object.
3. The Session is used to create MessageProducer and MessageConsumer objects, which are based on Destinations.
4. MessageProducers are used to produce messages that are delivered to destinations.
5. MessageConsumers are used to either poll or asynchronously consume (using Message-Listeners) messages from producers.

Figure 5.10 presents example Java code for creating a queue, publishing a message to the queue, and then consuming the published message from the queue.

5.7.3.1 JMS Pub/Sub Model

The JMS pub/sub model defines how JMS clients publish messages to, and subscribe to messages from, a well-known node in a content-based hierarchy. JMS calls these nodes *topics*.

A topic can be thought of as a mini message broker that gathers and distributes messages addressed to it. By relying on the topic as an intermediary, message publishers are kept independent of subscribers and vice versa. The topic automatically adapts as both publishers and subscribers come and go.

A Topic object encapsulates a provider-specific topic name. It is the way a client specifies the identity of a topic to JMS methods. Many pub/sub providers group topics into hierarchies and provide various options for subscribing to parts of the hierarchy. JMS places no restrictions on what a Topic object represents. It might be a leaf in a topic hierarchy, or it might be a larger part of the hierarchy (for subscribing to a general class of information).

The organization of topics and the granularity of subscriptions to them is an important part of a pub/sub application's architecture. JMS does not specify a policy for how this should be done. If an application takes advantage of a provider-specific topic grouping mechanism, it should document this. If the application is installed using a different provider, it is the job of the administrator to construct an equivalent topic architecture and create equivalent Topic objects.

```
import javax.jms.*;
mport javax.naming.*;
import java.util.*;

public class HelloWorldMessage {
  public static void main(String[] args) {
    try {
      ConnectionFactory myConnFactory;
      Queue myQueue;

      myConnFactory = new com.sun.messaging.ConnectionFactory();
      Connection myConn = myConnFactory.createConnection();
      Session mySess = myConn.createSession(false, Session.AUTO_ACKNOWLEDGE);
      myQueue = new com.sun.messaging.Queue("test");  // Create a queue

      MessageProducer myMsgProducer = mySess.createProducer(myQueue);
      TextMessage myTextMsg = mySess.createTextMessage();
      myTextMsg.setText("Test message");
      System.out.println("Sending: " + myTextMsg.getText());
      myMsgProducer.send(myTextMsg);   // send test message to queue

      MessageConsumer myMsgConsumer = mySess.createConsumer(myQueue);
      myConn.start();

      Message msg=myMsgConsumer.receive();  // Receive a message from the queue

      if (msg instanceof TextMessage) {
        TextMessage txtMsg = (TextMessage) msg;
        System.out.println("Received: " + txtMsg.getText());
      }
      mySess.close();
      myConn.close();
    } catch (Exception jmse) {
      System.out.println("Exception occurred: " + jmse.toString());
      jmse.printStackTrace();
}}}
```

Figure 5.10   Example JMS code for publishing and consuming events.

5.7.4   Message Delivery

Messages can be consumed in either of two ways:

- Synchronously. A subscriber or a receiver explicitly fetches the message from the destination by calling the receive method. The receive method can block until a message arrives or can time out if a message does not arrive within a specified time limit.
- Asynchronously. A client can register a message listener with a consumer. A message listener is similar to an event listener. Whenever a message arrives at the destination, the JMS provider delivers the message by calling the listener's onMessage method, which acts on the contents of the message.

A JMS message is not successfully consumed before it is acknowledged. The successful message consumption requires that the client receives the message, the client processes the message, and that the message is acknowledged. Acknowlegment is initiated either by the JMS provider or the client.

The basic JMS mechanisms pertaining to reliable message delivery are:

- Message acknowledgment policies.
- Message persistence: Messages can be persistent and stored to a disk.
- Message priority levels that affect the order in which messages are processed and delivered.
- Message expiration. A message specific expiration time determines when the message becomes obsolete.
- Temporary destination are valid only for the duration of the connection in which they are created.

Messages are acknowledged automatically in the transactional mode, however, if a session is not transacted there are three possible options for acknowledgement: lazy acknowledgment that tolerates duplicate messages, automatic acknowledgement, and client-side acknowledgement. In persistent mode delivery is once-and-only-once, and in nonpersistent mode the semantics are at-most-once.

5.7.5 Transactions

JMS supports transactions in which any series of operations are grouped into an atomic unit of work called a transaction. If any of the operations fails, the transaction can be rolled back to the state before the transaction was started. If the operations succeed, the transaction is successful and can be committed. The JMS API Session interface contains the commit and rollback methods. The invocation of the commit method results in the delivery of all produced messages and the acknowlegment of all consumed messages. The rollback method results in the dropping of the produced messages, and the recovery and redelivery of unexpired consumed messages.

5.7.6 Advanced Issues

The JMS specification does not consider advanced load balancing and failover issues. These can be supported by a JMS implementation. Figure 5.11 illustrates JMS usage in a cluster environment. There are three servers in the cluster. Two servers are responsible for a single queue, and one server is responsible for two queues. The queue Q1 and its corresponding destination D1 are replicated on two servers. There are two consumer queues on the third server, namely Q2 and a temporary queue QT. The replication of Q1 increases fault tolerance of the system.

5.7.7 JMS in Java EE and Implementations

The Java Enterprise Edition (Java EE) platform uses the JMS API in the following cases:

- *Enterprise JavaBeans (EJB)* components, Web components, and application clients use the API for sending and receiving JMS messages.

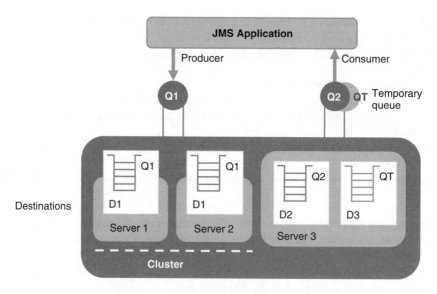

Figure 5.11   Example of a JMS cluster.

- Message-driven beans support the asynchronous processing of messages.
- Message send and receive operations can be used in distributed transactions. Transactions result in atomic operation and they support the rollback feature.

A message-driven bean can consume messages from a queue or a durable subscription. The messages can be sent by any JMS client and Java EE component. The message-driven bean contains an onMessage method that is invoked when a message arrives. A message-driven bean class must implement the javax.jms.MessageListener interface and the onMessage method.

There are several JMS implementations, for example: Fiorano MQ, Sun Java System MQ, WebSphere MQ, WebLogic Application Server, and Apache ActiveMQ. The last, ActiveMQ, is the only Open Source JMS server of the mentioned products.

5.8   TibCo Rendezvous

TibCo Rendezvous is a product that provides messaging APIs for application developers. The basic system is subject-based and it uses multicast to disseminate messages. The message passing consists of the following phases:

- A message has a single subject composed of elements that are separated by periods. The message is sent to a single Rendezvous Daemon component that are responsible for sending the message to interested listeners and Daemons in the distributed environment.
- A listener informs its interests to a Daemon. The listener can use wildcards to select interesting subjects. Messages matching the interests are then delivered by the Daemon to the listener.

The system offer also different enterprise services, such as fault tolerance and reliability, that build on the basic message passing system. The messages are typed name-value or number-value fields.

5.9   COM+ and .NET

Standard Windows COM and OLE support asynchronous communication and the passing of events using callbacks, however, these approaches have their problems. Standard COM publishers and subscribers are tightly coupled. The subscriber knows the mechanism for connecting to the publisher (interfaces exposed by the container). This approach does not work very well beyond a single desktop. Now, the components need to be active at the same time in order to communicate with events. Moreover, the subscriber needs to know the exact mechanism the publisher requires. This interface may vary from publisher to publisher making this difficult to do dynamically (ActiveX and COM use the IconnectionPoint mechanism for creating the callback circuit; an OLE server uses the method Advise on the IoleObject interface).

The COM+ event service is an operating system service that provides the general infrastructure for connecting publishers and subscribers. The service is a Loosely Coupled System, because it decouples event producers from event subscribers using the event service and a catalog for storing available events and subscription information. In this architecture, an event is a method in a COM+ interface called the event method, and it contains only input parameters.

The following steps are required to produce an event (Figure 5.12):

1. An event Class is registered.
2. Subscriber registers for an Event.
3. Publisher creates an Event Object at runtime.
4. Publisher fires the Event by calling the method in the Event Object.
5. Event Object reads the Subscription List from the Event Store.
6. The system delivers the event to the subscriber by calling the appropriate method.

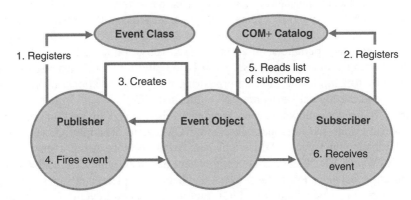

Figure 5.12   Overview of the COM+ Event Service.

The event service keeps track of which subscribers want to receive the calls, and mediates the calls. The event class is a COM+ component that contains interfaces and methods. A subscriber needs to implement the interfaces in order to receive the event, and a publisher calls the methods to fire events. Event classes are stored in a COM+ catalog that is updated either by the publishers or by the administration. Subscribers register their wish to receive events by registering a subscription with the COM+ event service.

A subscription is a data structure that contains the recipient, event class, and which interface or method within that event class the subscriber wants to receive calls from. Subscriptions are also stored in the COM+ catalog either by the subscribers or by the administration. Persistent subscriptions survive restarting the operating system whereas transient subscriptions will be lost on restart or reset. The publishers use the standard object creation functions to create an object of the desired event class. This event object contains the event system's implementation of the requested interface. The publisher then calls the event method that it wants to fire. The event system implementation of that interface searches the COM+ catalog and finds all the subscribers who have expressed interests in that event class and method. The event system then connects to each subscriber, using direct creation, monikers, or queued components, and calls the specified method. Event methods return only success or failure. Any COM+ client can become a publisher and any COM+ component can become a subscriber.

This event system has several limitations. The subscription mechanism is not itself distributed and there is no support for enterprise wide repository. The delivery time and effort increases linearly with the number of subscribers, which means that the system is not scalable to firing events to many subscribers. However, client-side disconnection is supported with queued components. COM+ supports components that record a series of method invocations (event occurrences) and are able to play them back in the recorded order. These components can be distributed using messages. Since the event object may be defined as queueable, a disconnected client may play back the desired event object upon reconnection. COM+ Events can be extended to support filtering, which needs to be implemented either on the publisher side or on the subscriber side. If an event is filtered by a component on the publisher side, it is never delivered to the event service. If an event is filtered on the subscriber side the event service will make the decision of whether to deliver the event to a particular subscriber.

Filtering on the publisher side is done by attaching a filter object to the event object interfaces (which correspond to events). The filter may query the subscription information and, for example, change the firing order for a set of subscribers. The subscriber-side filtering is done using parameter filtering for each subscription and method invocation. Parameter filtering evaluates the subscription FilterCriteria property against the parameters of the event method. The filter criteria string recognizes relational operators, nested parenthesis, and the logical keywords AND, OR, and NOT.

The .NET framework supports events at many levels. There is support for programming-language-level events and interoperability with COM events. The interoperation of Visual Basic .NET code and legacy COM component events is done using a runtime callable wrapper (RCW). In VB.NET listeners create event handlers, which are added to sources. The connection between events and event handlers is implemented by special objects called delegates. The benefit of the .NET runtime is that the events from components written in different languages, say C# and VB, are interoperable.

Microsoft's messaging infrastructure is called *Microsoft Message Queuing (MSMQ)*. MSMQ has been available for developers since 1997 and it provides the messaging framework for the *Windows Communication Foundation (WCF)*. MSMQ is a message oriented middleware platform that supports secure and reliable message transport.

5.10   Websphere MQ

IBM's MQSeries, currently known as Websphere MQ, is one of the most popular message oriented middleware products for electronic business. The product supports heterogeneous any-to-any communication between many different platforms. MQ is compatible with JMS and integrates with Java Beans (EJB), SOAP, REST, and .NET. MQ also supports SOAP for Web service creation. A JMS compliant embedded JMS provider supports point-to-point and publish-subscribe messaging [7].

5.10.1   Overview

The main features that WebSphere MQ provides is the assured *one-time delivery* of messages across a wide variety of platforms. The product emphasizes reliability and robustness of message traffic, and ensures that a message should never be lost if MQ is appropriately configured. WebSphere MQ provides facilities for reliable message queuing via *queue managers*. The queue managers maintain the queues of the message queuing infrastructure, and all of the messages that reside on those queues waiting to be processed or routed. Queue managers are tolerant to failures, maintaining the integrity of the business-critical data flowing through the message queuing infrastructure. The queue managers within the infrastructure are connected with *channels*. Messages automatically flow across these channels, from the initial producer (i.e. publisher) of a message to the eventual consumer (subscriber) of that message, based on the configuration of the queue managers in the infrastructure.

A queue manager manages queues that store messages, and applications communicate with their local queue manager. Remote queues are owned by remote queue managers, and each message that is inserted into a remote queue gets transmitted over the network. The queue manager may support a local queue, in which case the client is capable of supporting asynchronous communication. If no local queue is present, the client is bound to synchronous communication. Another configuration option is whether the client supports bridges and is capable of exchanging messages with other queue managers.

To support efficient and scalable message queuing implementation, WebSphere MQ allows (i) a single queue manager with local applications accessing the service, (ii) a single queue manager with remote applications accessing the service as clients, (iii) hub and spoke WebSphere MQ architectures, and (iv) MQ cluster. The hub and spoke architecture efficiently supports the distributed scenario with headquarters and branch office. Instead, a more flexible approach is to join many queue managers together in a dynamic logical network called a queue manager cluster. This allows multiple instances of the same service to be hosted through multiple queue managers.

As a JMS provider, WebSphere MQ implements the Java Message Service standard API. In addition, it has its own proprietary API, known as the Message Queuing Interface. Similar to JMS specification, there are two basic types of messaging style in WebSphere

MQ, point-to-point (PTP) and Publish/Subscribe. The concepts of PTP and Pub/Sub are similar to those in JMS.

5.10.2   Pub/Sub in WebSphere MQ

Applications that produce information about a particular subject are referred to as publishers. Applications that consume this information are referred to as subscribers. The information and subjects are managed by WebSphere MQ Publish/Subscribe (Pub/Sub). The subject is referred to as a topic. The information equates to WebSphere MQ messages.

Subscribing applications register their intention to receive information from particular topics with WebSphere MQ Pub/Sub. Publishing applications then send information about topics to WebSphere MQ Pub/Sub. The management and distribution of the information to registered subscribers is the responsibility of WebSphere MQ Pub/Sub. This decoupling of publisher and subscriber applications allows for greater scalability and a more dynamic network topology.

A *topic* refers to the subject on which publishers provide information. Subscribers interested in information about this topic can either subscribe to a topic object or a topic string to receive these publications.

The topic string is the central concept in WebSphere MQ Publish/Subscribe, and it provides the logical association between publishers and subscribers. Publishers can publish to a topic string and subscribers can subscribe to the publications using the topic string. A topic string can be up to 10,240 characters long and is case sensitive. A new data-type called the variable-length string (MQCHARV) has been introduced in WebSphere MQ in order to support the long string requirement.

The structure and semantics of the topic string are controlled by the slash (/). For example, there can be a high-level topic called deli to represent a delicatessen, which might be divided into separate subtopics relating to different categories of products that the deli sells, and further layers of subtopics beneath that to further qualify the product. Figure 5.13 shows an example of topic strings.

Topic strings imply a sense of hierarchy in the topic structure. The hierarchy is represented as the topic tree, as depicted in Figure 5.13. The topic tree has a root node that corresponds to the topic object SYSTEM.BASE.TOPIC.

Topic strings support wildcard characters. Subscribers can use two wildcards, hash (#) and plus (+), to subscribe to a range of topics. Both provide methods of topic-level

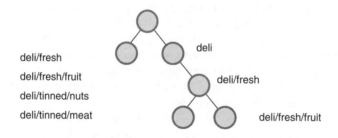

deli/fresh
deli/fresh/fruit
deli/tinned/nuts
deli/tinned/meat

deli

deli/fresh

deli/fresh/fruit

Figure 5.13   MQ Topics.

substitution. The hash can substitute for multiple levels in the topic hierarchy, whereas the plus can substitute for a single level in the topic hierarchy.

Topic strings are used to match information from a publisher to subscribers who are interested in that information. Topic strings do not have to be predefined. They come into existence dynamically when subscribing and publishing applications use them. Consider a publisher application publishing to a topic string called deli/fresh/fruit and no administrative topic objects have been defined at this time. The nodes on the corresponding tree, as shown in Figure 5.13, are referred to as nonadministrative topics. It is possible to define an administrative topic object for any node on this subtree (for example, /fresh/fruit) only if there is a need to associate specific attributes' settings with that particular node that are not the same settings as inherited from the parent node.

5.11   Advanced Message Queuing Protocol (AMQP)

The Advanced Message Queuing Protocol (AMQP)[1] is an open standard application layer protocol for messaging and pub/sub. The protocol has its roots in the financial application domain; however, the protocol is general and can be used to implement various message oriented applications. The protocol is currently being standardized at OASIS. The key features of the protocol are message orientation, queuing, routing reliability and security.

The AMQP specification defines the behaviour of the messaging provider and client as well as the broker component in order to achieve interoperability. The specification defines a wire-level protocol that contrasts many of the other standards, such as JMS, that define only an API. The protocol specification thus contains the description of the data packets that are sent across the network. AMQP messages consist of two parts, namely the envelope of properties and the content part. Message contents are binary blobs. Messages are passed between clients and brokers using the protocol commands and the wire format.

AMQP defines a set of basic messaging patterns that can then be used to build complex interactions. The basic messaging pattern are:

- request-response that defines an interaction between a requestor and a responder, and
- pub/sub that defines a message delivery process from a sender to one or more receivers while meeting the subscribers' criteria.

These patterns can be easily combined, for example sending a notification about the availability of data with pub/sub and then allowing interested receivers to request the data with the request-response interaction.

The core AMQP model defines four key entities:

- Message broker is a server to which clients connect using the protocol. There can be many brokers and they can be federated, but this is outside the scope of the basic specification.
- User is an entity that can connect to a broker.

---

[1] http://www.amqp.org/.

- Connection is a physical connection that is associated with a user.
- Channel is a logical connection that is associated to a connection. Communication over a channel is stateful.

These four entities form the basic model of the protocol.

Communication with the protocol happens through exchanges that are message routing processes implemented at brokers. The standard exchange simply routes messages from senders to receivers: a message is routed to queues where it is stored on behalf of recipients. There are different message routing techniques that can be implemented by an exchange: one-to-one, one-to-many, etc. The exchanges are configured through rules specified in bindings. The bindings determine how the exchange works and also what kind of selection of messages is used to accept messages into a queue. The AMQP protocol allows arbitrary exchange semantics with custom exchanges.

A specific message property called the routing key is used to implement versatile message forwarding behaviours. An exchange will deliver at most one copy of a message to a queue if the routing key in the message matches the routing key in a binding. There are many ways for defining the message matching process:

- A direct exchange matches when the routing key and the key of the binding are identical.
- A fanout exchange always matches.
- A topic exchange matches the routing key a message on binding key words. Words are strings which are separated by dots. The string matching also supports prefixes and postfixes.
- A header based exchange matches on the existence of keys as well as keyvalue pairs in the message header.
- Other exchange types are also possible; however, they are vendor specific and not supported by the specification.

Figure 5.14 illustrates how publishers send their messages through the exchanges to the consumer. The exchanges forward the incoming messages based on their rules to queues. Consumers then retrieve messages using the protocol from the queues. The lower part of the figure illustrates a fanout exchange that forwards the message to two queues.

There is no inherent semantics for the standardized AMQP exchanges for message storage. The exchanges will route the messages to queues and these store them for the recipient usage. An exchange is configured according to a set of bindings (rules) which vary from passing all messages into queue to a comprehensive examination of the message contents before queuing it. Furthermore, arbitrary exchange semantics is permitted in AMQP. This is accomplished using custom exchanges specifically designed for the application: messages can be created, queued, consumed and routed in desired ways.

Well-known AMQP implementations include the following:

- OpenAMQ is an open-source implementation of AMQP written in C by iMatix.
- RabbitMQ is an open-source implementation acquired by VMware in 2010.
- Apache Qpid is a project part of the Apache Foundation.

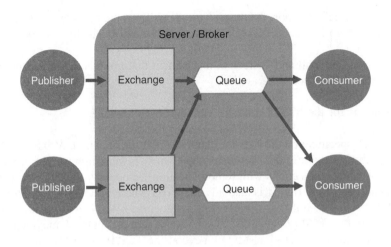

Figure 5.14   Example of AMQP.

5.12   MQ Telemetry Transport (MQTT)

There is a certain demand for lightweight broker-based pub/sub messaging protocols. *MQ Telemetry Transport (MQTT)* is such a protocol that is also open, simple and relatively easy to implement [8]. It is optimized for constrained environments but it is not limited to them. MQTT is advantageous in cases where:

- The network is expensive, with low bandwidth or is outright unreliable.
- It is used by an embedded device with limited processor resources or memory.

Main features of the MQTT protocol are:

- MQTT uses the pub/sub message pattern. This provides one-to-many message distribution and effective application-level decoupling.
- MQTT is a topic-based pub/sub protocol that supports hierarchical topics and wildcards in matching.
- MQTT applies a messaging transport essentially agnostic to the payload content.
- MQTT utilizes TCP/IP for basic network connectivity.
- MQTT has three different message delivery service qualities:
  - At-most-once allows messages to be delivered using the best effort of the TCP/IP network. This leads, however, to increased possibility of message loss or duplication. The quality is used with ambient sensor data where the loss of an individual reading is irrelevant because the next one will be published soon.
  - At-least-once meaning that message delivery is assured but with possible duplicates.
  - Exactly-once guarantees that message arrives exactly once. This quality is usable in billing systems intolerant to duplicates or lost messages leading to possible incorrect charging.

- MQTT has a small transport overhead. This stems from the fixed-length header being 2 bytes long. Therefore the protocol exchanges will be automatically minimized to reduce network traffic.
- MQTT has a Last Will and Testament mechanism that notifies the interested parties of an abnormal disconnection of a client.

Specifications for the MQTT protocol include the following:

- MQTT v3.1 specification. This is the main MQTT specification. MQTT v3.1 enables the pub/sub messaging model in an manner which is very lightweight, a very useful feature for many connections with remote locations, especially those with requirements for a small code footprint. It is also advantageous where network bandwidth scarce.
- MQTT-S v1.1 is meant for wireless sensor networks (WSN) and targeted at devices embedded in non-TCP/IP networks. This protocol supports pub/sub messaging protocol and aims to extend the original MQTT beyond the TCP/IP infrastructures.

Furthermore there are a some MQTT implementations of client APIs. Also a number of MQTT server implementations exist, from Open Source to commercial products.

5.13 Summary

In this chapter, we examined well-known standards based solutions for message oriented middleware and pub/sub. Figure 5.15 presents a summary of the examined standards and products. The table presents the differences and similarities in terms of system structure, state management, filtering capability, QoS attributes, and the operating environment. System structure pertains to the organization of the messaging and pub/sub functionality, for example the event channel pattern, the observer pattern, or a message queue. State management describes where subscription state is stored, for example in the event channel, message queue, or the subscribed resource. The structure and employed patterns determine how state management is realized. The filtering capability determines how message selection is performed. Message filtering capabilities differ between the systems. QoS support also differs among the examined solutions with some offering QoS attributes, such as reliability and timeliness. Finally, the environment is about the operating context, such as the CORBA environment, SIP, and Java.

Figure 5.16 considers the real-time support of the frequently employed solutions. Real-time support is a typical requirement for industrial and tactical systems, in which it is crucial that an asynchronous event can be delivered within certain time-frame to subscribers. Web services, Java, JMS, and CORBA do not typically offer very good real-time support. Java can be extended for soft real-time and hard real-time support; however, it is not specified how this should be reflected in JMS, which can at best cope with soft real-time requirements. For CORBA technology, there is the real-time RT CORBA extension. DDS, on the other hand, can cope with various real-time requirements, even the case of the hard real-time requirement. DDS is suitable for flexible QoS-aware data dissemination to many nodes in dynamic environments.

	Structure	State	Filtering	QoS Attributes	Environment
CORBA ES	Event Channel	Event Channel	No	No	CORBA
CORBA NS	Event Channel	Event Channel	Yes	Yes	CORBA
OMG DDS	Object Space	Subscribed Object	Topic-based with SQL queries	Yes	Various
SIP Event Framework	Pub/sub methods for SIP	Subscribed resource	Extension	Extension	SIP
Java Distributed Event Model	Observer pattern	Subscribed resource	Extension (adaptor)	Extension (Adaptor)	Java RMI
JMS	Point-to-point and pub/sub topic-based messaging	Message queue	Yes, SQL	Yes	Java
TibCo Rendezvous	Message queuing system	Message queue	Subject-based with wildcards	Yes	Various languages supported
COM+ and .NET	Observer pattern	Event store	Extension	Yes	Microsoft
Websphere MQ	Message queuing system	Message queue	Yes	Yes	Various languages and transports supported
AMQP	Message queuing system	Message exchange	Subject-based with wildcards	Limited	Various languages supported, standard wire protocol
MQ Telemetry Transport	Message queuing system	Broker	Topic-based (hierarchical), wildcards	Yes, failure notification	Sensor networks, embedded systems

Figure 5.15 Summary of the standards and products.

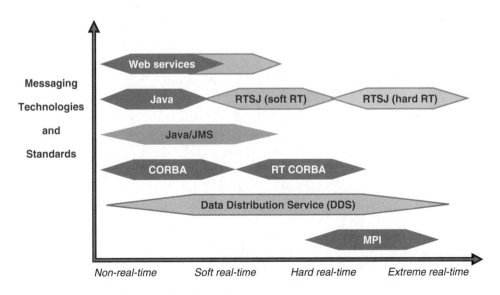

Figure 5.16 Messaging and real-time requirements.

AMQP can be used as a message transport protocol for DDS and JMS implementations. The protocol provides the basic facilities for a brokered communication architecture. The key features include broker interfaces and header based routing. DDS, on the other hand, provides a more peer-to-peer architecture, and topic/content based routing. DDS has been designed for scalability and real-time requirements. Both AMQP and DDS provide an interoperable wire protocol for pub/sub.

References

1. Object Computing, Inc. (2001) *CORBA Event Service Specification v.1.1*. OCI.
2. Object Computing, Inc. (2001) *CORBA Notification Service Specification v.1.0*. OCI.
3. Object Computing, Inc. (2001) *Management of Event Domains Specification*. OCI. http://www.omg.org/cgi-bin/ doc?formal/2001-06-03.
4. Object Computing, Inc. (2007) *Data Distribution Services, V1.2*. OCI.
5. Roberts S and Byous J (2001) *Distributed Events in Jini Technology*. Sun Microsystems. Available at: http://java.sun.com/developer/technicalArticles/jini/JiniEvents/
6. Sun (2002) *Java Message Service Specification 1.1*.
7. Davies S et al. January 2009 Websphere mq v7.0 features and enhancements. *IBM Redbooks*.
8. Hunkeler U, Truong HL and Stanford-Clark A (2008) Mqtt-s – a publish/subscribe protocol for wireless sensor networks. *COMSWARE*, pp. 791–98.

6

Web Technology

In this chapter, we examine key Web technologies for implementing pub/sub solutions. The key Web technologies pertain to basic enablers, such as W3C DOM Events, REST, AJAX, SOAP, and pub/sub systems built on top of them such as RSS and Atom, XMPP, and WS-Eventing and WS-Notification standards.

6.1 REST

Web application developers are typically faced with the question that how to define and structure their APIs for offering services provided by Web servers. The *REST (Representational state transfer)* [1] is a popular architectural model for structuring the APIs and the interactions that they offer. This model was introduced and defined by Roy Fielding in his doctoral dissertation in 2000. REST is essentially a request-response protocol that builds the API functions on a small well-defined set of core operations as well as a uniform resource locator scheme.

In REST the clients initiate requests to servers and the servers return responses after processing the requests. The system of requests and responses is essentially built around the idea of transferring resource representations. A resource is any coherent and meaningful addressable concept and the representation of a resource can be, for instance, a document capturing a resource's current or intended state. The process starts when the client sends requests to indicate that it is ready to make the transition to a new state. The client is considered to be in transition if one or more requests are outstanding, The representation of an application state contains a set of links which are usable when the client decides to initiate a new state transition.

REST was initially designed to function in the Web and HTTP context. In this case, HTTP method types are the basic API operations and the URI scheme is used to identify resources. Figure 6.1 illustrates how HTTP methods and URIs are used by a REST based application. The HTTP protocol is used to exchange data. The model is not limited to one specific protocol and other protocols can form a base for RESTful architectures as well. The requirement is that the protocols provide a rich and uniform vocabulary for applications built using the transfer of meaningful representational state.

Publish/Subscribe Systems: Design and Principles, First Edition. Sasu Tarkoma.
© 2012 John Wiley & Sons, Ltd. Published 2012 by John Wiley & Sons, Ltd.

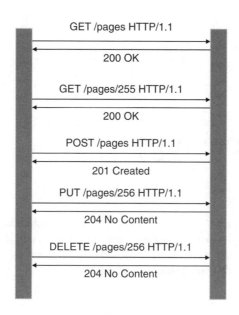

Figure 6.1   Example of REST.

6.2   AJAX

Asynchronous JavaScript and XML (AJAX) is based on JavaScript code executed by the Web browser. AJAX issues HTTP requests and thus essentially employs a poll-based communication model. Web applications can exploit AJAX when they try to create an interactive and less page-oriented user experience. AJAX is asynchronous because it decouples the request and response on the programming level. AJAX can support features like form validation, saving data on a server, fetching suggestions for autocompletion from the server, etc. In an AJAX application, JavaScript code sends requests from the browser to the server. The incoming response message (e.g. XML or JSON, a popular JavaScript-based data structure) is handled by a callback method which dynamically updates the browser window. This is important for the user convenience because the JS code does not wait for the server response. Instead it will postpone the response to a later time when the call back method is called. Thus the browser is available to process user input and possibly also to execute other scripts resident in the page.

It should be noted that AJAX is not fully asynchronous because it uses the request-response paradigm. For that reason the server is not permitted to send an HTTP response without a prior request. This is a limitation which is relaxed in a variation of AJAX, called Comet that can overcome this limitation [2].

The Comet system is based on an AJAX-like mechanism of callback functions to manage responses from the server. Figure 6.2 illustrates the key interactions of this scheme. Comet utilizes HTTP requests to keep the server connection open and thus establishes a long-lived connection to send and receive event data. Before the connection timeout, the client-side Comet automatically closes the connection sending out a new request. This technique essentially sets the server free to send responses to the client at any time. Comet

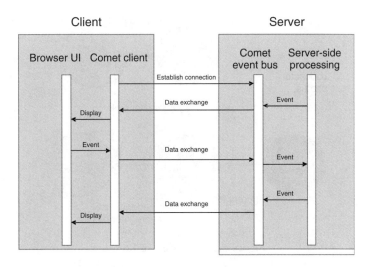

Figure 6.2 Example of the Comet system.

mechanism is also effective when implementing Web-based instant messaging and other applications that require asynchronous interaction.

For example, instant messaging needs a push type mechanism to deliver messages to the user. Until recently this was difficult to implement in browsers without the relying on various plugins. Today the common approach is to implement the push mechanism with long-lived connection techniques, such as the Comet.

In the near future, HTML5 and the new WebSocket communications option may become the way to create Web-based push solutions. WebSocket provides bidirectional, full-duplex communication channels for Web pages and applications over a single TCP socket. The WebSocket API has been designed for the Web environment and standardized by the W3C, and the protocol has been standardized by the IETF as RFC 6455.

WebSocket can be used to overcome the port number blocking problem in ordinary TCP connections where only port 80 is basically allowed for outside connections. Furthermore, WebSocket allows multiplexing several WebSocket services over a single TCP port, thus providing increased functionality with some additional protocol overhead.

6.3 RSS and Atom

Really Simple Syndication (RSS) is a family of specifications for the definition of Web-based information feeds using XML. RSS is essentially a simple pub/sub system that is based on polling the URL that identifies a feed and then determining if information has changed. RSS builds on existing Web standards, namely HTTP and XML, and it has become ubiquitous. RSS is used to disseminate updates, for example, pertaining to blog entries, news, video and audio resources.

The feeds based on RSS are nonpush-based and they require a simple pull-based mechanism to be used by the RSS reader implementations. The reader downloads the RSS feed periodically or when the user decides to reload manually.

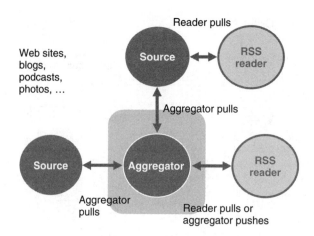

Figure 6.3   Overview of RSS.

An RSS document defines a feed that includes textual description of the elements and related metadata such as URIs and author information. RSS feeds complement the current Web sites, because they allow a standardized way to quickly disseminate updates and aggregate them. RSS is supported by a number of Web browsers and custom feed reader programs. The client program then checks the RSS feed URL regularly for any new information.

Figure 6.3 illustrates the key entities participating in the dissemination of Web feeds. The sources are Web servers that publish RSS feeds as XML-files. The feeds pertain to news pages, blogs, podcasts, photo sites, etc. For example, for a news site an RSS feed would contain the news items: titles, short descriptions, and links to the actual articles. The RSS readers are clients that pull RSS documents from the Web servers and display their contents to end users. Typically, a reader periodically tests whether or not the RSS document has been updated. The RSS document can contain information about the refresh cycle of the document to instrument the polling process. There are also aggregator services that pull RSS documents from various sites and combine their contents for a custom feed. RSS readers then pull content from this aggregator service. It is also possible to implement push from the aggregator to the RSS reader with the help of a long lived HTTP connection or a custom protocol. This is needed to achieve near real-time updates to the feeds.

The RSS 2.0 specification introduces an interesting element, namely the Cloud element that provides a mechanism for RSS document publishers to inform cloud resources that their content has been updated. Basically this allows cloud resources to subscribe content updates called pings. After receiving a ping, a subscriber must then pull the content from the site that sent the ping. This is necessary in order to build an efficient pub/sub system on top of RSS. A similar technique was introduced also in the Pubsubhubbub specification that we will investigate later in this book.

Atom is an alternative syndication protocol defined in RFCs 4287 and 5023. Atom is similar to RSS and uses HTTP and XML to realize Web feeds. Atom specifications emphasize internationalization support, modularity, and security features.

The *Cobra (Content-Based RSS Aggregator)* system has been proposed for efficient distributed RSS feed processing. The system filters and matches published RSS documents against client subscriptions and supports personalized document selection [3]. Cobra is based on a cluster of servers processing the documents, and it does not leverage peer-to-peer overlay technology. Therefore this solution is suitable for datacenters and cluster environments.

The Cobra system consists of a three-tiered network of crawlers, filters, and reflectors. Crawlers visit Web feeds, such as Web sites and blogs, and fetch new documents. Filters are applied to the recently crawled documents in order to find documents of interest to the subscribers. An index-based case-insensitive matching algorithm is used to find the documents of interest. The matching documents are then pushed to reflectors, which present personalized RSS feeds to the users that can be read using standard RSS readers. A reflector caches the last k matching elements and require that the user polls the feed periodically to obtain all matching articles.

This system has been evaluated in a cluster environment as well as on the large-scale Planetlab testbed. Experimental results suggest that the system is able to scale well in terms of the number of feeds and users. In order to cope with high data rates from upstream services, the system features a simple congestion control scheme that applies backpressure when a service is unable to cope with the incoming data rate. The backpressure is implemented with a 1 MB buffer for each upstream service. When the buffer fills up then the send attempts at the upstream services will block until the downstream service can catch up.

6.4 SOAP

We have mentioned SOAP in Chapter 2 in the messaging contexts. SOAP is a W3C specified and standardized one-way message exchange primitive. SOAP very flexible and

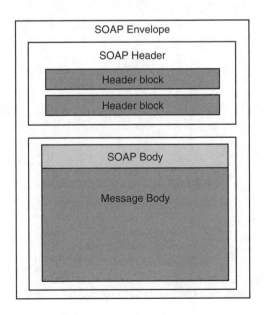

Figure 6.4 A SOAP message.

it can support various interactions which are built on the SOAP primitive. SOAP is utilized with many message transport protocols, such as HTTP and SMTP.

Figure 6.4 illustrates the structure of a SOAP message. A SOAP message has two parts: a header and a body. The header may include attributes related to the SOAP interaction. Certain parts of the header can be hallmarked for message processing intermediaries. SOAP is wholly encryptable and signable; this ensures that the primitive has the crucial security properties of confidentiality, integrity and authenticity.

The following list introduces some key SOAP terminology:

- A SOAP message path is consists of the nodes passing through a SOAP message.
- The initial SOAP sender originally launches the message.
- A SOAP intermediary is a both receiver and sender. It is targetable within a SOAP message. A SOAP intermediary will process those SOAP header elements targeted to it. Then it forwards the message towards the ultimate SOAP receiver.
- The ultimate SOAP receiver is the final destination of a SOAP message. It handles the processing of the message and any header blocks targeted at it.

Figure 6.5 illustrates SOAP routing with intermediaries and the key terminology. It shows us an example: a marketplace with buyers and sellers. In the middle part of the figure a marketplace engine is positioned connecting the buyers and sellers. A component of every SOAP sender and receiver is the SOAP protocol stack. This includes the necessary XML processing elements. The buyer has a SOAP application for sending the buying requests to the SOAP receiver host of the marketplace. This functions as the

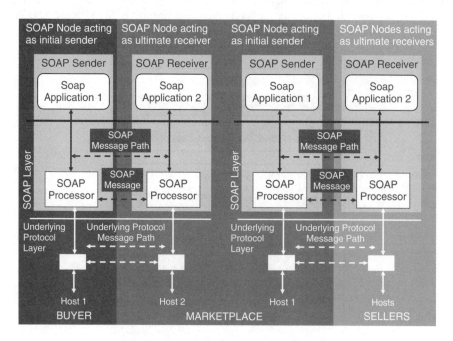

Figure 6.5   Example of SOAP routing.

ultimate SOAP receiver for the messages. Here the relevant messages are forwarded to sellers by sending new SOAP messages to the receivers.

The flexible header mechanism can be used so that the marketplace mediates messages between buyers and sellers without being the ultimate destination. A buyer could send a direct message through the marketplace to the buyer's SOAP receiver. As an intermediary, the marketplace can also log and validate the request. As to the security, both the buyer and seller can protect their messages and thus no intermediary can read the protected content.

6.5   XMPP

XMPP (Extensible Messaging and Presence Protocol) [4] (RFC 3920 based on the Jabber protocol) is an instant messaging protocol extendable to utilize many different message based communication models, such as:

- publish/subscribe mechanisms,
- presence and status updates,
- alerts,
- feature negotiation,
- service discovery.

XMPP implementations exist in two varieties; peer-to-peer and client-server interaction models. The following mechanism is typical:

- A client opens a connection to a server opening an XMPP stream.
- Connection parameters are negotiated, for example authentication and encryption. This enables both the client and the server to send XML stanzas over the connection.
- There are three main stanza types having each their own semantics:
  - The email-like message stanza which is pushed from entity to another entity.
  - The presence stanza which is the basic mechanism for broadcast and publish/subscribe. Thereby information can be received by multiple entities if they have previously subscribed to it.
  - The Info/Query stanza works as the basic request-response mechanism, similar to HTTP. It allows an entity to make requests and receive responses.
- A stanza may be sent by the client to a node which is not in the server database. Then the server and the foreign domain will negotiate using a server-to-server protocol.
- The message will be sent to the corresponding foreign server for delivery to the ultimate destination.

XMPP is rapidly becoming a very popular general API protocol on the Internet for companies like, for example Google, Twitter, and Facebook. Figure 6.6 illustrates a simple distributed XMPP case with multiple servers. XMPP clients utilize the client-server protocol with XMPP servers. The servers then use server-to-server protocol in forwarding content. The servers can interoperate with other XMPP compatible systems, such as Googletalk servers.

XMPP is based on incremental parsing of XML because an XMPP server manages the XML streams and XML stanzas. In the model XML streams rely on long-lived TCP

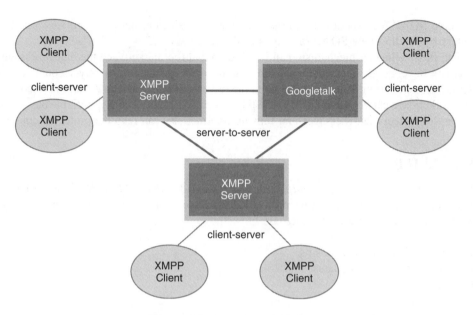

Figure 6.6   Overview of XMPP.

connections and the XML stanzas are sent over the stream as first-level child elements. Furthermore, XMPP complies with RFC 3920 which requires a server implementation to support *Transport Layer Security (TLS)* for channel encryption, *Simple Authentication and Security Layer (SASL)* for authentication, and DNS SRV records for port lookups. Bogus messages are prevented by XMPP as the XML stanzas are stamped by servers with validated source addresses. Furthermore services such as storage of contact lists, management of IM session, management of presence subscription, and management of block lists (RFC 3921) are supported by XMPP instant messaging and presence servers.

6.6   Constrained Application Protocol (CoAP)

Constrained Application Protocol (CoAP) is a Web protocol designed for the constrained Internet of Things environment [5]. The protocol realizes the REST architecture for limited and embedded nodes and thus is closely resembles the HTTP protocol. CoAP is typically used within nodes in constrained networks and also between nodes of a contained networks and nodes in the Internet. CoAP interoperates with HTTP and can be easily translated to the format for integrating data from constrained networks with Web resources. The typical application areas of CoAP include Internet of Things and *machine-to-machine (M2M)* communications.

The main features of CoAP are:

- Designed for constrained environments and M2M communications.
- Asynchronous messaging.
- UDP binding with optional reliability for unicast and multicast requests.

- Low header overhead and parsing overhead.
- URI and Content-type support.
- Simple proxy and caching capabilities.
- A stateless HTTP mapping that allows proxies and gateways to provide HTTP access to CoAP resources and CoAP access to HTTP resources.
- Security binding to *Datagram Transport Layer Security (DTLS)*.

CoAP provides a lightweight reliability mechanism that includes the following features:

- Stop-and-wait retransmission reliability with exponential back-off for messages that require an acknowledgement ("confirmable" in CoAP terminology).
- Duplicate detection of messages.
- Multicast support.

CoAP can be used to implement the observer design pattern introduced in Chapter 4. The pattern is realized as an extension of the basic protocol with the GET request that is sent by a client to a server. The request identifies the subscribed resource and includes a Lifetime Option that determines the duration of the subscription. The technique is very similar to the long polling discussed already in this chapter. The server will then monitor the resource identified in the request and then notify the client regarding changes to the resource.

The CoAP protocol provides a request/response interaction model between application endpoints that follows the Web conventions such as URIs and content types. The protocol supports built-in resource discovery, and is based on UDP. In general, CoAP complies with the requirements of constrained environments and provides features like simplicity, low overhead, and acceptance of sleeping nodes. CoAP operates over IP networks. However, an example of operating environments specifically suitable for CoAP is *6LoWPAN (IPv6 over LoW power Wireless Personal Area Network)*.

6.7 W3C DOM Events

W3C's Document Object Model Level 2 Events is a platform- and language neutral interface that defines a generic event system [6]. The event system builds on the DOM Model Level 2 Core and on DOM Level 2 Views. The system supports registration of event handlers, describes event flow through a tree structure, and provides contextual information for each event. The specification provides a common subset of the current event systems in DOM Level 0 browsers.

For example, the model is typically used by browsers to propagate and capture different document events, such as component activation, mouse overs, and clicks. The two propagation approaches supported are capturing and bubbling. Capturing means that an event can be handled by one of the event's target's ancestors before being handled by the event's target. Bubbling is the process by which an event can be handled by one of the event's target's ancestors after being handled by the event's target. The specification does not support event filtering or distributed operation. The Document Object Model Events Level 3 is currently being standardized by W3C and it builds on the DOM level 2 model [7].

6.8   WS-Eventing and WS-Notification

Two key mechanisms have been proposed for realizing pub/sub for Web services, namely WS-Eventing and WS-Notification. The former was submitted to W3C in 2006 as member submission and the latter was standardized by OASIS in 2006.

The Web Services Eventing (WS-Eventing) specification[1] describes a protocol that allows Web Services to subscribe or to accept subscriptions for event notifications. An interest registration mechanism is specified using XML Schema and WSDL. The specification supports both SOAP 1.1 and SOAP 1.2 Envelopes.

The key aims of the specification are to specify the means to create and delete event subscriptions, to define expiration for subscriptions, and to allow them to be renewed. Figure 6.7 presents the subscription message format defined in the specification. The format supports different filtering languages and specifically supports XPath 1.0 and 2.0.

The WS-Eventing specification relies on other specifications for secure, reliable, and/or transacted messaging. The specification supports filters by specifying an abstract filter element that supports different filtering languages and mechanisms through the Dialect attribute.

WS-Notification version 1.3 was standardized by OASIS in 2006 These specifications provide a way for a Web service to disseminate information to a set of other Web services. WS-Notification consists of the following specifications [8]:

- The WS-Base Notification specification defines the pub/sub interfaces for Web services.
- The WS-Topics specification defines a mechanism to organize items of interest, called topics. The specification defines an XML model for metadata and several topic expression dialects for filtering.
- The WS-Brokered Notification specification defines the Web services interfaces for notification brokers. This specification supports the creation of distributed pub/sub topologies based on WS-Notification.
- The WS-Notification Policy specification defines a set of policy statements that can be used with the other specifications.

```
[Action] http://www.w3.org/2011/03/ws-evt/Subscribe

[Body]
<wse:Subscribe ...>
 <wse:EndTo> endpoint-reference </wse:EndTo> ?
 <wse:Delivery ...> xs:any* </wse:Delivery>
 <wse:Format Name="xs:anyURI"? > xs:any* </wse:Format> ?
 <wse:Expires BestEffort="xs:boolean"? ...>
    (xs:dateTime | xs:duration)
 </wse:Expires> ?
 <wse:Filter Dialect="xs:anyURI"? ...> xs:any* </wse:Filter> ?
 xs:any*
</wse:Subscribe>
```

Figure 6.7   WS-Eventing subscription message format.

---
[1] W3C Proposed Recommendation 27 September 2011.

6.9   Summary

In this chapter, we examined the key Web standards and specifications that support pub/sub systems. The Web and Internet do not have a basic pub/sub communication primitive at the moment, but promising candidates include RSS and XMPP. We will later consider more advanced distributed pub/sub services for the Web, such as the Pubsubhubbub system. In all, the basic protocols such as REST and SOAP support the creation of interoperable solutions. Interoperability is a key requirement in order to create a global communication system.

The emerging HTML5 and WebSocket standards may become the new industry solutions for implementing push and asynchronous notification in Web applications.

References

1. Fielding RT and Taylor RN (2002) Principled design of the modern web architecture. *ACM Trans.Internet Technol*. **2**: 115–50.
2. Mahemoff M (2006) *Ajax Design Patterns*. O'Reilly Media, Inc.
3. Rose I, Murty R, Pietzuch P, Ledlie J, Roussopoulos M and Welsh M (2007) Cobra: content-based filtering and aggregation of blogs and rss feeds *Proceedings of the 4th USENIX Conference on Networked Systems Design And Implementation*, pp. 3–3 NSDI'07. USENIX Association, Berkeley, CA.
4. Saint-André P (2004) *RFC 3920: Extensible Messaging and Presence Protocol (XMPP): Core Internet Engineering Task Force*.
5. Shelby Z, Hartke K, Bormann C and Frank B (2011) *Constrained Application Protocol (CoAP)* IETF. Internet Draft (work in progress).
6. W3C (2000) *Document Object Model (DOM) Level 2 Events Specification, version 1.0*. W3C Recommendation.
7. W3C (2011) *Document Object Model (DOM) Level 3 Events Specification*. W3C Working Draft.
8. Niblett P and Graham S (2005) Events and service-oriented architecture: the oasis web services notification specifications. *IBM Syst. J*. **44**(4), 869–86.

7

Distributed Publish/Subscribe

In this chapter, we examine distributed pub/sub systems, in which the subscribers and publishers are located at different physical or logical locations in the communication environment. In this setting, the pub/sub system needs to find an information delivery strategy that takes the communications cost and requirements into account. The choice of the event model, routing algorithm, overlay network, and network protocols play a crucial role in determining the overall distributed solution.

7.1 Overview

Figure 7.1 illustrates the key components of a distributed pub/sub system. Starting from the bottom layers, the key components are the following:

- Network protocols that are used to send and receive notification messages. The employed solution can utilize network-level primitives such as broadcast and multicast in order to efficiently disseminate the message. On the other hand, primitives such as multicast and broadcast are not globally available, which has resulted in ubiquitous use of TCP (transport) and HTTP (application layer) protocols. Nevertheless, network primitives are useful in specific settings, such as ad hoc and mesh networks.
- Overlay networks. As discussed in Chapter 3, overlay networks provide useful communication and data processing primitives on top of the network layer. Thus overlay networks are a frequently used building block for distributed pub/sub systems. Overlay networks are useful in distributing functions over a large set of network nodes in a scalable and robust manner. Overlay networks come in many forms and the typical examples are structured, unstructured, and fixed overlays.
- The routing algorithm is the core component of a distributed pub/sub system. The routing algorithms can be classified into three groups, namely flooding, gossip, and filter-based subspaces.
- An event model determines the expressiveness of the pub/sub system. The three well-known examples are topic, type, subject, and content based systems.

Publish/Subscribe Systems: Design and Principles, First Edition. Sasu Tarkoma.
© 2012 John Wiley & Sons, Ltd. Published 2012 by John Wiley & Sons, Ltd.

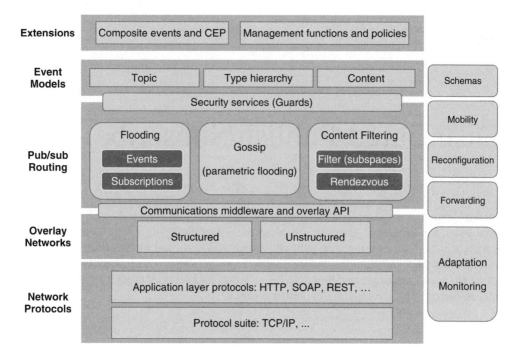

Figure 7.1   The components of a distributed pub/sub system.

The figure illustrates several important components and additional services that are needed. Some of these components span several layers in the figure. The adaptation and monitoring component is required in order to cope with networking failures and changes. The pub/sub network needs to adapt to the underlying overlay network and its under-lay. Event and message forwarding are crucial functions of a pub/sub system, and thus a forwarding component is needed. In order to separate concerns, the routing and for-warding tables should be separate structures. The reconfiguration component operates in the pub/sub routing layer and interoperates with the lower layer adaptation and monitor-ing component. The reconfiguration component needs to configure the pub/sub routing topology, and as a consequence routing and forwarding tables, in such a way that a good mapping of the pub/sub to the network topology is achieved. The mobility compo-nent is responsible for managing mobile subscribers and publishers, and for configuring the pub/sub routing topology to reflect the current locations of them. Finally, the schema component is responsible for the event types and their structures. In addition to these com-ponents, also various management and security functions and policies and user interfaces are needed for a complete pub/sub solution. Security solutions need to consider each layer of the solution stack in order to fully protect the subscribers, publishers, brokers, and the pub/sub network in general. The composite event service can be seen as an application of the basic pub/sub system; however, parts of the composite event detection can and should be distributed across nodes. This requires interaction with the underlying components.

The event system is a logically centralized component that may be a single server or a number of federated servers. In a distributed system consisting of many servers, there are two approaches for connecting sources and listeners:

- The event service supports subscription of events, and it routes registration messages to appropriate servers (for example, using a minimum spanning tree). One optimization to this approach is to use advertisements, messages that indicate the intention of an event source to offer a certain type of event, to optimize event routing.
- Some other means of binding the components is used, for example, a lookup service.

In this context, by event listener we mean an external entity that is located on a physically different node on the network than the publisher. However, events are also a powerful method to enable interthread and local communication, and there may be a number of local event listeners that wait for local events.

Event routing requires that store-and-forward type of event communication is supported within the network. This calls for intermediate components called event routers or brokers. Each event source is connected to at least one router. Each router needs to know a suitable subset of other routers in the domain. In this approach the request, in the worst case, is introduced at every router to get a full coverage of all message listeners. This is not scalable, and the routing needs to be constrained by locality or by hop count. Effective strategies to limit event propagation are zones, the tree topology used in JEDI, or the four server configurations addressed in the SIENA architecture. SIENA broadcasts advertisements throughout the event system; subscriptions are routed using the reverse-path of advertisements, and notifications are routed on the reverse-path of subscriptions. IP Multicast is also a frequently used network-level technology for disseminating information and works well in closed networks, however, in large public networks multicast or broadcast may not be practical. In these environments universally adopted standards such as TCP/IP and HTTP may be better choices for all communication.

A number of different router topologies have been proposed in event literature. Well-known router topologies include:

- *centralized*,
- *hierarchical*,
- *acyclic*, *cyclic*, and
- *rendezvous point (RP)-based* topologies.

Centralized routers represent the trivial case for distributed operation, in which subscribers and producers use a client-server protocol for sending and receiving event messages and invoke the interest-registration service provided by the router.

In hierarchical systems each router has a master and a number of slave routers. Notifications are always sent to the master. Notifications are also sent to slaves that have previously expressed interest in the notifications. The basic hierarchical design is limited in terms of scalability, because one master router is the root of the distribution tree and will receive all the notifications produced in the system. The hierarchical topology was used in the JEDI system [1, 2], and an acyclic topology with advertisements in Rebeca [3–5].

For acyclic and cyclic topologies routers employ a different, server-to-server, protocol to exchange interest propagation information and control messages. Acyclic topologies allow more scalable configurations than hierarchical topologies, but they lack the redundancy of cyclic topologies. On the other hand, topologies based on cyclic graphs require techniques, such as the computation of minimum spanning trees, to prevent loops and unnecessary messaging. The SIENA project investigated and evaluated the topologies with different interest propagation mechanisms [6, 7]. In general, the acyclic and cyclic topologies have been found to be superior to hierarchical topologies [1, 6, 8].

The rendezvous point model differs from acyclic and cyclic topologies, because the routing of a specific type of event is constrained by a special router, the RP. The RP serves as a meeting point for advertisements and subscriptions and avoids the flooding of advertisements throughout the system. Rendezvous-based systems limit the propagation of messages using the RP and thus attempt to address scalability limitations presented by the flooding of subscriptions or advertisements. Typically, an RP is responsible for a predetermined event type. RPs may be used to create a type hierarchy. In this case, a message needs to be sent to the proper RP and any super-type RPs, which may increase messaging cost and limit scalability [9].

7.2 Filtering Content

Pub/sub systems may be classified into five categories based on their expressiveness: channel-based, topic and subject-based, type-based, header-based, and content-based. Topic and channel-based systems make the content delivery decision based on channel names that have been agreed by the communicating participants. Type-based systems make the content delivery decision based on a type hierarchy that has been agreed beforehand by the subscribers and publishers. Subject-based systems make the delivery decision based on a single field. Header-based systems use a special header part of the message in order to make the delivery decision. We can consider the SOAP system as an example, SOAP supports header-based routing of XML-messages [10]. Content-based systems use the content of the message in making the decision [11]. Next, we describe the four well-known categories of message routing systems.

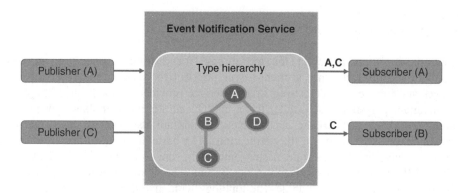

Figure 7.2 Type-based routing.

Channel/topic-based. Delivery decision is made based on the channel on which the event is published. A channel is a discrete communication line with a name. Named channels are also called topics, and they represent an abstraction of numeric network addressing mechanisms. Usually with channel-based messaging, new channels need to be added to the address space, because the producers and consumers must agree on a channel. Channel-based messaging allows the use of IP multicast groups [12]. The channels can be allocated to multicast addresses.

Type-based. The type hierarchy is shared by the publishers and subscribers. The publishers publish events under one or more categories in the hierarchy. The service then matches the published events to the subscribers. A subscriber will receive all events under the subscribed hierarchies. For example, subscriber for category A will receive events for all categories under A in the hierarchy. Subscriber for category B will similarly receive events under that part of the hierarchy, namely the event published in the C category. Type hierarchy allows efficient routing of the messages and ensures that type safety is maintained. On the other hand, this kind of routing is not very expressive and it may be difficult to change the type hierarchy. Figure 7.2 illustrates an event notification service that supports type hierarchies.

Subject-based. Delivery decision is made based on the subject of the event. Subject-based routing is more expressive than channel-based routing. On the other hand, a single field may not be enough to properly describe the content of a message.

Header-based. Delivery decision is made based on a number of fields in the message header. In header-based routing the message has two distinct parts: the header and the body. Only fields in the header are used for making routing decisions. Header-based routing is more expressive than subject-based and has performance advantage to content-based routing, because only the header of a message is inspected.

Content-based. Delivery decision is made based on the content of a message, for example strongly typed fields in the event message. Content-based routing is the most expressive of the four types.

Pub/sub systems typically allow subscribers to specify the delivery constraints in the form of a filter. The filtering model is restricted by the system type, for example header-based systems restrict filters to header fields only, whereas content-based systems allow filters to evaluate statements on the whole message content.

Figure 7.3 illustrates the key concepts of pub/sub in a three-dimensional space defined by the level of aggregation, correlation, and filtering. Aggregation denotes the level of compactness of events and filters. For example: in many cases it is possible to combine two events into a single event without losing information. Similarly, two filters can be combined, or merged, into a single filter given that certain merging conditions are maintained. Correlated filtered events pertains to the detection of patterns and sequences based on events. Typically event correlation is provided by a composite or compound event detection subsystem. Event correlation is the basic building block of CEP systems.

Figure 7.4 illustrates an example event schema and an event that conforms to the schema. The schema defines the structure of the event. Typically the schema defines the event type and can be used to enforce type safety that is necessary for interoperability. The schema can be defined in many ways, for example with XML schemas, an IDL language, or as a set or sequence of typed tuples that define the attribute names and their

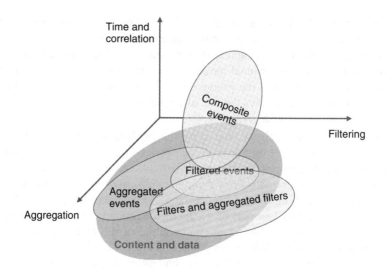

Figure 7.3 Three key dimensions for filtering.

Name	Type	Allowed values
Resource_name	String	ANY
Address	URL	ANY
Resource_type	String	Web page, image, stylesheet, audio file, video file, misc

Event schema

Name	Value
Resource_name	CS Department's home page
Address	www.cs.helsinki.fi
Resource_type	Web page

Event

Figure 7.4 An example event schema and its instance.

types. Our example illustrates the latter in which an event consists of a set of strongly typed tuples. In this case, the schema defines the allowed values for the attributes. The event instance conforms with the schema.

7.3 Routing Function

The main functions of a pub/sub router are to match notifications for local clients and to route notifications to neighbouring routers that have previously expressed interest in the notifications. The routing function is an important part of the distributed system and heart of the routing algorithm. The desirable properties for the routing function are small routing table sizes and forwarding overhead [8], support for frequent updates, and high performance.

There are many ways to realize the routing function and how it builds and maintains the routing tables in the distributed environment. It is evident that a fair amount of

Event Message Flooding

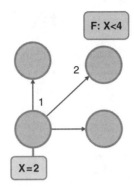

Figure 7.5 Event flooding-based routing.

control messages are needed to be able to connect the publishers and subscribers across the network. A simple and straightforward approach would be to flood all events to all subscribers or to match all subscriptions against the event one by one. The former strategy is illustrated by Figure 7.5 and each subscriber evaluates all published events for a potential match. In the latter, the subscriptions are flooded and each publisher evaluates whether or not a publication should be delivered to a subscriber. These approaches do not scale well and more sophisticated techniques are needed.

Assuming that many events are published and there are more events than subscriptions, the basic strategy is to selectively flood subscription messages in the network in order to reach all possible publishers of requested data. This strategy is typically implemented with reverse path routing so that the published notifications are sent on the reverse path of the subscription messages. This type of routing has different variants depending on the expressiveness.

In *simple routing*, each router in the network follows every subscription in the system. Figure 7.6 illustrates this simple approach that requires the flooding of subscriptions to the whole network. This can be achieved with the following strategy.

Upon receiving a message subscribe (F,X), where X can be either a neighboring router or a local client, the current router R performs the following actions:

- The pair (F, X) is added to the routing table.
- R sends a new message, subscribe (F,R), to all neighbouring routers except the one it received the subscription from X.

Unsubscription can be handled in a similar fashion. This approach is reasonable for small networks or for networks where subscriptions are relatively static. The approach lends itself well to acyclic networks; however, for cyclic networks an additional broadcasting function is needed to prevent loops. With large networks the signalling cost can be excessive.

Identity-based routing is a minor improvement over simple routing, in which the routing tables may contain redundant entries. In identity-based routing, a subscription is not forwarded to a neighboring router if an identical subscription has been forwarded to that

Subscription Flooding

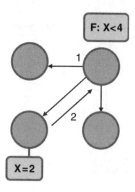

Figure 7.6 Subscription flooding-based routing.

router before [4]. This removes duplicate entries from routing tables and thus reduces unnecessary forwarding of subscriptions, but it complicated the handling of unsubscriptions. An unsubscription is sent to neighboring routers only if no local client has an outstanding subscription with an identical filter. An inverse operation to subscription processing is used to propagate the unsubscriptions.

In *covering-based routing* a covering test is used instead of an identity test. This results in the propagation of the most general filters that cover more specific filters. On the other hand, unsubscription management becomes more complicated because previously covered subscriptions may become uncovered due to an unsubscription.

Merging-based routing allows routers to combine exiting routing entries in order to reduce the number of propagated filters (in other words, routing table elements). Merging-based routing may be implemented in many ways and combined with covering-based routing [8]. Merging-based routing has more complex unsubscription processing when a part of a previously merged routing entry is removed.

Typically, the set of neighbours for a given pub/sub router is not static, but can change over time with new neighbours joining and existing ones leaving. This requires support for dynamic neighbour set. This involves that for a new neighbour, the routing table is scanned for entries that should be sent to the new neighbour. In addition, new subscriptions are accepted from the new neighbour and then first updated to the local routing table and then sent to other neighbours. Care must be taken in order to ensure that loops are not created. Typically, the routing updates are propagated with the help of shortest spanning trees that ensure that the network is acyclic. When a neighbour leaves or fails, routing table entries for that neighbour are removed. This routing state removal is done first locally, and then necessary updates are sent to other neighbours so that they do not send unnecessary notifications to the current router. The dynamic routing table process differs between the filtering model and routing scheme, but the basic idea is the same for the different forms of pub/sub routing.

So far we have discussed subscription driven distributed operation. It is also possible to take into account explicit publisher announcements in the form of advertisements. With advertisement semantics the routers first propagate advertisements. The subscriptions are then connected on the reverse path of advertisements. Notifications are forwarded on the reverse path of subscriptions as before. Advertisements may be used with various routing mechanisms. Advertisements typically have their own routing table and they are managed using the same algorithms as subscriptions. The removal of an advertisement causes a router to drop all overlapping subscriptions for the neighbour that sent the unadvertisement message. Similarly, an incoming advertisement requires that overlapping subscriptions are forwarded to the neighbour that sent the advertisement message. The use of advertisements considerably improves the scalability of the event system [1, 6, 8].

One of the first wide-area pub/sub system based on these two semantics with optimizations was the SIENA system that used covering relations between filters to prevent unnecessary communications. The SIENA system developed the notion of covering for three different comparisons: matching a notification against a filter, covering relation between two subscription filters, and overlapping between an advertisement filter and a subscription filter. Covering and overlapping relations have been used in many later systems, such as Rebeca [5] and Hermes [9, 13]. These systems are examined in more detail in Chapter 9.

7.4 Topic-Based Routing

Topic-based publish/subscribe is a communication paradigm that supports selective message dissemination among a set of data producers and consumers. Messages are associated with topics and are selectively routed to destinations with matching topic interests. Figure 7.7 illustrates topic-based pub/sub with a centralized notification service.

Many industry solutions (such as TIBCO, IBM WebSphere MQ) supported the topic-based pub/sub. In the topic-based pub/sub, a publication message belongs to one of a set of topics (equally groups, subjects, or channels). Users define subscriptions by specifying one or multiple topics, and will receives all publications that are associated with such topic(s).

Brokers maintain channels (or queues) of messages, and decouple publishers and subscribers. Typically, the topic name is associated with a channel, and used to create the channel. Every channel is then identified by a unique name (or with other alias). The unique name is frequently used as the argument of the functions of publish() and subscribe().

In terms of the semantics of the topics, a topic is simply a string that represents a name, a subject, or a topic according to which messages are classified. As a simple semantic model, all topics are *flat*, and users define their content of interest directly by a topic. Besides that, the *hierarchical* relationship of topics is useful to orchestrate topics. Via the hierarchical topics, brokers organize topics according to containment relationships. Based

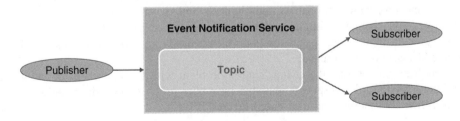

Figure 7.7 Topic-based routing.

on the hierarchical topics, a subscription made to some topic in the hierarchy implicitly subscribe to all the subtopics under that topic. The hierarchy of topics offers many flexible manners for subscribers and publishers, and is widely adopted by most industry vendors.

The matching problem of the topic-based pub/sub can be solved over the topic space, and is much simpler than its content-based counterpart.

7.4.1 Mechanisms

Topic-based publish/subscribe systems are distinguished by the qualities of service (QoS) the system offers to its clients, such as various degrees of reliability, messaging persistence, message ordering constraints, message delivery guarantees, and message delivery latencies constraints, etc. Since the QoS of topic-based pub/sub can follow the design for QoS of the general pub/sub system.

In this section, we mainly focus on the unique mechanism related topic-based pub/sub: how to maintain topics in topic-based pub/sub. First, The channelization problem studied the minimum cost to maintain channels typically inside a broker of the topic-based pub/sub. Second, given an overlay network to connect distributed topic-based pub/sub with minimal edges while ensuring for each topic t, a message published on t could reach all the nodes interested in t. Finally, based on real RSS news dissemination applications, in a Peer-to-Peer scenario, a novel clustering algorithm continuously adapts clusters of topics to minimize the maintenance of topics.

Following the classic survey [14], we do not strictly differentiate the topic-based pub/sub with channel-based pub/sub. It is because publishing a message to a channel is similar to associating a message with a topic and a topic could be the name or identity of the channel.

7.4.2 Channelization Problem

The channelization problem was first studied in [15] in the context of traffic flows and a fixed number of channels. In large scale data dissemination applications, for example, topic-based pub/sub systems, a large number of publications are delivered to a large number of subscribers. However, because of differences of user interests, not all receivers are interested in all publications. The channels provide opportunities to deliver users with publications of interest. The maintenance of channels, however, requires resources and management overhead. Given a very large number of topics or publications, it is impossible to maintain a channel for each topic (or each publication).

Thus, given a limited number of channels with a large number of information sources (i.e. publishers) and consumers (i.e. subscribers), this problem defines the map (i) between information publishers and channels and (ii) the map between information consumers and channels. Based on the maps, users are expected to receive all needed information without false exclusion and meanwhile, the amount of unneeded data received by users is minimized.

It has been proven that the channelization problem with constrained or unconstrained conditions is NP-hard and several heuristics have been proposed including random assignment, flow-based merge, and user-based merge [15].

7.4.3   Distributed Overlay with Many Topics

Differing from [15] that maintains channels of a broker for topic-based pub/sub, [16, 17] studied the problem of designing a scalable overlay network to support decentralized topic-based pub/sub communication. They proposed a new problem, namely the *Minimum Topic-Connected Overlay (Min-TCO)*, to capture the tradeoff between the scalability of the overlay (in terms of the fanout parameter) and the message forwarding overhead incurred by the communicating parties. That is given a collection of nodes and their subscriptions, connect the nodes using the *minimum possible number of edges* so that for each topic t, a message published on t could reach all the nodes interested in t by being forwarded by only the nodes interested in t.

It has been proved that the decision version of Min-TCO is NP complete [16] and there is a polynomial algorithm that approximates the optimal solution within a logarithmic factor with respect to the number of edges in the constructed overlay. It has been further proved that this approximation ratio is almost tight by showing that no polynomial algorithm can approximate Min-TCO within a constant factor (unless $P = NP$) [16].

7.4.4   Dynamic Clustering in Topic-Based Pub/Sub

With the popularity of RSS news syndication there is a need for scalable and real-time document distribution over the Internet. By treating each RSS URL as a topic, the dissemination of RSS news can be treated as an application of the topic-based pub/sub. The dissemination of events in each topic incurs two main costs: (1) the actual transmission cost for the topic events, and (2) the maintenance cost for its supporting structure. When a pub/sub system supports a large number of topics with moderate event frequency, such maintenance overhead becomes particularly dominating.

To overcome the maintenance overhead, a dynamic clustering method for reducing this maintenance overhead to the minimum has been proposed [18]. The distributed clustering algorithm utilized correlations between user subscriptions to dynamically group topics together, into virtual topics (called topic-clusters), and thereby unifies their supporting structures and reduces costs. Such a technique continuously adapted the topic-clusters and the user subscriptions to the system state, and incurs only very minimal overhead.

7.4.5   Summary

In this section, we introduced the basic concept of topic-based pub/sub, and gave industry examples that have adopted the topic-based pub/sub model. Finally, we present three recent research systems on topic-based pub/sub. Such systems mainly focus on the maintenance of topics, especially on reducing the maintenance cost when the number of topics is very large.

7.5   Filter-Based Routing

Figure 7.8 illustrates the filter propagation technique, in which a filter is selectively flooded in the network and matching event messages are delivered either directly or on the reverse

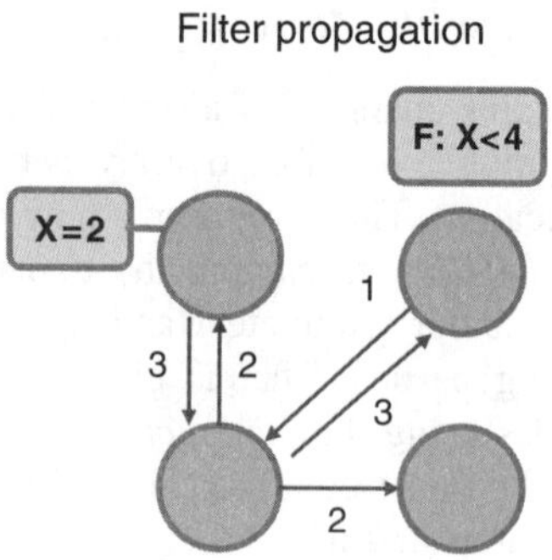

Filter propagation

Figure 7.8   Filter propagation-based routing.

path to the subscriber. This contrasts the naive subscription flooding discussed earlier in this chapter. Subject-based, header-based, and content-based routing are examples of routing with filters. This approach lends itself well to filter aggregation that can be used to reduce routing table sizes and matching cost. This filter based strategy is frequently employed by many of the solutions presented in this book. Indeed, the more elaborate content-based routing model builds on this.

Algorithm 7.1 gives an example of a simple filtering router that maintains two structures: the routing and matching structures. The routing structure is for remote subscriptions and the matching structure for local clients. The algorithm does not handle duplicate detection or filter containment. In addition, it does not handle the case of dynamic neighbours and the code for managing new neighbours and removing existing ones would need to be added for dynamic environments.

Dynamic neighbours can be handled with the following process description:

- For a new neighbour, it is important to ensure that no loops will be created in the network. If the network is based on an acyclic graph topology, then this is guaranteed. Otherwise, additional mechanisms are needed to ensure that a loop is not created.
- The new neighbour will receive those filters that must be subscribed from it. The new neighbour will receive the current routing table (possibly in aggregated form). The new neighbour then sends its own routing table. If the routing topology is cyclic, the exchanged routing table data needs to processed in such a way that cycles are not created.
- After this the new neighbour has been connected to the pub/sub network (or two separate networks have been combined).
- When a neighbour leaves, the filters that have been received from the departing neighbour are simply dropped, and they are also removed from other neighbours by unsubscription, if necessary.

This general process for adding and removing neighbours generalizes to the many forms of filter-based routing, for example rendezvous-based and content-based variants.

Algorithm 7.1 Filtering-based routing

Data: *Op* denotes the operation for the router, *m* denotes the data of the operation,
  x denotes the node that sent the operation.
begin
  SUB is the set of local subscriptions.
  EXT is the set of external subscriptions.
  switch *Op* **do**
    case *Publish*
      $matched \longleftarrow match(m, SUB)$
      trigger notify(m) for *matched*
      $fwd \longleftarrow match(m, EXT)$
      forward data m to $fwd \setminus \{x\}$
    end
    case *Subscribe*
      if *x is a client* **then**
        $SUB \longleftarrow SUB \cup \{m\}$ /* Update matching structure. */
      else
        $EXT \longleftarrow EXT \cup \{m\}$ /* Update routing structure. */
      end
      Send (Op, m) to $neighbours \setminus \{x\}$
        end
    case *Unsubscribe*
      if *x is a client* **then**
        $SUB \longleftarrow SUB \setminus \{m\}$ /* Update matching structure. */
      else
        $EXT \longleftarrow EXT \setminus \{m\}$ /* Update routing structure. */
      end
      Send (Op, m) to $neighbours \setminus \{x\}$
    end
  end
end

7.6 Content-Based Routing

In the content information model, the users subscribe to information based on their interests. A network of content-based routers or brokers is then responsible for delivering messages that match the interests to the proper subscriber. Content-based routing is typically implemented above the network level. This content-based model contrasts the typical networking model, in which data is sent to a recipient address (unicast) or to a set of addresses (multicast). The classical content-based pub/sub system was the SIENA system developed in the late 1990s. In this section, we focus on this model.

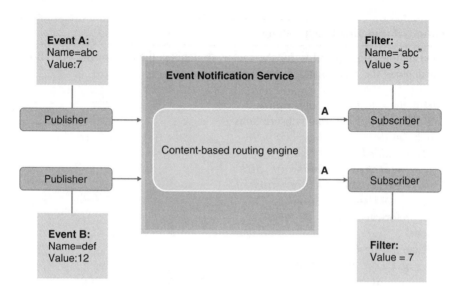

Figure 7.9   Content-based routing.

Figure 7.9 illustrates the content-based routing scheme with a simple one broker example. Two subscribers have connected with the content-based router and installed active filters. Two publishers are also connected and they have published two events that are then matched by the routing service to the active filters. As a result of the matching, the subscribers are then notified. Event B does not have any matching filters, but event A is forwarded to both subscribers. The two filters together define the content-based subnetwork of this router.

7.6.1   Addressing Model

Carzaniga and Wolf defined a content-based addressing model by considering the predicates that define subscriptions as the destination addresses [19]. In this model, datagrams are implicitly addressed to a node by their content. Various predicate models can be developed. Typically, the predicate model is based on a set of Boolean functions that capture the receiver interests.

In a content-based network, each node (client or router) advertises a *receiver predicate* (*r-predicate*) that defines the set of datagrams the node is interested in receiving. A node may also advertise a *sender predicate* (*s-predicate*), which defines the datagrams the node intends to send. The *content-based address* of a node n is its r-predicate p_n [7, 20].

The content-based address of a node is implicit and very different from the standard notion of a network address. The key differences of content-based addresses to IP addresses include the rate at which a content-based of a node address changes, the abstraction of the sending and receiving host by the address and the network, and the fact that many nodes can have the same address. The change rate can be several orders of magnitude faster than with traditional addresses creating challenges for network scalability. Nodes do not have unique addresses in a content-based network; however, unique addresses are typically used on the lower layers [7, 20].

Content-based routing requires an algorithm that uses a forwarding table. The forwarding table is responsible for mapping receiver predicates to interfaces in such a way that given a message the proper destination interfaces can be determined. The forwarding system typically needs to take into account the network topology and prevent forwarding loops. An incoming datagram is forwarded only to nodes that are on a broadcasting path from the source. The broadcasting path is defined by the broadcast function; for example, it may be determined by a spanning tree rooted at the source of the datagram.

In networking, a *subnet* is a set of nodes that are topologically close to each other under a single subnetwork address. The fact that subnets are identified by a single address allows routers to maintain compact routing tables, which improves scalability. Subnetworks can also be aggregated by routers with *supernetting*, in which multiple routing able entries are combined into a single entry. These two principles can also be applied in the context of content-based networking. Let filters F_1 and F_2 denote filters of routers p and q, respectively. If $F_2 \sqsupseteq F_1$, then p is a *content-based subnet* if q and q is a *content-based supernet* of p.

7.6.2   Propagating Routing Information

The content-based routers propagate routing information by selectively flooding subscriptions and optionally advertisements in the network. The content-based model lends itself well to subscription and advertisement covering, which happens by determining the containment relationships of the predicates (filters). Covering relations and related data structures are examined in Chapter 8.

Figure 7.10 illustrates how the next-hop neighbours are selected in a content-based pub/sub system implementing advertisements and the covering optimization when processing an incoming subscription message. The set of destinations for any message

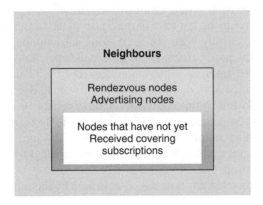

Figure 7.10   Routing table structure.

is constrained by the neighbours. The neighbour set is then constrained by the received advertisements from neighbours and possibly by rendezvous related information. In the latter, the network has assigned specific content specific rendezvous points that constrain the propagation of messages. After removing those neighbours that do not have matching advertisements, a subscription is sent to the neighbours who have not yet received covering subscriptions.

7.6.3 Routing Behaviour: Subscriptions

We consider how to implement the routing and forwarding functions of a content-based router by examining the workings of the SIENA system. SIENA defines a specific routing table structure called the poset that stores filters based on their covering relations. The idea is simple – more general filters cover, or contain, more specific filters. This way the structure can aggregate filters for routing purposes. We return to the poset structure in Chapter 8, where the detailed algorithms are presented. Now, we assume that there is a poset structure for storing the filters, and with the information stored by the poset, we calculate the *forwards* set that contains the next-hop neighbours for an incoming subscription or unsubscription message.

In distributed operation based on an acyclic graph router topology, the SIENA server defines the set *forwards(f)* as presented in the equation

$$forwards(f) = neighbours - NST(f) - \bigcup_{f' \in P_s \wedge f' \sqsupseteq f} forwards(f'). \tag{7.1}$$

The *neighbours* set contains the event brokers connected to the current broker (one application-level hop distance). The functor *NST (Not on any Spanning Tree)* means that the propagation of f must follow the computed spanning trees rooted at the original subscribers of f. With acyclic topologies *NST* contains the neighbour that sent f. With the *NST* term, the system can support different topologies, including cyclic ones.

P_s denotes the subscription poset. Using the equation, f is never forwarded to the neighbour that sent it. Due to the last term of the equation the subscription is not forwarded to any routers that have already been sent a covering subscription. We observe that only the two first levels of the poset structure can have nonempty *forwards* sets.

Because X is removed from all subscriptions covered by f, an intermediary pub/sub router does not know which subscriptions should be forwarded due to unsubscription. This information is essentially lost by this optimization; however, the origin of the subscriptions has this information and propagates any subscriptions due to the unsubscription in the same message, which is applied atomically by other servers. The *unsubscribe(X,f)* removes X from the *subscribers* set of all subscriptions that are covered by f. Filters with empty subscriber sets are removed.

Algorithm 7.2 gives an overview how the poset is used by the subscribe and unsubscribe message handlers.

Algorithm 7.2 Message handlers for subscription processing

Data: *Op* denotes the operation for the router, f denotes the input filter, x denotes
     the node that sent the operation.
begin
    P_s is the set of subscription poset.
    switch *Op* **do**
        case *Subscribe*
            Add (f,x) to P_s.
            Forward subscription message using *forwards*(f).
        end
        case *Unsubscribe*
            Remove (f,x) from P_s.
            Let F_O denote the old forwards set and F_N a newly computed forwards
            set for f after the subscriber x has been removed from the *subscribers* set.
            if *subscribers set is empty* **then**
                $F_N = \emptyset$.
            end
            The unsubscription is forwarded to $F_O \setminus F_N$.
            The set may be empty if there are subscriptions from other neighbours
             that cover f.

            The *forwards* sets of subscriptions covered by f may change, which may
            require the forwarding of new subscriptions.
            Any uncovered subscriptions in P_s are forwarded with the
            unsubscription message.
            An uncovered subscription is such that its forwards set gains
            an additional element due to the removal of a covering filter.
        end
        end
end

7.6.4 *Routing Behaviour: Advertisements*

The basic subscription semantics may be optimized by using advertisements. In this
model, advertisements are propagated to every node, and subscriptions are propagated
only towards advertisers that have previously advertised an overlapping filter. The idea
is to use the additional advertisement information to prevent subscription flooding. The
model uses two poset data structures, one for each type of message.

In advertisement semantics a second poset P_a is used for advertisements [21]. The sets
advertisers(a) and *forwards*(a) are needed for each advertisement $a \in T_A$, where T_A is the

set of all advertisements in the poset. Instead of forwarding subscriptions to a global set *neighbours*, a set constrained by advertisements is used as presented by the equation

$$neighbours_s = \bigcup_{a \in T_A : a \simeq s} advertisers(a) \cap neighbours. \qquad (7.2)$$

In this case, Equation 7.2 uses the *neighbours*$_s$ set instead of the *neighbours* set. An advertisement may thus result in a number of subscriptions being forwarded to the sender of the advertisement. The process of unadvertisement is similar to unsubscription.

Algorithm 7.3 Advertisement extension

Data: Op denotes the operation for the router, m denotes the data of the operation, x denotes the node that sent the operation.
begin
    EXT is the set of external subscriptions.
    ADV is the set of advertisements from neighbours.

    switch Op **do**
        case *Advertise*
            $ADV \longleftarrow ADV \cup \{m\}$ /* Update adv structure.   */
            Add m from x to ADV. Forward m to $neighbours \setminus \{x\}$
        end
        case *Subscribe*
            $EXT \longleftarrow EXT \cup \{m\}$ /* Update remote subsc. structure. */
            $A \longleftarrow getMatchingADV Destinations(m, x)$
            /* Received advertisements constrain the routing */.
            Send (Op, m) to $(neighbours \setminus \{x\}) \cap A$
        end
        case *Unsubscribe*
            Forward unsubscription and uncovered subscriptions.
        end
        case *Unadvertise*
            Forward unadvertisement and uncovered advertisements.
        end
    end
end

7.6.5 Routing Tables

The desirable characteristics for a content-based routing table are efficiency, small size, support for frequent updates, and extensibility and interoperability. The routing table data structure should be generic enough to support a wide range of filtering languages.

    The routing tables of content-based routers are typically represented as sets and the mechanisms for inserting and removing filters are left unspecified. For example, JEDI [2] and Hermes [9] keep filters in a simple table, and Rebeca uses sets and a counting algorithm for finding covering filters and mergeable filters [4]. Two counting-based algorithms are needed for routing. One to determine the covered filters, and one to determine the covering filters. We investigate matching algorithms and structured in Chapter 8.

The filters poset data structure was used in the SIENA system to store filters by their covering relations and manage information related to forwarded messages. The filters poset can be thought of as the routing table for a SIENA router. The poset stores filters by their generality and may also be used to match notifications against filters by traversing only matching filters in the poset starting from the most general filters. We call the set of most general filters that covers other filters the *root set* of the data structure in question.

The filters poset is a generic data structure and may be used with various filter semantics. The poset may also be used for various interest propagation mechanisms, such as subscription and advertisement semantics. On the other hand, this generality has a performance drawback. One of the findings in SIENA was that the filters poset algorithm limits the performance of routers and more efficient solutions are needed [20]. The two-layer CBCB design examined in Chapter 9 was presented as a solution to the scalability challenge.

7.6.6 Forwarding

The forwarding function is responsible in determining the set of next-hop destinations for a datagram that has arrived at a router [19]. The set of next-hop destinations may contain both adjacent routers and client nodes connected to the router. The router computes the next-hop destinations based on the datagram content and its *forwarding table*. The forwarding table is an internal data structure that is compiled and updated by the routing function.

Conceptually, content-based forwarding and routing tables are different structures. The former may be optimized for fast matching while the latter is optimized for efficient update and remove operations. The forwarding table is then periodically updated or rebuilt by the routing function, based on the routing state updates it has received. On the implementation level, however, routing and forwarding tables may be implemented by the same data structure. Figure 7.11 illustrates an example of a content-based forwarding table.

More formally, the forwarding table can be seen as a map from the set of interfaces to the set of content-based addresses. The interfaces of the router represent the neighbouring nodes. The local client nodes can be treated as a single *local interface* I_0 [7, 20]. The task of the forwarding function is finding the set of next-hop interfaces for an incoming datagram or message.

Interface	Node ID	Address
I_0	local	$(p < 400) \vee (al = \text{FINNAIR})$
I_1	R2	$orign = \text{HEL}$
I_2	R3	$destination = \text{SFO}$
I_3	R4	$(p < 200.00) \vee (p < 300.00)$

Figure 7.11 A content-based forwarding table maintained by a router.

7.6.7 Performance Issues

Throughput of the forwarding function determines the throughput of the router. Thus is is vital to optimize the performance of the forwarding function. This optimization involves both forwarding and routing functions, because the routing function provides the data for the forwarding function. This optimization has been an active area of research and development, and many efficient algorithms are known for various data models. In addition to router throughput also other metrics need to be considered, such as the following:

- overall performance;
- size of the routing table;
- signalling overhead of the routing protocol;
- accuracy of the forwarding function (false positives and negatives);
- fluctuations in time (burstiness);
- reconfiguration time when topology changes.

Early research pertaining to the forwarding function was carried out in the context of event matching in centralized pub/sub systems [7, 20]. We review the history and current state of the art for matching and forwarding algorithms in Chapter 8.

7.6.8 A Generalized Broker with Advertisements

A content-based broker or router based on advertisement semantics is depicted in Figure 7.12. The router is responsible for connecting advertisers and subscribers, and forwarding notifications. The router operates on messages, which may be subscriptions, advertisements, notifications, or control messages that are sent to interfaces and received from interfaces. An interface I represents a particular neighbour of a router. The top part of Figure 7.12 presents a simple configuration with four routers.

The middle part of the figure presents the router functionality in more detail. For message forwarding functionality each input interface maps to all output interfaces except itself. The routing core is responsible for maintaining routing state and mapping messages to interfaces. The routing state is typically maintained using a poset data structure examined in Chapter 8. The router is also responsible for queue management and prioritization of traffic.

We may now formulate the assumptions on the behaviour of the content-based router with advertisements.

- Any message from input interface I is not forwarded to output interface I.
- A notification must be forwarded to an output interface if the interface has a matching input subscription.
- A subscription must be forwarded to an output interface if the interface has an overlapping input advertisement and a covering subscription has not been forwarded.
- An advertisement must be forwarded to an output interface if a covering advertisement has not yet been forwarded.
- An unsubscription message must be forwarded to an output interface if the interface has subscribed the message and there are no covering subscriptions and local

subscriptions. If there are active subscriptions that are covered by the unsubscription they are forwarded with the unsubscribe message and installed at the other router using an atomic operation.

- An unadvertisement must be forwarded to an output interface that has previously forwarded the advertisement and there are no covering advertisements or local advertisements. If there are active advertisements that are covered by the unsubscription they are forwarded with the unadvertise message and installed at the other router using an atomic operation.

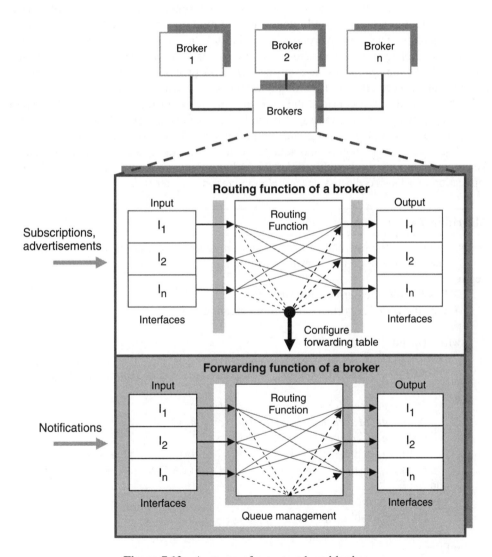

Figure 7.12   Anatomy of a content-based broker.

7.7   Rendezvous-Based Routing

Rendezvous points (RP) are used in many overlay routing systems [22–24] to reduce communication costs and realize nonfixed indirection points. In essence, RPs constrain the propagation of advertisements and subscriptions, and they are a coordination mechanism. RPs can be uniformly distributed over the addressing space or placed using some specific scheme. The placement RPs using uniform distribution is motivated by the fact that the event types are disjoint from the viewpoint of matching. On the other hand, event traffic distribution may well be nonuniform, which should also be taken into account. The problem with nonuniform traffic distribution is that an RP may be located on the other side of the network. The RP may then become a performance and scalability bottleneck.

Figure 7.13 illustrates the rendezvous based routing, in which an RP is used to coordinate signalling. This form of routing is similar to filter based routing, but now the RP restricts the propagation of messages. The key idea is that the subscriptions (filters) and the published event messages meet at the RP or on their way to the RP. This form of routing can be more efficient than simply selectively flooding subscriptions.

In the following, we consider the case that the RP constrains the basic filter-based routing by only propagating messages towards the rendezvous point. Algorithm 7.4 gives an outline of this kind of routing with advertisements. As illustrated in Figure 7.10, the neighbours, rendezvous points, and advertisements all constrain subscription propagation. There are many variations of the rendezvous scheme depending on the role of the RP and replication of RPs. We discuss several alternatives in Chapter 9.

Algorithm 7.4 Covering extension

Data: Op denotes the operation for the router, m denotes the data of the operation,
        x denotes the node that sent the operation.
begin
   EXT is the set of external subscriptions.
   ADV is the set of advertisements from neighbours.

   switch Op **do**
      case $Subscribe$
          $EXT \longleftarrow EXT \cup \{m\}$ /* Update remote subsc. structure.
          */
          $R \longleftarrow getRendezvousDestinations(m, x)$
          $A \longleftarrow getMatchingADVDestinations(m, x)$
          $C \longleftarrow getCoveredDestinations(m, x)$
          /* Both advs and rendezvous constrain the routing */.
          Send (Op, m) to $(neighbours \setminus (C \cup \{x\})) \cap A \cap R$
      end
      case $Unsubscribe$
          Forward unsubscription.
      end
   end
end

Rendezvous routing

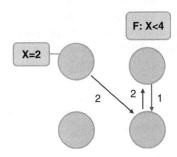

Figure 7.13   Rendezvous-based routing.

7.8   Routing Invariants

Different invariant system properties are useful in characterizing networks and ensuring that certain important requirements are not violated. For example, we are interested in showing that no published events are lost even though the routing system changes over time.

In this section, we first examine the notion of configuration and invariants for a general routing algorithm, and then examine pub/sub specific invariants.

7.8.1   Configurations

The first step towards defining useful routing and system invariants is the notion of a configuration, which is a certain snapshot of the distributed system state. The configuration contains the nodes of the network and their current links. Typically the configuration is represented as a graph with the nodes and the links modelled as edges between nodes.

Now, we can define the notions of soundness, completeness, and stretch for routing configurations and algorithms [25]. Soundness for a configuration means that all the nodes that are reported to be reachable are reachable, and for a routing algorithm that all such nodes are reachable when update activities have ended. A reachability function $f_u(w)$ reports whether or not node w is reachable at u. The function $\phi(u, w)$ gives the path between nodes u and w if it exists.

Definition 7.1 *A configuration is sound if for all nodes u and w, $f_u(w) \neq NONE$ implies $\phi(u, w)$ is a path from u to w. A routing algorithm is sound if it produces a sound configuration after the network becomes quiet.*

Definition 7.2 *A configuration is loop-free if for all u and w, $\phi(u, w)$ is finite. A routing algorithm is loop-free if it produces a loop-free configuration after the network becomes quiet.*

The difference between a sound and a loop-free configuration is that in the latter, a node only needs to know that forwarding to its next hop will not cause a loop (but the packet could be dropped somewhere down the path), while in a sound configuration, forwarding to the next hop must actually reach the destination.

The easiest way to achieve soundness is for every node to pretend everyone is unreachable by setting $f_u(w) = NONE$ for all destinations w. Clearly this is a undesired configuration, so $f_u(w)$ should be NONE only if w is unreachable from u. This property is called completeness.

Definition 7.3 *A configuration is complete if for all distinct u and w, if w is reachable from u then $f_u(w) \neq NONE$. A routing algorithm is complete if it produces a complete configuration after the network becomes quiet.*

Together the soundness and completeness properties say that all nodes are reachable by forwarding; however, they do not say anything about the optimality of the paths. Thus we need the notion of stretch that defines how the optimal are the paths given by the routing algorithm when compared with the best possible paths. Stretch is one frequently used metric for evaluating the goodness of routing algorithms.

7.8.2 Pub/Sub Configurations

The above notions of soundness and completeness are generic and only concern the basic reachability of nodes in the network. Soundness and completeness are useful invariants for a pub/sub network; however, additional definitions are needed to ensure that events are properly delivered and that no events are lost when network topology changes occur.

We are interested in ensuring that a routing configuration is valid so that the pub/sub system does not allow illegal operations. In order to formally define the notion of a valid routing configuration, we need to define the possible operations that may take place. A trace is a sequence of operations, such as subscribe, notify, and unsubscribe in causal order that determine the execution of the system. Now, we can formalize the constraints that define a valid routing configuration. One frequently used technique for defining the constraints is the *Linear Temporal Logic (LTL)*. LTL formulas are used to define a specification and a system is correct when it exhibits only traces allowed by the specification. The key constraints are defined with: □ which denotes "always", ◊ which means "eventually", and ○ which applies to the next operator in the trace. N denotes the set of all notifications [4], and $N(F)$ the notifications selected by the filter F.

Any valid routing configuration must satisfy the following constraints on traces defined using the operators of the LTL. Property 7.1 gives the liveness constraint for the pub/sub system with subscription semantics. This property defines when a notification should be delivered and ensures that they are eventually delivered.

Property 7.2 gives the safety constraint that ensures that incorrect events are not processed and delivered. Note that this property does not specify any particular delivery order for notifications, and we need additional constraints to introduce, for example, causal ordering using Lamport's happened-before relation [26] discussed in Chapter 2.

The properties are from the definitions in [4] with minor changes in presentation, which also contains the definitions for the safety and liveness properties for advertisement semantics and proofs are given for the correctness of content and merging-based routing.

Property 7.1 *Liveness:*

$$\Box[Sub(A, F) \Rightarrow [\Diamond\Box(Pub(B, n) \wedge n \in N(F) \Rightarrow \Diamond Notify(A, n))] \vee [\Diamond Unsub(A, F)]],$$

specifies that a subscription with filter F and the publication of an event n that matches the subscription will lead to an eventual notification of subsequent publications of that event unless the subscription is invalidated by unsubscription.

Property 7.2 *Safety:*

$$\Box[Notify(A, n) \Rightarrow [\circ\Box\neg Notify(A, n)] \wedge [n \in Published] \wedge [\exists F \in Subs(A) : n \in N(F)]],$$

specifies that a notification is delivered only once, it has been published previously, and that the recipient has a matching subscription. Published is the set of published events, and the set Subs gives the subscriptions for each client.

7.8.3 False Positives and Negatives

In order to understand the model and instances based on the model we need useful metrics to characterize the system and the event flows. The two most important metrics are the number of false positives and negatives, and even more so proofs for the existence of false negatives or positives in a particular system. False positives are events that are delivered but were not subscribed, and similarly false negatives are events that were subscribed but were not delivered upon publication. Clearly, the presence of false negatives indicates a serious error in any event system. Therefore, we are interested in proving that a candidate event system does not manifest this erroneous behaviour. Other interesting and useful metrics are end-to-end latency in relevant time units, number of intermediate brokers for each flow (end-to-end path length), and size of routing tables.

False positives are easy to define, but false negatives require that we examine the duration in which a subscription is active. For example, the current formulation tolerates any delay and the issuance of the unsubscription forgets any possible false negatives that have not yet been delivered.

7.8.4 Weakly Valid Routing Configuration

The weakly valid routing configuration ensures the delivery of events to subscribers whose subscription update process has terminated in a static topology. Thus a routing algorithm based on this configuration that ensures updates process terminates satisfies Definitions 7.4 and 7.3 ([4]).

We call all update procedures that have ended successfully as complete in the topology, and observe that completeness can be used to characterize the pub/sub network. Completeness can thus be applied on multiple levels in the system. In pub/sub, we are also interested in reachability, but not the reachability of particular node, but in the reachability of content from publishers to subscribers. Thus a reformulation of the previous notion of completeness for pub/sub and for content delivery is needed. The completeness of subscriptions and advertisements in given by Definition 7.4. Note that Definition 7.4 imposes new requirements for the topology.

Definition 7.4 *An advertisement A is complete in a pub/sub system P S iff there does not exist a router* $s \in S_B$ *with an overlapping subscription that has not processed A. Similarly, a subscription S is complete in P S iff there does not exist a router* $s \in S_B$ *such that s has an an advertisement that overlaps with S and S is not active on s.*

7.8.5 Mobility-Safety

In distributed pub/sub systems it is clear that a subscription may require an arbitrary number of network hops before it is connected with publishers. Events published during this configuration period may be lost. The mobility-aware weakly valid routing configuration tolerates these false negatives due to the reconfiguration caused by stationary subscribers and publishers.

Mobile clients change this scenario, because the flow endpoints are mobile and brokers transfer endpoints using a handover process. False negatives that occur during mobile clients are not tolerated.

A mobility-safe pub/sub system can be defined as follows [27]:

Definition 7.5 *A pub/sub system is mobility-safe if starting from an initial configuration* C_0 *at time* T_0 *and ending in a final configuration* C_e *at time* T_e *handovers (mobile clients) will not cause any false negatives.*

The notion of subscription completeness and mobility-safety are important in engineering efficient solutions for pub/sub mobility and topology reconfiguration. We investigate these solutions in Chapter 11.

7.8.6 Stabilization and Eventual Correctness

It may be difficult to maintain the above properties of safety and liveness in dynamic pub/sub systems, and therefore they may be relaxed. A self-stabilizing pub/sub system ensures correctness of the routing algorithm against the specification and convergence [4]. The safety property may be modified to take self-stabilization into account by requiring eventual safety. Well-known stabilization mechanisms include periodic heartbeats and subscription leases to detect failures.

These two solutions address differing challenges in the distributed system:

- Periodic heartbeats address the failure of brokers and routers in the network that process and propagate subscription and advertisement messages. When a failed broker is detected, the subscriptions and advertisements received from it are cleared.
- Subscription and advertisement leases address the failure of subscribers and publishers, for example endpoints of the content-based system. The expiration of a lease results in the removal of the subscription or advertisement from the network. With leases, the pub/sub API does not necessarily need to have specific unsubscribe and unadvertise operations.

Periodic heartbeats are another necessary mechanism for coping with failures as demonstrated by the Hermes system [9]. Typically pub/sub routers and brokers send heartbeat messages in order to prove that they are still available and functioning properly. If a broker has not sent a heartbeat message for several rounds it can be suspected of a failure. Eventually failed brokers are removed from routing tables and the subscriptions they have sent removed as well.

Leases are a frequently employed technique in order to stabilize a distributed system when failures can occur. In this case, the expiration lease always leads to unsubscription. Thus it can be ensured that even with failures the subscriptions of failed subscribers are eventually removed from the distributed system. On the other hand, this requires that the network maintains state in order to be able to unsubscribe an expired subscription. Thus leases complicate optimizations such as the covering and merging optimizations discussed in this chapter. Leases also require periodic resubscriptions in order to keep the lease from expiring. In a distributed system this increases the signalling cost of the solution. The key parameter is the refresh time (lease time) of the subscriptions that needs to be set carefully to minimize unnecessary signalling [28].

These solutions alleviate the hard state nature of many traditional content-based systems. A hard state system assumes that the state set by subscriptions and advertisements will be eventually removed when unsubscription or unadvertisement happens. Unfortunately, if subscribers, publishers, and brokers fail, state is not removed. As a consequence, the brokers' routing tables will eventually become polluted by obsolete entries and the content routing topology may become very unoptimal.

Soft state [29] pub/sub systems contrast hard state systems by achieving eventual correctness with stabilization mechanisms such as heartbeats and leases. These systems can be made to automatically recover from subscriber, publisher, and broker failures.

7.8.7 Soft State

In the soft state solution of the XSIENA system, the publishers and subscribers determine the lease time of periodically published subscription and advertisement messages [28]. These messages need to be renewed before their leases expire or otherwise they are removed from the network. The system adaptively calculates the lease time in such a way that no prior network information is required such as the diameter of the network. The underlying model for the soft state system is the *Timed Asynchronous Distributed System Model* [30] that has been shown to be suitable for modelling loosely coupled systems. The model allows the development of adaptive systems that react to violations of the real-time requirements.

A major challenge with leases is the estimation of the transmission and processing delay across the network. It is important to have a good estimate of the delay in order to set the lease time in such a way that the current lease does not expire while a refresh message is still being processed in the network having not yet reached all the brokers. This estimation is made difficult by not knowing the pub/sub overlay topology, which is a blackbox for our purposes, and that typically there is no clock synchronization between the brokers and nodes. Variable delays may lead to the refresh messages arriving late and resulting in the expiration of a lease.

The main idea is to adaptively estimate the propagation times of messages. The propagation time for a message is the sum of the transmission and processing delays it experiences in the network until received by the recipient. The transmission delay is defined as the actual time spent by the message in transit between two nodes in the network. The processing time is the time spent by the message at the given node from its reception until its dispatching.

The lease time calculation involves determining a posteriori uncertainty regarding the propagation times of the messages. This uncertainty is estimated as the difference between the minimum and maximum propagation times for every message within a certain window defined either in time or number of refresh messages. The estimated uncertainty is then added to the lease time of every message in order to factor in the varying propagation delays. The uncertainty is determined based on the one way propagation delay of messages estimated with the upper bound message propagation delay technique [28].

The upper upper bound for the message transmission delay is based on feedback from the destination with helper messages [31]. The idea is simply to calculate the time difference based on round trip communications and local clocks. This gives a good estimation of the round trip delay given that the local clock drifts are bounded. This is a simple scheme, and it is non-trivial to apply it to the context of a large-scale pub/sub system, in which the destinations are not known and the diameter of the network is unknown.

7.9 Summary

In this chapter, we examined the basics of distributed pub/sub, in which subscribers and publishers are located at different physical or logical locations and connected through the logically centralized event service. We covered many different routing strategies and mechanisms in order to be able connect subscribers and publishers. The choice of the event model, routing algorithm, overlay network, and network protocols play a crucial role in determining the overall distributed solution.

The routing algorithm is the core component of a distributed pub/sub system. The routing algorithms can be classified into three groups, namely flooding, gossip, and filters. An event model determines the expressiveness of the pub/sub system. The three well-known examples are topic, type, and content based systems. We focused on topic- and content-based systems and examined their distributed operation. Different invariant system properties are important in characterizing networks and ensuring that certain important requirements are not violated. We examined the safety and liveness properties for pub/sub systems as well as stabilization and soft state.

Figure 7.14 illustrates the key solutions considered in this chapter that are the basis of distributed pub/sub systems. Reverse path routing is frequently used to build multicast trees. The idea is very simple: published events follow the reverse path of matching subscriptions. This can be implemented directly on top of the network layer or on top of an overlay routing substrate such as a DHT network. Advertisements are also a frequently used optimization that work together with subscriptions and reverse path routing. With this extension, subscriptions are sent on the reverse path of matching advertisements. Typically this matching is based on overlap between a subscription and an advertisement.

The subscription and advertisement models can be optimized with rendezvous points that coordinate their propagation. Rendezvous points have been shown to compare favourably to the basic routing and also offer benefits with mobility support. We will investigate mobility and topology reconfiguration later in this book in Chapter 11.

Solution	Examples	Description
Reverse path routing	Multicast, SIENA	A routing scheme that forwards subscriptions towards destination brokers (flooding, advertisements, rendezvous, and other strategies can be used) and uses the reverse path of these messages to deliver content. A frequently used scheme that was first proposed in the context of IP-multicast.
Advertisements	SIENA	Optimizes subscription propagation by taking into account explicit advertisements from neighbouring routers. An advertisement expresses capability to publish content in the future. Subscriptions are only sent towards publishers that have advertised the subscribed content.
Rendezvous routing	Rebeca, multicast	Routes messages towards a logically centralized rendezvous node that is responsible for the message type or contents. Thus rendezvous nodes partition the event space.
Gossip with unstructured overlay	REDS	Gossip with nearby nodes to deliver message towards the destination. This scheme is suitable for dynamic environments.
DHT or structured overlay network	Hermes, Scribe	This routing scheme is implemented on top of a DHT structure and it leverages DHT naming, addressing, and message routing functions. DHT offers certain fault tolerance and scalability features for the higher level pub/sub routing system.
Event Space Partitioning / channelization	Meghdoot, Fuego	This technique partitions the event space into disjoint or overlapping subspaces and assigns each subspace to a specific part of the network, for example, a broker. The technique can be used to implement load balancing and channelization of the traffic.

Figure 7.14   A summary of pub/sub solutions.

As mentioned above, DHT and structured networks are frequently used as the basic routing substrate for pub/sub. They abstract many network specific issues and faults from the pub/sub middleware. For example, churn, IP address changes and failed routers can be handled by the overlay.

Gossiping has been also considered for pub/sub, and it is suitable for dynamic environments. Gossiping is frequently combined with infrastructure based overlays in order to realize a scalable multilayer system.

In addition to rendezvous, event space partitioning and channelization can be used to distribute subscriptions and events in the pub/sub network in order to minimize costs. We will consider these in more detail in Chapter 11.

Figure 7.15 presents a matrix that considers how the above techniques and optimizations work together. It turns out that in many cases they can be combined in order to meet given application requirements and constraints of the operating environment. Some of the techniques can be applied in a recursive manner, for example DHT structures can be used recursively or hierarchically, for example, in order to support organizational boundaries. Gossip does not work well with techniques that rely on fixed infrastructure solutions that place state in the pub/sub network, such as the commonly used reverse path routing technique.

	Reverse path	Advertisements	Rendezvous	Gossip	DHT	ESP
Reverse path		Yes	Yes	Not effective	Yes	Yes
Advertisements	Yes		Yes	Yes	Yes	Yes
Rendezvous	Yes	Yes	Yes (recursive)	Not effective	Yes	Yes
Gossip	Not effective	Yes (depends)	Not effective		Yes (Dynamo, and others)	No
DHT	Yes	Yes	Yes	Yes (Dynamo, and others)	Yes (recursive, hierarchical)	Yes
ESP	Yes	Yes	Yes	No	Yes	Recursive

Figure 7.15 A matrix of pub/sub solutions.

References

1. Bricconi G, Tracanella E, Nitto ED and Fuggetta A (2000) Analyzing the behavior of event dispatching systems through simulation. *HiPC*, pp. 131–40.
2. Cugola G, Di Nitto E and Fuggetta A (1998) Exploiting an event-based infrastructure to develop complex distributed systems *Proceedings of the 20th International Conference on Software Engineering*, pp. 261–70. IEEE Computer Society.
3. Fiege L, Gärtner FC, Kasten O and Zeidler A (2003) Supporting mobility in content-based publish/subscribe middleware *Middleware*, pp. 103–22.
4. Mühl G (2002) *Large-Scale Content-Based Publish/Subscribe Systems*. PhD thesis. Darmstadt University of Technology.
5. Mühl G, Ulbrich A, Herrmann K and Weis T (2004) Disseminating information to mobile clients using publish-subscribe. *IEEE Internet Computing* 8: 46–53.
6. Carzaniga A (1998) *Architectures for an Event Notification Service Scalable to Wide-area Networks*. PhD thesis. Politecnico di Milano Milano, Italy.
7. Carzaniga A, Rosenblum DS and Wolf AL (2001) Design and evaluation of a wide-area event notification service. *ACM Transactions on Computer Systems* 19(3): 332–83.
8. Mühl G, Fiege L, Gärtner FC and Buchmann AP (2002) Evaluating advanced routing algorithms for content-based publish/subscribe systems. In: *The Tenth IEEE/ACM International Symposium on Modeling, Analysis and Simulation of Computer and Telecommunication Systems (MASCOTS 2002)* (Boukerche A, Das SK and Majumdar S eds), pp. 167–76. IEEE Press, Fort Worth, TX.
9. Pietzuch PR (2004) *Hermes: A Scalable Event-Based Middleware*. PhD thesis. Computer Laboratory, Queens' College, University of Cambridge.
10. W3C 2003 *SOAP Version 1.2*. W3C Recommendation.
11. Carzaniga A, Rosenblum DS and Wolf AL (2000) Content-based addressing and routing: A general model and its application. Technical Report CU-CS-902-00, Department of Computer Science, University of Colorado.
12. Opyrchal L, Astley M, Auerbach J, Banavar G, Strom R and Sturman D (2000) Exploiting IP multi-cast in content-based publish-subscribe systems. *Middleware '00: IFIP/ACM International Conference on Distributed Systems Platforms*, pp. 185–207. Springer-Verlag New York, Inc., Secaucus, NJ.
13. Pietzuch PR and Bacon J (2002) Hermes: A distributed event-based middleware architecture. *ICDCS Workshops*, pp. 611–18.
14. Eugster PT, Felber P, Guerraoui R and Kermarrec AM (2003) The many faces of publish/subscribe. *ACM Comput. Surv*. 35(2), 114–31.
15. Adler M, Ge Z, Kurose JF, Towsley DF and Zabele S (2001) Channelization problem in large scale data dissemination *ICNP*, pp. 100–9.
16. Chockler G, Melamed R, Tock Y and Vitenberg R (2007) Constructing scalable overlays for pub-sub with many topics *PODC*, pp. 109–18.

17. Chockler G, Melamed R, Tock Y and Vitenberg R (2007) Spidercast: a scalable interest-aware overlay for topic-based pub/sub communication *DEBS*, pp. 14–25.

18. Milo T, Zur T and Verbin E (2007) Boosting topic-based publish-subscribe systems with dynamic clustering *SIGMOD Conference*, pp. 749–60.

19. Carzaniga A and Wolf AL (2003) Forwarding in a content-based network. *Proceedings of ACM SIGCOMM 2003*, pp. 163–74, Karlsruhe, Germany.

20. Carzaniga A, Rutherford MJ and Wolf AL (2004) A routing scheme for content-based networking. *Proceedings of IEEE INFOCOM 2004*. IEEE, Hong Kong, China.

21. Carzaniga A, Rosenblum DS and Wolf AL (1999) Interfaces and algorithms for a wide-area event notification service. Technical Report CU-CS-888-99, Department of Computer Science, University of Colorado. Revised May 2000.

22. Rowstron AIT and Druschel P (2001) Pastry: Scalable, decentralized object location, and routing for large-scale peer-to-peer systems *Middleware 2001: Proceedings of the IFIP/ACMInternational Conference on Distributed Systems Platforms Heidelberg*, pp. 329–50. Springer-Verlag, London.

23. Stoica I, Morris R, Karger D, Kaashoek F and Balakrishnan H (2001) Chord: A scalable peer-to-peer lookup service for internet applications. *Computer Communication Review* **31**(4): 149–60.

24. Zhao BY, Kubiatowicz JD and Joseph AD (2002) Tapestry: a fault-tolerant wide-area application infrastructure. *SIGCOMM Comput. Commun. Rev.* **32**(1): 81.

25. Levchenko K, Voelker GM, Paturi R and Savage S (2008) Xl: an efficient network routing algorithm. *SIGCOMM '08: Proceedings of the ACM SIGCOMM 2008 Conference on Data Communication*, pp. 15–26. ACM, New York, NY.

26. Lamport L (1978) Time, clocks, and the ordering of events. *Communications of the ACM* **21**(7): 558–65.

27. Tarkoma S and Kangasharju J (2007) On the cost and safety of handoffs in content-based routing systems. *Computer Networks* **51**(6): 1459–82.

28. Jerzak Z (2009) *XSiena: The Content-Based Publish/Subscribe System*. PhD thesis. TU Dresden Dresden, Germany.

29. Raman S and McCanne S (1999) A model, analysis, and protocol framework for soft state-based communication. *SIGCOMM Comput. Commun. Rev.* **29**: 15–25.

30. Cristian F and Fetzer C (1999) The timed asynchronous distributed system model. *IEEE Trans. Parallel Distrib. Syst.* **10**(6): 642–57.

31. Casimir A, Martins P, Rodrigues L and Veríssimo P (2001) Measuring distributed durations with stable errors *Proceedings of the 22nd IEEE Real-Time Systems Symposium*, pp. 310–19. RTSS '01, IEEE Computer Society, Washington, DC.

8

Matching Content Against Constraints

This chapter investigates content matching techniques and efficient filtering solutions. Content matching is a basic requirement for a pub/sub system and thus it should be efficient and scalable as well as support different filtering constraints. We investigate well-known techniques including counting based algorithms and the poset and forest algorithms as well as several probabilistic matching techniques based on Bloom filters.

8.1 Overview

Filtering reduces the number of events sent from the sources to the listeners by matching events against a template. Those events that match the template are forwarded to the listeners. Matching is usually done on single events, but may be also performed on composite events. Filtering improves the scalability of the system. Also, the location of the filtering of events affects the scalability of the framework. Here we face two separate issues: the filtering of simple events and the filtering of composite events. We focus on the filtering of simple atomic events. Composite event detection can be implemented on top of the basic filtering capability.

Both kinds of event filtering can be done at several locations:

- At the subscriber when events are flooded.
- At the event source with the observer pattern.
- In the infrastructure (event routers or brokers) with the event notifier pattern.

Simply flooding all events and performing filtering at the subscribers is not feasible due to the excessive traffic and filtering overhead with high publication rates. Event source-side filtering reduces the burden on infrastructure, but faces scalability challenges when the number of subscribers grows. Hence distributed infrastructure based event service is typically utilized when scalability is a desired system property.

Publish/Subscribe Systems: Design and Principles, First Edition. Sasu Tarkoma.
© 2012 John Wiley & Sons, Ltd. Published 2012 by John Wiley & Sons, Ltd.

We examined various distributed filtering strategies in Chapter 7. In this chapter, we focus on the basic problem of matching an event against active filters. Typically the active filters are stored in a routing or forwarding table. The internal organization of the matching system is very important when considering expressiveness and scalability of the system.

Event filtering is used in most current event systems. The CORBA Notification Service uses the extended Trader Constraint Language (TCL) [1]. The Java Messaging Service (JMS) supports a subset of SQL-92 for event filtering [2]. These two specifications do not define any particular way of doing distributed event delivery although distributed filtering may be implemented based on them.

Research efforts such as JEDI [3], Elvin [4], Rebeca [5], Gryphon [6], and SIENA [7] have investigated distributed event filtering. Wide-area scalability of event filtering was investigated in the SIENA architecture and they define filter relationships formally using covering relations. Filter covering is used in many systems to find the noncovered set of filters, or minimal cover set, that is propagated by event routers. Although filter covering was observed to have favourable properties, this filtering approach has been observed to have limited scalability to wide-scale environments due the overhead of filter-based routing and forwarding. This observation has led to many proposals for optimizing filter-based routing systems. We will investigate data structure improvements, namely the forest structure, separation of routing and forwarding in the XSIENA system, and fast forwarding structures (counting, tree, and hybrid algorithms). We will later in Chapter 9 consider the CBCB strategy that divides the filter-based routing environment into two parts, namely the broadcast layer and then the content-based layer.

In general, filter matching is done by counting attributes using the counting algorithm [8–13], counting and clustering [14], using a tree-based data structure [15], or a *binary decision diagram* (BDD) [16]. Fast matching algorithms combine client-side processing, caching, and filter clustering [17]. In addition to exact event matching also approximate matching has been proposed based on fuzzy logic [18]. The Elvin [4] filtering language is based on Lukasiewicz's tri-state logic with values *true*, *false*, and *undecidable*.

8.2 Matching Techniques

Figure 8.1 presents a summary of well-known content matching techniques. The naive basic approach for matching incoming messages against filters and continuous queries is to simply iterate the filters and test each filter separately for a potential match. The counting algorithm takes a different approach and divides the filters and events into attributes and attribute filters, respectively. This technique tests each event attribute against relevant attribute filters, and for each match increments counters in all filters associated with the matching attribute filter. Filters whose counters are equal to the number of attribute filter match the event. This way it is not necessary to test all the filters. One of the earliest counting algorithms was proposed by Yan and Garcia-Molina [13]. Attribute counter-based algorithms for finding the set of covering filters for a given input filter and the set of mergeable filters were presented in [11].

The SIFT algorithm is similar in principle to the counting algorithm and it creates an inverted index of query terms that can be used to quickly map query terms to matching documents, and documents to queries, respectively [19]. The inverted index is very similar

Year		Structure	Notes
–	Sequential scanning	Naive method	Linear time and linear space. Does not scale
–	Counting algorithm	Each event attribute is tested against query constraints and the matching constraints are counted to find matching queries	Structural assumptions, fast if redundant query constraints
1995	SIFT Algorithm	Directory and Inverted lists	Each term in a document are tested with term specific inverted lists. Key idea is to index the queries
1998, 2005	Poset and forest	DAG / forest	Suitable for covering optimizations, general technique, scalability issues when matching events
1999	Gryphon	Tree	Sublinear matching, linear space, preprocessing needed
2000-2002	XFilter and YFilter	Finite state automata (XFilter) and Non-deterministic finite automata (YFilter) for matching XML with XPath and XQuery	Compiles an automata based on the queries
2003	SIENA fast forward algorithm	Directory with attribute specific matchers	Assumptions about the structure
2009	Inverted index for Boolean expressions (VLDB 2009)	Inverted index structure for Boolean expressions over high-dimensional multivalued attribute space	
2010	BE-tree (SIGMOD 2010)	Tree structure for Boolean expressions with adaptive properties, suitable for high dimensional workloads	

Figure 8.1   Content matching techniques.

to the counting algorithm, because the event attributes, in many cases keywords, are mapped to an index of query terms that then maps to the actual subscriptions.

The poset algorithm was proposed in the SIENA system as a content-based routing table structure. The idea is to determine the covering relations between the input filters and then create and maintain the poset structure based on the relations. This allows compact representation of the routing table, but has a relatively high cost in terms of management operations. The poset-derived forest was later proposed as a simple and more efficient structure.

Many matching mechanisms do not take the distribution and selectivity of filters into account. The poset structure takes selectivity into account through the covering relation; however, also matching rate can be used to optimize the filtering process. Efficient selectivity-based filtering has been examined in [20]. Selectivity-based filtering evaluates the most general filters first that have the highest selectivity. A high selectivity can be estimated based on different information: the distribution of events, the distribution of subscriptions, or both.

The R-tree [21] and its variants, such as R* and R+-tree, are efficient structures for determining containment, intersection, and overlapping of multidimensional data, such as geographical data. They require that minimum bounding rectangles can be computed for the nodes in the trees. These structures can be used as specialized matching components in counter based designs.

The SIENA fast forward algorithm was developed to mitigate the high cost of forwarding with the poset algorithm. This structure is based on the counting algorithm and uses specific efficient components to implement support for a particular attribute filter type, such as ranges, strings, etc. This algorithm assumes that the structure of the filters and events are known and that efficient matching components are available for the attribute filters. The algorithm also uses a selectivity table for removing unmatchable predicates [8].

A generic tree matcher was proposed in the Gryphon project. The idea is to organize the filters into a tree in such a way that each layer presents a test on the input event. Thus an event is tested first on the root node and then on subsequent matching nodes progressing depth-wise towards the leaves of the tree. The leaf nodes encountered are then said to match the event. This structure is quite efficient; however, requires a two-phase approach that first builds the tree and then uses it for matching.

W3C is specifying and working on two XML query languages: XPath [22] and XQuery [23], which may also be used in the routing of events that are represented using XML. Efficient XPath filtering is an active research topic. Most XPath query evaluation implementations run in exponential time to the size of input queries [24]. XPath query covering and merging are computationally demanding, which motivates simpler schemes. Tree pattern aggregation is a recent research area and covering algorithms and a minimization algorithm have been presented for conjunctive tree queries [25].

The XFilter and YFilter were developed for the matching of structured XML documents against XPath and XQuery filters. The key idea is to create a finite state automata (XFilter) or a nondeterministic finite automata (YFilter) of the filters and then utilize this automata for matching an incoming XML event. These structures are efficient in matching the structured documents, but require that the automata is first created based on the queries.

Newer structures include an extension of the inverted index for arbitrarily complex Boolean expressions[1] over high-dimensional multivalued attribute space. Prior solutions have supported the indexing of only a subset of conjunctive and/or disjunctive expressions. The approach can return the top-N matching Boolean expressions making it suitable for determining the top subscriptions matching a given event [26].

The BE-Tree is a another new tree data structure for matching Boolean expressions over a high-dimensional discrete space. This structure adjusts itself in order to adapt to the workload [27].

8.3   Filter Preliminaries

We follow the common typed pub/sub notification and filtering model. Formally, a filter is a stateless Boolean function

$$F : N \rightarrow \{true, false\},$$

where N is the set of all notifications [28].

---

[1] Conjunctive Normal Form and Disjunctive Normal Form.

A notification $n \in N$ is said to *match* F if $F(n) = true$. Further, we let $N(F)$ denote the set $\{n \in N | F(n) = true\}$, that is, the notifications that match F. Filters F_1 and F_2 are *identical*, written $F_1 \equiv F_2$, if and only if $N(F_1) = N(F_2)$.

In the commonly used data model, a filter is a conjunction of attribute constraints, called attribute filters. A notification matches a filters if its attributes satisfy every attribute constraint of the filter. A filter is not satisfied if it contains an attribute not present in the evaluated notification. Typically it is not necessary that all the attributes of a notification match an attribute filter of a filter in order for the filter to match the notification. In this case, it is considered that the filter implicitly accept the attribute.

A filter can be seen as a subspace of the content space, in which notifications are points. A filter therefore selects points belonging to the subspace its defines. Filter covering can be visualized by considering the containment of subspaces. If a subspace is contained in a larger subspace, it said to be covered by this space.

Next, we examine four filters that illustrate the filtering model. The first three filters have two attribute filters, and the fourth filter has a single attribute filter.

$$F_1 = (\text{price} < 200 \wedge \text{price} > 300),$$

$$F_2 = (\text{price} < 100 \wedge \text{price} > 400),$$

$$F_3 = (\text{price} < 350 \wedge \text{price} > 500),$$

$$F_4 = (\text{price} \notin [200, 300]).$$

A filter F_1 is said to cover filter F_2, denoted by $F_1 \sqsupseteq F_2$, if and only if all notifications that are matched by F_2 are also matched by F_1, that is, $N(F_1) \supseteq N(F_2)$. In other words, F_1 is *more general* than F_2. If $F_1 \not\sqsupseteq F_2$ and $F_2 \not\sqsupseteq F_1$, the filters F_1 and F_2 are said to be *unrelated* or *incomparable*. The covering relation $\sqsupseteq$ is reflexive, transitive and antisymmetric and thus defines a *partial order* over the set of all filters [29]. A set with a partial order is called a *partially ordered set* or *poset*.

Filters F_1 and F_2 *overlap*, written $F_1 \sqcap F_2$, if and only if $N(F_1) \cap N(F_2) \neq \emptyset$. It is possible that two filters may overlap and that they are unrelated.

Next, we consider examples illustrating relationships between filters based on the previous example. In our example, these statements are true:

$$F_1 \equiv F_4,$$

$$F_1 \sqsupseteq F_2, F_4 \sqsupseteq F_2,$$

$$F_2 \sqcap F_3, F_1 \sqcap F_3,$$

$$F_2 \not\sqsupseteq F_3 \wedge F_3 \not\sqsupseteq F_2.$$

8.4 The Counting Algorithm

The counting algorithm (also called the inverted index) is a frequently employed design for implementing fast filter-based routing and forwarding engines. The idea is simply to count the number of matched attribute filters or components of a filter. When all the components of a filter have been matched and each of them satisfies the applied matching operation, it is said that the whole filter matches with the input event.

8.4.1 Overview

The counting algorithm is based on the fact that filters are conjunctions of attribute filters. Typically, matching with the counting algorithm is divided into a preliminary elimination phase in which unmatchable filters and interfaces are removed, and a counting phase. If the counter of a filter becomes equal to the number of attribute filters in the filter, the filter matches the input notification and the corresponding interface is added to a set of output interfaces. The counting algorithm returns either the identifiers of matching filters or a set of output interfaces. Optimized matchers use efficient data structures for different predicates, for example hashtable lookup for equality tests and interval trees [30] for range queries.

The counting algorithm has been extended by Carzaniga and Wolf to support various attribute filters [8] illustrated by Figure 8.2. In this model, filters are disjunctions of conjunctions, and thus a filter can be added to the results set if any of conjunctions in the filter is satisfied. The extended algorithm uses the forwarding table as a dictionary-type data structure optimized for searches. We note that typically the counting structure have not been optimized for dynamic updates of the structure.

The data structure is indexed by attributes that are present in the filters in the forwarding table. For each of the attributes in an incoming event, the index is searched for constraints that match the attribute. The count of satisfied constraints is incremented for each satisfied constraint found. Based on the satisfied constraints it is then easy to determine the set of matching filters that should be in the result set.

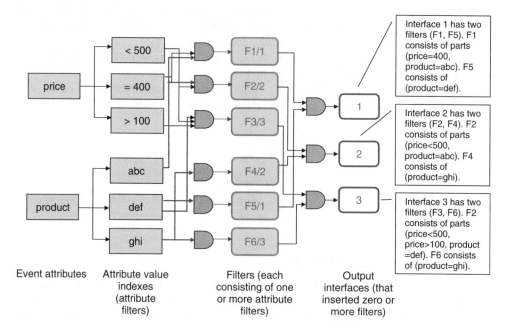

Figure 8.2 The SIENA fast forward structure.

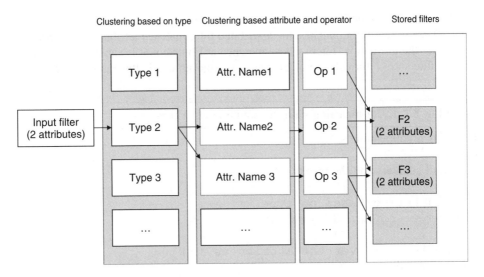

Figure 8.3 Structures for the counting algorithm.

8.4.2 Algorithms

Figure 8.3 illustrates this design that involves the decomposition of filters into their constituent parts. In this section, we assume that a filter consists of constraints, called attribute filters, which are evaluated against the attributes of a notification. Each attribute filter specifies an attribute and at least one operation that is evaluated on the attribute. In addition, each filter has a type, which determines the structural properties of the filter.

The figure shows how the filter can be decomposed by first clustering attribute filters based on their type and then further clustering the attribute filters based on their properties. The type-based clustering can also take the number of attribute filters into account. It is clear that a filter cannot match an event with fewer attributes than the number of attribute filters. Each attribute filter maps to one or more filters, and each filter has an associated count of the number of attribute filters it contains.

8.4.2.1 Add and Remove a Filter

Filter insertion is a simple operation for the counting design. The filter is decomposed into its constituent parts and each part is stored in a separate data structure. Bookkeeping structures are used to store the relations of the components so that the filter can be reconstructed if necessary.

Filter deletion is also a simple operation, which involves removing the components of the filter from the data structures into which they were stored.

8.4.2.2 Matching a Notification

A notification can be matched against the stored filters by simply first finding the set of attribute filters that can potentially match with the notification based on the notification type. Then each attribute filter of the notification is evaluated against the stored attribute filters. If an attribute filter matches, the associated filter matched attribute counts are increment. After the evaluation, filters whose matched attribute count is equal to the number of attribute filters are returned as matching filters. Algorithm 8.1 illustrates this technique.

Algorithm 8.1 Event matching with the counting algorithm

Data: n is a notification.
begin
    Init all counters and the results set

    /* Clustering based on type   */
    $cluster \longleftarrow getClusterByType(n.type)$

    /* We iterate over the attributes in n */
    for $nf \in n$ **do**
        /* We iterate over the AF set. A more efficient
        structure can also be used. Get candidate AFs based
        on the attribute nf and the cluster   */

        $AFSet \longleftarrow getAFset(cluster, nf)$

        for $af \in AFSet$ **do**
            if af matches with nf **then**
                $FSet \longleftarrow getFSet(af)$
                for $f \in FSet$ **do**
                    $f.counter \longleftarrow f.counter + 1$
                    if $f.counter$ is equal to the number of attribute filters **then**
                        Add f to results set
                  end
                end
            end
        end
    end
end
return results set

8.4.2.3 Covering and Covered Filters

Finding the covered and covering filters is a more challenging task with the counting algorithm than matching a notification. For covered and covering filters, the containment relationships of the components of the input filter with the counting structure need to be determined.

Algorithm 8.2 determines the covered filters given the input filter f. The idea is to iterate the stored attribute filters that have greater than or equal number of attribute filters than the input filter. For each covered attribute filter, the algorithm increments the counts of the associated filters. Finally, the algorithm returns filters for which the attribute count is equal to the number of attribute filters. The algorithm assumes that the covering relation can be computed for any two given attribute filters. This can be done easily for elementary constraints, such as ranges, inequalities, and substring matching.

Algorithm 8.2 Finding covered filters in a counting structure

---

Data: f is a filter.
begin
    Init all counters and the results set

    /* Clustering based on type   */
    $cluster \longleftarrow getClusterByType(f.type)$

    /* We iterate over the attributes in f */
    for $af \in f$ **do**
        /* We iterate over the AF set. A more efficient
        structure can also be used. Get candidate AFs based
        on the attribute af and the cluster   */

        $AFSet \longleftarrow getAFset(cluster, af)$

        /* Optimization: Attribute filters associated only with
        filters that have less attribute filters than f can be
        removed from consideration.   */

        for $af' \in AFSet$ **do**
            if af covers af' **then**
                $FSet \longleftarrow getFSet(af')$
                for $f' \in FSet$ **do**
                    $f'.counter \longleftarrow f'.counter + 1$
                    if $f'.counter$ *is equal to the number of attribute filters* **then**
                        Add f' to results set
                  end
                end
            end
        end
    end
end
return results set

The algorithm for finding the covering filters for a given input filter f is similar to the previous Algorithm 8.2 with the difference that now the direction of the relation is reversed (af is covered by af') and that only stored filters that have an equal or less than number of attribute filters need to be considered. It is clear that filters that have a greater number of attribute filters than the input filter cannot cover the input filter.

8.4.2.4   Discussion

The counting algorithm can be summarized with the following observations:

- Filter insertion and deletion are typically very fast operations.
- Scalability is limited by the size of the candidate attribute sets. In the worst case performance may be linear to the number of attributes unless specialized matching structures are used for the attribute filter sets.
- Allows the clustering of the attributes and operations. Custom data structures can be used, for example R-trees for ranges.
- Assumes typed structure and thus it is not a generic technique for arbitrary filtering languages.
- Can be used for detecting covering and covered filters; however, it is more difficult to determine the immediate predecessors and successors.
- There are many variations of the counting algorithm.

8.5   Matching with Posets

In this section, we consider the poset based matchers, namely the SIENA poset and the poset-derived forest. First, we present present some details pertaining to the filtering model, and then proceed to the algorithms. The main idea of the algorithms is that they are organized based on filter containment, and thus capture the selectively of filters within the data structures.

Figure 8.4 illustrates the two structures. The poset is a directed acyclic graph (DAG) structure that maintains the immediate predecessor and successor sets for each filters. Next, we define these sets in more detail. The poset-derived forest keeps a subset of the these sets and thus only stores a forest representation of the partial order.

Figure 8.5 gives example configuration of content-based routers based on posets and forests. The poset can be easily combined with a forest in such a way that the local clients are managed by the forest and the poset stores the remote subscriptions as well as the most general subscriptions of the local clients. Another configuration involves the forest for storing the local subscriptions and an efficient matcher structure, such as the counting

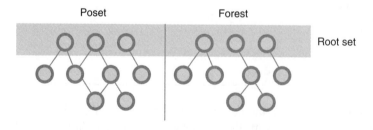

Figure 8.4   Example of a poset and a forest.

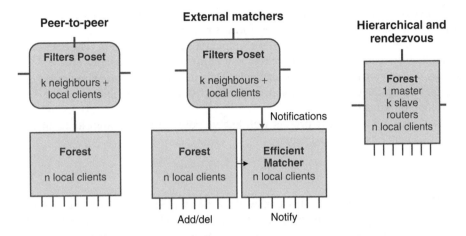

Figure 8.5 Posets and forests for routing table organization.

algorithm, for high speed matching events to local clients. A yet another variant would also utilize a fast matcher to realize a fast forwarding plane for external routers as well. The forest is suitable for standalone use in hierarchical and rendezvous configurations.

The poset and forest can be seen as structures for the slower routing state configuration process, and the counting algorithm and other efficient matchers are more suitable for the fast forwarding path. Observation that the basic SIENA has limited scalability, partly due to the poset, led to the development of the more efficient CBCB model examined in Chapter 9.

One optimization is that although the poset and forest typically operate on filters, they can also be applied on the attribute filter level. Thus attribute filters can be organized based on the covering relation, and a counting algorithm can be then run over this structure creating a more efficient design. The XSIENA system examined later in this chapter aims to improve forwarding performance by creating a fast forwarding structure based on the poset and forest designs.

8.5.1 Poset Preliminaries

Now, we extend the basic filtering model so that it is possible to organize the filters based on their relationships into a routing table. In order to do this, we need to define the more general and less general filters for a given filter.

Let $\mathcal{F}$ denote a set of filters. Now, the *predecessors* of a filter $G \in \mathcal{F}$ are defined as the set $Pred(G) = \{F \in \mathcal{F} | F \neq G \wedge F \sqsupseteq G\}$. The *successors* of G are defined similarly: $Succ(G) = \{F \in \mathcal{F} | F \neq G \wedge G \sqsupseteq F\}$. If $F \sqsupseteq G$ and there is no $F' \in \mathcal{F}$ such that $F \sqsupseteq F'$ and $F' \sqsupseteq G$, we say that F *immediately covers* G and denote it with $F \succ G$.

The *immediate predecessors* and *immediate successors* of a filter G are defined as $ImPred(G) = \{F \in \mathcal{F} | F \neq G \wedge F \succ G\}$ and $ImSucc(G) = \{F \in \mathcal{F} | F \neq G \wedge G \succ F\}$.

A *cover* for a set of filters $\mathcal{F}$ is defined to be a set $\mathcal{G} \subseteq \mathcal{F}$ such that for each $F \in \mathcal{F}$ there exists a $G \in \mathcal{G}$ for which $G \sqsupseteq F$. This cover is *minimal* if it does not contain a proper subset that is also a cover of $\mathcal{F}$. It is clear that $\sqsupseteq$ cannot hold between two members of a minimal cover. We call this minimal set for the ground set $\mathcal{F}$ as the *root set*. Figure 8.4 illustrates the root set for the poset and forest structures.

Cover can be efficiently determined for simple typed attribute filters. The data structures assume that the pair-wise cover checking can be made in a reasonably efficient manner, and they do not consider unions of elements. Thus the algorithms are simple and efficient from the viewpoint of computational complexity. It is also possible to devise schemes based on approximate cover.

In the following we briefly present well-known poset and forest algorithms that are the basis of content-based routing tables.

8.5.2 SIENA Poset

The SIENA filters poset data structure was used in the SIENA distributed pub/sub system for maintaining covering relations between filters. The structure is an example of a routing table that stores filters. The SIENA system did not separate between local and external subscriptions and thus all filters were stored in the same structure. Later research has shown that it is more efficient to divide the internal organization of the content-based router into separate parts and have separate routing tables for local clients and external routers. The filters poset uses a DAG-based algorithm for finding and maintaining the direct predecessor and successor sets [7].

A two-phase algorithm for partial order maintenance for strings and string subpattern containment was presented in [31]. The SIENA algorithm is similar to this, but it does not keep a sorted linked list of the node set, but rather iterates the structure using the direct successor sets. In the general case, the node set cannot be sorted unless an additional heuristic, such as string length, is employed.

In SIENA's peer-to-peer architectures the poset stores additional information for each subscription that is inserted into the poset. The *subscribers(f)* set gives the set of subscribers for the given subscription filter f, and similarly, *forwards(f)* contains the subset of peers to which f needs to be sent. A subscription *subscribe(X,f)* where X is the subscriber and f is the filter representing the subscription proceeds as follows [7]:

1. If a filter f' is found for which $f' \sqsupseteq f$ and $X \in subscribers(f')$ then the procedure terminates, because f for X has already been subscribed by a covering filter.
2. If a filter f' is found for which $f' \equiv f$ and $X \notin subscribers(f')$ then X is added to *subscribers(f')*. The server removes X from all subscriptions covered by f. Also, subscriptions with no subscribers are removed.
3. Otherwise, the filter f is placed in the poset between two possibly empty sets: immediate predecessors and immediate successors of f. The filter f is inserted and X is added to *subscribers(f)*. The server removes X from all subscriptions covered by f, and subscriptions with no subscribers are also removed.

Next, we examine the algorithms for inserting and deleting a filter from the poset. Algorithm 8.3 presents the insertion process that accepts as input the filter and the root set of the structure. The insertion determines the immediate predecessors *ImPred* for the new filter, and then positions the filter between the immediate predecessors and successors.

The *Disconnect(p,s)* operation removes s from the immediate successors of p, and s from the immediate predecessors of p. The *Connect(p,c)* operation adds c to immediate successor set of p, and p to the immediate predecessors of c. The insert and delete algorithms need to also update the root set for those nodes that do not have any predecessors.

Algorithm 8.3 Poset filter insert

Data: s is a filter, R is the root set.
Result: Poset with the new element s.
begin

    $ImPred \longleftarrow findImPred(s, R)$

    if $ImPred$ is empty **then**

        s is a root node.

        Connect s with direct successors.

        return

    end

    for $p \in ImPred$ **do**

        $ImSucc \longleftarrow findImSucc(p, s)$

        for $c \in ImSucc$ **do**

            /* Connect the proper immediate successors.   */

            $Disconnect(p, c)$

            $Connect(s, c)$

        end

        /* Connect the proper immediate predecessors.   */

        $Connect(p, s)$

    end

end

Algorithm 8.4 presents the filter deletion algorithm that disconnects the filter to be deleted from its predecessors, and then connects its immediate successors with the predecessors. A predecessor p is connected with a successor s if there does not exist a successor of p that is in relation with s.

Both the list and DAG-based poset algorithms require $O(n)$ comparisons for insertion and additional bookkeeping for the visited edges, and edges to be connected. Deletion is more complicated and the DAG-based algorithm first disconnects the direct predecessor and successor sets, and then connects necessary elements that are in direct relation.

Algorithm 8.4 Poset filter delete

Data: s is a filter, R is the root set.
Result: Poset without the element s.
begin

    $ImPred \longleftarrow findImPred(s)$

    if $ImPred$ is empty **then**

        Move appropriate immediate successors of s as root nodes.

        return

    end

```
    for p ∈ ImPred do
        /* Disconnect s.                                                     */
        Disconnect(p, s)
        /* Get the immediate successors of s.                               */
        ImSucc ⟵ findImSucc(s)
        /* Connect predecessors and successors.                            */
        for i ∈ ImSucc do
            /* Iterate over all the children of the predecessor.
            */
            if ¬indirectSuccessorTest(p, i) then
                Connect(p, i)
            end
        end
    end
end
```

The insert and delete operations utilize two important sets, namely the *ImPred* and *ImSucc*. Algorithms 8.5 and 8.6 outline how these sets are determined. The *ImPred*(x) set is found by starting from the root set and descending towards nodes for which the ⊒ relation holds. Some bookkeeping is needed to keep track of visited nodes. The algorithm finds *Succ*(x) by simply starting from the successors of *ImPred*(x).

Algorithm 8.5 Find immediate predecessors

Data: s is a filter, R is the root set.
Result: Immediate predecessor set for s.
begin

```
    visit_list ⟵ R
    /* Find immediate predecessors and keep track of visited
    nodes.                                                               */
    for p ∈ visit_list do
        direct ⟵ true
        clist ⟵ getChildren(p)
        for c ∈ clist do
            if c covers s then
                /* p is not an immediate predecessor.                   */
                direct ⟵ false
                if c ∉ visit_list then
                    visit_list ⟵ visit_list ∪ {c}
                end
            end
        end
        /* Test if an immediate predecessor was found.                  */
        if direct then
            ImPred ⟵ ImPred{∪p}
```

        end
     end
     return *ImPred*
end

Algorithm 8.6 Find immediate successors

Data: s is a filter, *ImPred* is the immediate predecessor set for s, R is the root set.
Result: Immediate successor set for s.
begin
    ImSucc $\longleftarrow \emptyset$
    ImSuccTemp $\longleftarrow \emptyset$
    if $p = \emptyset$ **then**
        visit_list $\longleftarrow R$
    else
        visit_list $\longleftarrow$ *ImPred*
    end
    /* Find immediate successors and keep track of visited nodes.   */
    for $c \in$ *visit_list* **do**
        if c *not yet visited* **then**
            if s *covers* c **then**
                ImSuccTemp $\longleftarrow$ *ImSuccTemp* $\cup \{c\}$
            else
                visit_list $\longleftarrow$ *visit_list* $\cup$ *ImSucc(c)*
            end
        end
    end
    /* Finalize the result set.   */
    for $c \in$ *ImSuccTemp* **do**
        if c *is not covered by any other node in* *ImSuccTemp* **then**
            ImSucc $\longleftarrow$ *ImSucc* $\cup \{c\}$
        end
    end
    return *ImSucc*
end

8.5.3 *Poset-Derived Forest*

The poset-derived forest data structure is a similar structure to the SIENA filters poset; however, it is simpler and more efficient, because only a subset of the covering relation is stored [29]. Each node in the structure has at most one parent. The main drawback of the structure is that it becomes more difficult to find the direct predecessors and successors of a given node. The structure, on the other hand, computes the same root set as the poset and thus lends itself well to hierarchical and rendezvous-based routing systems as well as storing filters from clients. Indeed, the forest can be seen to be a building block for various filter-based pub/sub systems. The structure can be extended with the sets $subscribers(f)$, $forwards(f)$, $advertisers(a)$, and $forwards(a)$.

A pair $(\mathcal{F}, \succ')$ is a poset-derived forest, where $\mathcal{F}$ is a finite set of filters and $\succ'$ is a subset of the covering relation. It is convenient to imagine the roots of the trees as children of a virtual node not in $\mathcal{F}$.

Next, we outline the algorithm for the poset-derived forest [29].

Let $(\mathcal{F}, \succ')$ be a poset-derived forest. We define the following algorithms with inputs $\mathcal{F}$ and a filter x and output a poset-derived forest:

add($\mathcal{F}$, x): This algorithm maintains a *current node* during its execution. First, set the current node to be the imaginary root of $\mathcal{F}$.
1. If x is already in the forest then return.
2. Else if x is incomparable with all children of the current node, add x as a new child of the current node.
3. Else if x covers some child of the current node, move all children of the current node that are covered by x to be children of x and make x a new child of the current node.
4. Else pick a child of the current node that covers x, set the current node to this child and repeat this procedure from step 2.

del($\mathcal{F}$, x): Let C be the set of children of x and r be the parent of x. Then run add for each of the elements of C starting from step 2 and setting r as the current node. In this an element of C carries the whole subtree rooted at it with the addition.

8.5.4 Matching Events

The poset and forest structures have two interesting properties for matching that follow from the definition of the covering relation (Property 8.1 and 8.2).

Property 8.1 *If a node n_1 matches a notification then all the predecessors of n_1 must also match the notification.*

Property 8.2 *If a node n_1 does not match a notification then none of the descendants of n_1 matches the notification.*

Algorithm 8.7 presents the poset-based matching technique. The matching starts from the root of the structure and proceeds in breadth-first order towards matching nodes. Each node contains a filter and is a test on the input event. The algorithm is the same for the forest with the exception that the visited list is not needed.

Algorithm 8.7 Match an event with the poset

Data: p is an event, R is the root set.
Result: Matching filters for p
begin
    $visit\_list \longleftarrow R$
    $result \longleftarrow \emptyset$
    /* Iterate over the structure.   */
    for $f \in visit\_list$ **do**
        if f *not yet visited* **then**
            if f *matches* p **then**
                $result \longleftarrow result \cup \{f\}$
                $visit\_list \longleftarrow visit\_list \cup \{ImSucc(f)\}$
            end
        end
    end
    return $result$
end

The forest and poset also support approximate matching. For example, we may walk the forest with the notification breadth-first and define a time bound for matching. When this time expires the algorithm simply walks the remaining nodes and records the interfaces as matched. This is approximate, because it may result in false positives.

The interesting feature of the algorithm is that the matching mechanism does not know the details of the filtering language–it only assumes that there are covering relations between nodes. This makes the algorithm suitable for environments where the filtering language and the operators (predicates) are dynamic and change. In addition, adding new operators does not require complicated changes to the matching algorithm, such as creating new indexing structures.

8.6 Tree Matcher

The tree matcher structure differs from the counting algorithm, because it starts from the attribute constraints in the forwarding table rather than in the event. We briefly consider the algorithm due to Aguilera *et al.* [15]. A similar algorithm is due to Campailla *et al.* [16] that is based on binary decision diagrams.

The tree matcher algorithm consists of two phases:

- Pre-processing phase, in which the set of subscriptions is inserted into the data structure. The pre-processing assumes that subscriptions do not change frequently. Updates can be incrementally added to pre-processed data. The matching tree is created in this phase. In the tree, each node is a test on some of the attributes. Each edge starting at a node is a result of a test. Thus each lower layer is a refinement of the tests performed at higher levels. The subscriptions are at the leaves of the tree.

- The matching phase, in which each event is iterated with the tree structure to find the matching subscriptions. The process starts from the root of the tree. At each node the test at that node is conducted and matching edges are followed until a leaf node is encountered. The leaves are then finally mapped to the matching subscriptions.

The tree can have also special *-edges that represent the observation that subscriptions reachable through the edge do not care about the test result. These edges allow for a very flexible structure.

Algorithm 8.8 presents the tree matcher that takes the root node of the tree v and the the event to be matched p as arguments [15]. The algorithm then iterates the tree and at each node tests whether not to iterate that particular branch further. If the current node matches with p, then the algorithm proceeds deeper into that branch of the tree. Eventually, the iteration will end at a leaf node that will be part of the result set. This algorithm traverses the tree in depth-first order; however, other orderings can also be used such as breadth-first.

The time complexity of the tree matcher is sublinear in the number of subscriptions, and its space complexity is linear. The structure can be further optimized with static analysis of the subscriptions [15].

Algorithm 8.8 The recursive *visit* procedure for tree matching

---

Data: *Tree* denotes the structure, v is the root or a node of the tree, p is an event to be matched.
Result: Matching filters for p
begin
    if v *is a leaf node of Tree* **then**
        Output (v)
    else
        perform test by v on p
        if v *has an edge e with the result of the test* **then**
            visit(*Tree*, (child of v at endpoint of e in *Tree*), p)
        end
        if v *has a *-edge e* **then**
            visit(*Tree*,(child of v at the endpoint of * in *Tree*), p)
        end
    end
end

8.7 XFilter and YFilter

The XFilter project pioneered the use of event-based parsing and *finite state machines (FSM)* for XML document filtering [32]. XFilter supported XPath expressions that were converted to FSMs by mapping the locations steps of the expressions to machine states.

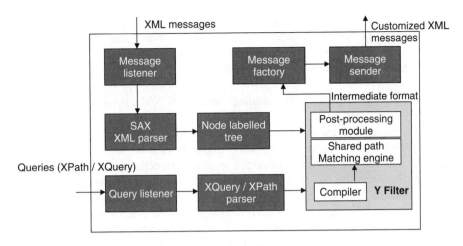

Figure 8.6   Overview of YFilter.

The incoming XML messages are then parsed with an event-based parser that drive the execution of the FSM. A query matches a document if the parsing runs into an accepting state. XFilter generates a FSM for each distinct path expression and thus does not take into account the commonality between the queries.

The YFilter project built on the XFilter concept and this system takes the query commonality into account by creating a single machine represented by a nondeterministic finite automata [33, 34]. The key goal of YFilter is to match XML documents to a large number of queries, and also to support transformation of the matched XML documents based on subscriber requirements.

Figure 8.6 presents an overview of the YFilter system that defines a general query processor for structured data. The system accepts incoming XML messages and documents through a message listener interface. The incoming XML documents are then parsed and a node labelled tree representation is generated that is then processed by the core YFilter system. XPath and XQuery filters are accepted through a query listener interface. The queries are parsed and then given to the core matching system. The queries are compiled into a non-deterministic finite state machine (YFilter). Incoming messages, namely the node labelled tree, is then matched against a combined state machine containing all the active queries. Matching documents are then post-processed and customized for subscriber and then finally sent to the subscribers [33, 34].

The main components of the YFilter architecture are:

- Query listener that accepts queries.
- XQuery parser that parses queries.
- Message listener for accepting incoming XML messages.
- XML parser for generating an internal representation of an XML message.
- YFilter, which is the main component of the system and consists of several subcomponents, namely the compiler, a path matching engine, and a post-processing module. The compiler constructs execution plans for the queries. The runtime system matches the given message with the queries. The system also extracts message contents for

the matched queries and organizes the contents in an intermediate format for efficient translation into the final customized output messages. When filtering without transformations, YFilter matching only results in a boolean result for a given query.

- Message factory is responsible for generating an output message for each matched query.
- Message sender is responsible for sending customized XML messages to users.

Filtering is performed using path expressions that address parts of an XML document. A path expression is a sequence of location steps. A location step consists of an axis, a node test, and zero or more predicates. The axis parameter specifies the hierarchical relationships between the nodes. YFilter focuses on two common axes: the parent-child operator /, and the ancestor-descendant operator //. A node test is an element name or a wildcard operator. Predicates are matched against text data, attributes, or the positions of the addressed elements. Predicates can also contain other path expressions. YFilter supports filtering and transformations through a subset of XQuery *FLWOR (for-let-where-order by-return)* expressions.

8.8   Bloom Filters

The Bloom filter is a space-efficient probabilistic data structure that supports set membership tests. The data structure was conceived by Burton H. Bloom in 1970 [35]. The structure offers a compact way to represent a set that can result in false positives, but never in false negatives. Bloom filters are useful for many different kinds of tasks that involve lists and sets [36]. They have been extensively applied in pub/sub routing and forwarding.

In this section,we give a brief overview of the Bloom filter structure, and then examine several applications for pub/sub. We consider briefly summary subscriptions, multicast forwarding, content-based forwarding, and multilevel filters for XML document matching.

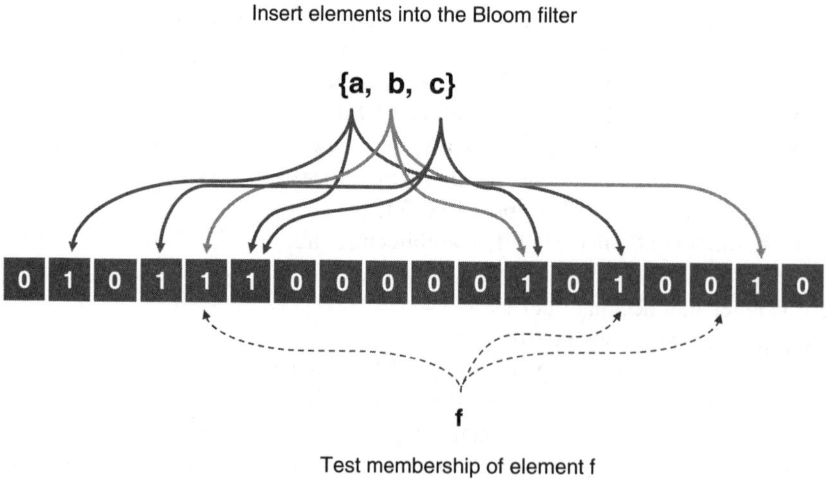

Insert elements into the Bloom filter

{a, b, c}

Test membership of element f

Figure 8.7   Overview of a Bloom filter.

8.8.1 Definition

A Bloom filter is a sequence of m bits for representing a set $S = \{x_1, x_2, \ldots, x_n\}$ of n elements. Initially all the bits are set to zero. The key idea is to use k hash functions, $h_i(x), 1 \leq i \leq k$ to map items $x \in S$ to random numbers uniform in the range $1, \ldots m$. The hash functions are assumed to be uniform. Figure 8.7 illustrates a Bloom filter with three elements, each hashed with three hash functions.

An element $x \in S$ is added into the filter by setting the bits $h_i(x)$ to one for $1 \leq i \leq k$. The element y is reported to be a member of S if the bits $h_i(y)$ are set, and guaranteed not to be a member if any bit $h_i(y)$ is not set.

Algorithm 8.9 presents the insertion operation and Algorithm 8.10 presents the membership test operation. The limitation of Bloom filters is the possibility for a false positive. False positives are elements that are not members of the set S, but are reported to be members by the Bloom filter.

Algorithm 8.9 Bloom filter insertion operation

Data: x is the object key.
Result: BFinsert(x)
begin
    for $j : 1 \ldots k$ **do**
        /* Iterate all k hash functions h_j */
        $i \leftarrow h_j(x)$
        if $B_i == 0$ **then**
            /* Set bit at position i */
            $B_i \leftarrow 1$
        end
    end
end

Algorithm 8.10 Bloom filter membership test operation

Data: x is the object key to be tested for membership.
Result: BFismember(x) returns true or false
begin
    $m \leftarrow 1$
    $j \leftarrow 1$
    while $m == 1$ *and* $j \leq k$ **do**
        $i \leftarrow h_j(x)$
        if $B_i == 0$ **then**
            $m \leftarrow 0$
        end
        $j \leftarrow j + 1$
    end
end
return m

A Bloom filter constructed based on S requires m bits of space and answers membership queries in $O(1)$ time. Given $x \in S$, the Bloom filter will always report that x belongs to S, but given $y \notin S$ the Bloom filter may report that $y \in S$ resulting in a false positive.

The false positive probability of Bloom filter is given by the equation $(1 - e^{-kn/m})^k$. The false positive probability decreases as the size of the Bloom filter, m, increases. The probability increases with n as more elements are added. False positive rate can be minimized by minimizing the equation with respect to k. Taking the derivative and equaling to zero gives the optimal value of k:

$$k_{opt} = \frac{m}{n} \ln 2 \approx \frac{9m}{13n}.$$

Increasing or decreasing the number of hash functions towards k_{opt} can lower false positive ratio while increasing computation in insertions and lookups. The cost is directly proportional to the number of hash functions. The size of the filter can be used to tune the space requirements and the false positive rate. A larger filter will result in fewer false positives. The size of the set that is inserted into the filter determines the false positive rate.

8.8.2 Summary Subscriptions

Bloom filters and additional predicate indices have been used to summarize subscriptions [37, 38]. The problem with Bloom filters is that they are not directly suitable for summarizing arithmetic operators other than equality. Thus two additional structures were developed to implement the summary subscriptions, namely an *Arithmetic Attribute Constraint Summary (AACS)* and a *String Attribute Constraint Summary (SACS)*. This approach is similar to the filter merging technique examined in Chapter 11, but it is not transparent to the routers, because they have to be aware of the structures. Moreover, the operators and attributes need to be known beforehand.

8.8.3 Multicast Forwarding

Bloom filters can also be used to implement efficient multicast systems. The LIPSIN system is one recent example that represents multicast trees in the form of Bloom filters, called zFilters, and includes these in data packets [39]. The system realizes source routing by making forwarding decisions based on the multicast tree links stored in the Bloom filter. This system is discussed more in Chapter 13.

8.8.4 Content-Based Forwarding

The XSIENA system uses a content-based forwarding engine based on Bloom filters [40]. Experimental results with the system indicate that it is efficient and retains the flexibility of the content-based routing concept. The key idea of the system is to determine a compact probabilistic representation of each filter and then utilize this representation in matching the event. This requires that such a compact representation can be derived and that the routers have synchronized routing tables so that the compact representation is used correctly in the forwarding process.

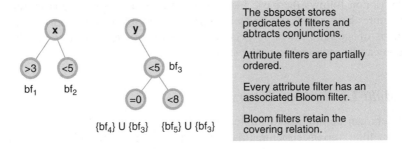

Figure 8.8   The sbsposet structure proposed in XSIENA.

This system is based on two new data structures, namely the sbsposet and sbstree. The sbsposet decomposes filters into their attribute filter components, and stores these in the structure grouped by their attribute name. Each attribute filter is represented by a Bloom filter derived with the structure. The sbsposet is used to derive a compact representation of each attribute filter that retains the relations of the poset. The tree is a simpler structure that stores the filters in conjunctive form based on their Bloom filter representation. The tree avoids the loss of precision due to the decomposition of the filters into attribute filters, because it takes the conjunctions of attribute filters into account when matching events. A variant of the sbstree structure was also developed based on the counting algorithm. The sbsposet can be seen to be a slow path structure whereas the sbstree is a fast path structure consulted during forwarding.

The sbsposet structure is illustrated by Figure 8.8. The sbsposet stores the predicates of filters and the associated Bloom filter representations. The Bloom filters follow the covering relation of the poset. If a filter covers another filter, then their Bloom filters retain this covering relation as well. This is achieved by applying the logical OR operation with the predecessor Bloom filters. This structure is responsible for the content-based matching of events at the first broker and determining the Bloom filter representation of a new filter. The sbstree stores a disjunction of conjunctions of Bloom filters and it is used for fast forwarding of events.

Content-based subscription is processed with the following steps:

1. A subscriber issues a filter and contacts a broker for constructing a set of Bloom filters that summarize the filter's content.
2. The filter along with the Bloom filter representation are disseminated in the pub/sub network using the identity based routing scheme presented in Chapter 7.
3. Each broker adds the filter and the Bloom filters into the sbsposet routing table and the Bloom filters into the sbstree routing table.

The forwarding of events involves the following steps:

1. A publisher publishes an event that is delivered to the nearest broker.
2. The broker matches the event with the sbsposet structure and extracts a Bloom filter that summarizes the subscriber's interest matched by the event. This matching with the sbsposet is performed only once.
3. The Bloom filter is then used with the sbstree structure to determine the output interfaces for the event. This is done with each intermediate broker.

Unsubscription follows the basic identity-based routing model. The sbsposet and sbs-forest structures have counters in order to cope with multiple identical attribute filters, and conjunctions of attribute filters, respectively. When a counter reaches zero, the element in question can be removed from the structure.

8.8.5 Multi-Level Bloom Filters

Bloom filters can be applied to implement efficient matching of structured documents. Two Bloom filter variant structures have been proposed for matching queries on XML documents. Koloniari *et al.* have defined the *Breadth Bloom Filter (BBF)* and *Depth Bloom Filter (DBF)* to support path expressions. These are based on deriving a long bitstring that represents the documents available on each node in the distributed network. Similarly, a bitstring can be derived that summarizes the local and external queries active at a node [41, 42]. These bitstrings can then be shared and compared to find matching XML documents for each node.

Let T be an XML tree with j levels with 1 as the root level. The Breadth Bloom Filter for an XML tree T with j levels is a set of Bloom filters. There is one Bloom filter for each level i of the tree. Each level filter contains the elements of all nodes at that level. An additional Bloom filter denoted by BBF0 contains all elements that appear in any node of the tree. The matching process distinguishes between path queries starting from the root and partial path queries. All elements of a query must be present in BBF0. If all elements are present, the process continues to evaluate the structure of the path.

The Depth Bloom filter provides an alternative way to summarize XML trees. Different Bloom filters store paths of different lengths. The Depth Bloom Filter for an XML tree T with j levels is a set of Bloom filters. In this construction, each path of length i has its own Bloom filter. The matching process consults the BBF0 first as before, and then checks whether whether the subpaths of the path query are contained. If a subpath is not contained, the algorithm reports a miss.

Experimental results with the structures indicate that they perform significantly better than simply using a single Bloom filter to summarize structured documents. The results suggest that the DBF has favourable properties compared to the BBF in terms of reducing the number of false positives [41, 42].

8.9 Summary

In this chapter, we examined a number of well-known techniques and algorithms for matching events and content to subscriptions. This filtering process is an essential part of pub/sub systems and necessary to ensure that relevant content is delivered to subscribers.

We examined the counting algorithm, poset structures, the tree algorithm, XFilter and YFilter, and Bloom filter based probabilistic matchers. There are many ways to design and implement a filtering system. The systems differ based on the design parameters and assumptions, for example:

- Evaluation can start with event contents or the constraints. The counting algorithm iterates the event structure whereas the tree algorithm and the poset walk the structure with the event.

- The structure can be probabilistic with a certain accuracy. Bloom filter based matchers have a false positive rate.
- The structure can be used as a routing table structure or a forwarding structure or both. The routing table should support dynamic operation and aggregation, whereas the forwarding structure should be very fast. The separation of these is evident in many recent system proposals.
- The matching structure can be very specific and only support a certain filtering language and event structure or it can be generic and support a variety of filtering languages. The counter and tree algorithms are specific to the event structure whereas the poset and forest are generic, but assume that the covering relation can be determined.
- The system may support structured content, for example XML documents. XFilter and YFilter are examples of such structures.
- Basic matching algorithms can be combined, for example XSIENA combines the poset, tree, and counting algorithm. This kind of combination can result in performance improvements.
- The matching structure may or may not be suitable for parallel operation and implementation with hardware. For example, the poset is not very suitable due to the many graph operations needed, whereas Bloom filter based matchers typically allow parallelization and can be implemented efficiently with hardware.

Figure 8.9 presents the frequently used solutions for improving the performance of the pub/sub matching subsystem. The poset and forest, trees, and tables for the counting algorithm are examples of routing table structures. Filter covering and merging are useful optimizations for the distributed environment. The poset and forest are examples of

Solution	Examples	Description
Routing table	SIENA, Gryphon, Rebeca, X-SIENA	A dedicated data structure for storing and processing subscriptions (filters). Classical examples are the SIENA poset, and tree-based structures.
Filter covering	SIENA, Rebeca, Fuego	Optimizes routing tables by propagating only the most general filters.
Filter merging	Rebeca, dynamic mergingin Fuego	Optimizes routing tables by aggregating two or more filters into a single filter. The result, a merger, can be either imperfect or perfect.
Fast forward algorithm	SIENA, Gryphon, others	A filter language and format specific algorithm that is able to determine matching output interfaces for a given event in an efficient manner. The algorithm makes assumption on the filterand event structure.
Approximate routing tables	X-SIENA	A scheme that allows approximate routing tables that summarize routing state or approximate summary information in event messages that is used in forwarding. This results in compact table size and fast matching; however, may result in false positives or false negatives, depending on the scheme. One example is Bloom-filter based routing.
Separating routing and forwarding	SIENA, X-SIENA	This solution separates the routing table structure and the forwarding structure.

Figure 8.9   Solutions.

structures that are based on the covering relation and thus implicitly support this optimization. Both can easily be extended with filter merging. We also considered how the counting algorithm can be extended to support the covering optimization. Many systems separate the routing and forwarding parts of the pub/sub router. We examined fast forwarding with the counting algorithm, tree algorithm, XFilter and YFilter, and the Bloom filter based algorithms. These can be combined with more general routing structures. We also discussed approximate routing tables with Bloom filter. The sbsposet used in XSIENA is an example of such a structure. XSIENA features poset based routing tables and tree based forwarding tables with approximate Bloom filter based forwarding.

References

1. Object Computing, Inc. (2001) *CORBA Notification Service Specification v.1.0*. OCI.
2. Sun (2002) *Java Message Service Specification 1.1*.
3. Cugola G, Di Nitto E and Fuggetta A (1998) Exploiting an event-based infrastructure to develop complex distributed systems. *Proceedings of the 20th International Conference on Software Engineering*, pp. 261–70. IEEE Computer Society.
4. Sutton P, Arkins R and Segall B (2001) Supporting disconnectedness-transparent information delivery for mobile and invisible computing. *CCGRID '01: Proceedings of the 1st International Symposium on Cluster Computing and the Grid*, p. 277. IEEE Computer Society, Washington, DC.
5. Mühl G and Fiege L (2001) Supporting covering and merging in content-based publish/subscribe systems: Beyond name/value pairs. *IEEE Distributed Systems Online (DSOnline)*.
6. IBM (2002) *Gryphon: Publish/subscribe over Public Networks*. (White paper) http://researchweb.watson.ibm.com/distributedmessaging/gryphon.html.
7. Carzaniga A, Rosenblum DS and Wolf AL (2001) Design and evaluation of a wide-area event notification service. *ACM Transactions on Computer Systems* **19**(3): 332–83.
8. Carzaniga A and Wolf AL (2003) Forwarding in a content-based network *Proceedings of ACM SIGCOMM 2003*, pp. 163–74, Karlsruhe, Germany.
9. Carzaniga A, Deng J and Wolf AL (2001) Fast forwarding for content-based networking. Technical Report CU-CS-922-01, Department of Computer Science, University of Colorado.
10. Mitidieri C and Kaiser J (2003) Attribute-based filtering: Improving the expressiveness while keeping the predictability in P/S systems. *Proceedings of the 2nd International Workshop on Distributed Event-Based Systems (DEBS'03)*.
11. Mühl G (2002) *Large-Scale Content-Based Publish/Subscribe Systems*. PhD thesis. Darmstadt University of Technology.
12. Pereira J, Fabret F, Llirbat F and Shasha D (2000) Efficient matching for web-based publish/subscribe systems. In *Cooperative Information Systems, 7th International Conference, CoopIS 2000, Eilat, Israel, September 6–8, 2000, Proceedings* (ed. Etzion O and Scheuermann P), vol. 1901 of *Lecture Notes in Computer Science*, pp. 162–73. Springer.
13. Yan TW and García-Molina H (1994) Index structures for selective dissemination of information under the boolean model. *ACM Trans. Database Syst*. **19**: 332–64.
14. Jacobsen HA, Ashayer G and Leung H (2002) Predicate matching and subscription matching in publish/subscribe systems. In *Proceedings of the 1st International Workshop on Distributed Event-Based Systems (DEBS'02)* (ed. Bacon J, Fiege L, Guerraoui R, Jacobsen A and Mühl G), pp. 539–46.
15. Aguilera MK, Strom RE, Sturman DC, Astley M and Chandra TD (1999) Matching events in a content-based subscription system. *PODC '99: Proceedings of the Eighteenth Annual ACM Symposium on Principles of Distributed Computing*, pp. 53–61. ACM Press, New York, NY.
16. Campailla A, Chaki S, Clarke E, Jha S and Veith H (2001) Efficient filtering in publish-subscribe systems using binary decision diagrams *ICSE '01: Proceedings of the 23rd International Conference on Software Engineering*, pp. 443–52. IEEE Computer Society, Washington, DC.
17. Fabret F, Jacobsen HA, Llirbat F, Pereira J, Ross K and Shasha D (2001) Filtering algorithms and implementation for very fast publish/subscribe. In *Proceedings of the 20th Intl. Conference on Management of Data (SIGMOD 2001)* (ed. Sellis T and Mehrotra S), pp. 115–126, Santa Barbara, CA.

18. Liu H and Jacobsen HA (2002) A-TOPSS – a publish/subscribe system supporting approximate matching. *Proceedings of the 28th VLDB Conference*, Hong Kong, China, pp. 1281–4.

19. Yan TW and García-Molina H (1999) The SIFT information dissemination system. *ACM Trans. Database Syst*. **24**, 529–65.

20. Hinze A and Bittner S (2002) Efficient distribution-based event filtering. In *Proceedings of the 1st International Workshop on Distributed Event-Based Systems (DEBS'02)* (ed. Bacon J, Fiege L, Guerraoui R, Jacobsen A and Mühl G), pp. 525–32.

21. Guttman A (1984) R-Trees: A dynamic index structure for spatial searching. In *SIGMOD'84, Proceedings of Annual Meeting, Boston, Massachusetts, June 18–21, 1984* (ed. Yormark B), pp. 47–57. ACM Press.

22. W3C (1999) *XML Path Language (XPath) 1.0*. [W3C Recommendation] http://www.w3.org/TR/xpath.

23. W3C (2010) *XQuery 1.0: An XML Query Language (Second Edition)*. [W3C Recommendation] http://www.w3.org/TR/xquery/.

24. Gottlob G, Koch C and Pichler R (2003) The complexity of XPath query evaluation. *Proceedings of the Twenty-Second ACM SIGMOD-SIGACT-SIGART Symposium on Principles of Database Systems*, pp. 179–190. ACM Press.

25. Chan CY, Fan W, Felber P, Garofalakis MN and Rastogi R (2002) Tree pattern aggregation for scalable XML data dissemination. *VLDB*, pp. 826–37.

26. Whang S, Brower C, Shanmugasundaram J, *et al*. (2009) Indexing boolean expressions. *PVLDB* **2**(1): 37–48.

27. Sadoghi M and Jacobsen HA (2011) Be-tree: an index structure to efficiently match boolean expressions over high-dimensional discrete space. *SIGMOD Conference*, pp. 637–48.

28. Mühl G (2001) Generic constraints for content-based publish/subscribe systems. In *Proceedings of the 6th International Conference on Cooperative Information Systems (CoopIS'01)* (ed. Batini C, Giunchiglia F, Giorgini P and Mecella M), vol. 2172 of *LNCS*, pp. 211–25. Springer-Verlag, Trento, Italy.

29. Tarkoma S and Kangasharju J (2006) Optimizing content-based routers: posets and forests. *Distributed Computing* **19**(1): 62–77.

30. Cormen TH, Leiserson CE and Rivest TL (2001) *Introduction to Algorithms*. The MIT Press.

31. Collin C and Levinson R (1989) Partial order maintenance. *SIGIR Forum* **23**(3–4): 59–88.

32. Altinel M and Franklin MJ (2000) Efficient filtering of XML documents for selective dissemination of information. *Proceedings of the 26th International Conference on Very Large Data Bases*, pp. 53–64. VLDB '00. Morgan Kaufmann Publishers Inc., San Francisco, CA.

33. Diao Y, Altinel M, Franklin MJ, Zhang H and Fischer P (2003) Path sharing and predicate evaluation for high-performance XML filtering. *ACM Trans. Database Syst*. **28**, 467–516.

34. Wu E, Diao Y and Rizvi S (2006) High-performance complex event processing over streams. *Proceedings of the 2006 ACM SIGMOD International Conference on Management of Data*, pp. 407–18. SIGMOD '06. ACM, New York, NY.

35. Bloom BH (1970) Space/time trade-offs in hash coding with allowable errors. *Commun. ACM* **13**(7): 422–6.

36. Tarkoma S, Rothenberg C and Lagerspetz E (2011) Theory and practice of bloom filters for distributed systems. *Communications Surveys Tutorials*, IEEE, pp. 1–25, preprint.

37. Triantafillou P and Economides A (2002) Subscription summaries for scalability and efficiency in publish/subscribe systems. In *Proceedings of the 1st International Workshop on Distributed Event-Based Systems (DEBS'02)* (ed. Bacon J, Fiege L, Guerraoui R, Jacobsen A and Mühl G).

38. Triantafillou P and Economides A (2004) Subscription summarization: A new paradigm for efficient publish/subscribe systems. *ICDCS*, pp. 562–71.

39. Jokela P, Zahemszky A, Esteve C, Arianfar S and Nikander P (2009) LIPSIN: Line speed publish/subscribe inter-networking. *SIGCOMM'09*, pp. 195–206.

40. Jerzak Z and Fetzer C (2008) Bloom filter based routing for content-based publish/subscribe. *Proceedings of the Second International Conference on Distributed Event-Based Systems*, pp. 71–81. DEBS '08. ACM, New York, NY.

41. Koloniari G and Pitoura E (2003) Bloom-based filters for hierarchical data. *5th Workshop on Distributed Data Structures and Algorithms (WDAS 03), Thessaloniki*, pp. 13–14.

42. Koloniari G and Pitoura E (2004) Filters for XML-based service discovery in pervasive computing. *Comput. J*. **47**(4): 461–74.

9

Research Solutions

In this chapter, we survey well-known pub/sub research systems and prototypes. The research prototypes are implemented as application layer overlay networks. Many of the first distributed pub/sub systems featured hierarchical and fixed overlay designs, in which the connections between routers are based on a configuration policy. Many of the newer systems are based on DHT structures presented in Chapter 3.

Typical examples of early systems are Gryphon, SIENA, JEDI, and Elvin that offer content-based routing support and utilize non-DHT-based underlays. These systems introduced many of the frequently used optimizations and extensions for distributed pub/sub systems, such as filter covering, filter merging, mobility support, clustering, load balancing, and security services. We also examine DHT-based solutions that leverage the structured overlay as an underlay for the higher level pub/sub routing. DHT-based solutions are able to support wide-area environments with clients joining and leaving the network.

The design principles, patterns, and solutions for the research systems have already been introduced in the previous chapters, and we revisit many of them when discussing the research systems. For example, the many of the solutions discussed in Chapters 7 and Chapter 8 have been proposed in the context of the research systems presented in this chapter.

We conclude the chapter by summarizing the key differences and similarities of the systems.

9.1 Gryphon

Gryphon is a Java-based pub/sub message broker developed at the IBM T.J. Watson Research Center. The aim of the system is to distribute data in real time over a large public network [1, 2]. The project was started in 1997 to develop the next generation Web applications and it has been tested several times with real-world use cases. For example, it was used to deliver information about the Tennis Australian Open to 50000 concurrently connected clients. Moreover, it has been used to distribute real-time sports scores in the Tennis US Open, Ryder Cup, and for monitoring statistics at the Sydney Olympics.

Gryphon supports both topic-based and content-based pub/sub, relies on TCP/IP and HTTP, and supports recovery from server failures and security. In Gryphon, the flow of

Publish/Subscribe Systems: Design and Principles, First Edition. Sasu Tarkoma.
© 2012 John Wiley & Sons, Ltd. Published 2012 by John Wiley & Sons, Ltd.

streams of events is described using an *information flow graph (IFG)*, which specifies the selective delivery of events, the transformation of events, and the creation of derived events as a function of states computed from event histories. Gryphon implements the JMS API that the clients can use to send and receive messages.

Gryphon extends the publish/subscribe one-to-many model with request-reply and solicit-response models. By using unique topics JMS users can use request-reply-style messaging. In the solicit-response model a client may make an advertisement to which one or several clients may respond privately. The basic unit of the Gryphon multibroker configuration is the cell, which is a group of fully connected servers. Cells may be further linked together for geographical scaling through link bundles. Link bundles provide redundant connections between cells, which includes load balancing and fault tolerance. The internal protocols and systems ensure that loops are avoided and messages are routed around failed nodes.

Event processing in Gryphon is based on the notion of an IFG. An IFG contains stateless event transforms that combine events from various sources, and stateful event interpretation functions that can be used to derive trends, alarms, and summaries from published events. Each event is a typed tuple. Stateful events depend on the event history. States are used to express the meaning of an event stream and the equivalence of two event streams.

The Gryphon model consists of information spaces, which are either event histories or states. Event histories grow monotonically over time as new events are published. Event sources and sinks are modelled as event histories. States capture certain relevant information about event streams, and they are typically not monotonic. Information spaces are defined using information schemas. Dataflows are directed arcs that connect nodes in the IFG, which needs to be acyclic.

Gryphon supports four types of dataflows (arcs in the IFG):

- The select arc connects two event histories that have the same schema. Each arc is a predicate on the attributes of the event type in the information space. All events that satisfy the constraint are delivered to the destination information space. This arc makes a subset of events flow from the source history to the target history.
- The transform arc connects any two event histories that may have different schemas. Each arc has a rule for mapping event types between the two spaces. This rule may include functions that transform particular event attributes.
- The collapse arc connects an event history to a state using a rule. The rule maps a new event and a current state into a new state.
- The expand arc is the inverse of collapse, and links a state to an information space.

When the state at the source of the arc changes, the destination space is updated in such as way that the sequence of events it contains collapses to the new state. This transformation is nondeterministic. Gryphon has two techniques for the implementation of systems based on IFGs:

- A flow graph rewriting optimization that allows stateless IFGs to be used with multicast technology.
- An algorithm for converting a sequence of events to the shortest equivalent sequence of events.

The information flow graph is abstract and separated from the physical topology of the network. The mapping of an IFG to a network of message brokers is nontrivial. Gryphon reduces an arbitrary IFG by rewriting it. All the select operations are moved together and closer to publishers and all the transform operations are also grouped together closer to the subscribers. Transform operations are done at the periphery of the network.

The Gryphon system detects failed brokers and reroutes traffic around failed nodes. Moreover, the system incorporates several security mechanisms, such as access control, and four authentication methods. Gryphon supports the JMS publish/subscribe API, and supports topic-based subscriptions. In addition, clients may specify filters using the WHERE-clause of SQL92 supported by JMS.

9.2   The Cambridge Event Architecture (CEA)

The CEA uses the publish-register-notify paradigm [3], in which the object publishes its interface, for example specified in *IDL (Interface Definition Language*, which is different from the IDL in CORBA). This interface includes the events of which it is capable of notifying. A client invokes the object synchronously and can register for events by indicating parameters (attributes) or wildcards. Wildcard matching is applied on the parameters of a notification, but it may not be applied on the event type. The template system provides rudimentary filtering by matching parameters one by one. The object accepts registrations and notifies the clients that match the registration template. The notification is performed when the event firing conditions and access restrictions are satisfied (Figure 9.1). The paradigm supports direct source-to-client event notification.

In CEA an object, if asked, publishes the events it is capable of notifying of in IDL. The object has a register method in its interface that has parameters for the type of event and wildcards. Event occurrences are objects of a specific type, and the set of types defines the level of event detection and notification granularity. CEA enforces access control upon registration, and authentication is based on a parameter value. CEA supports defining intermediate services, which are called event mediators in the architecture. Event mediators act as middlemen between primitive event sources and event clients, and provide the facilities for detecting more complex events. Moreover, if an event source cannot

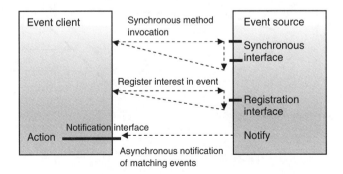

Figure 9.1   Overview of the CEA system.

afford the overhead of supporting template matching, it can send all its events to the mediator. The mediator then matches the template on behalf of the source.

The mediator is capable of providing equivalent functionality to the CORBA event service. The CORBA event service registers interests in all notifiable events with event sources and supports both a synchronous pull interface and an asynchronous push interface.

Composite events can be detected by giving mediators the capability to filter simple events of different types across different sources. The event composition is supported by the combination of event templates. Composite events are detected by monitors, which are busy until the event is detected and fired. A composite event specification language may be used to design a monitor that detects complex templates. The system has been demonstrated by implementing an active badge system that monitors badges within a building.

9.3 Scalable Internet Event Notification Architecture (SIENA)

SIENA is an event notification service that introduces many of the features of content-based systems. We have examined the basics of the SIENA model in Chapter 7 and introduced the notions of subscription and advertisement semantics as well as their optimizations. Its characteristics include content-based routing in a wide-area network and aims to balance scalability with expressive content delivery semantics [4].

SIENA architecture supports many network topologies such as hierarchical topology, acyclic peer-to-peer topology, and general peer-to-peer topologies. Also hybrid network topologies are supported. Overhead of routing table management is decreased because the servers know only the identities of their neighbours. A server-server protocol is used for communication of servers with their peers. Communication of servers with the subscribed notification clients is accomplished using a client-server protocol. Figure 9.2 illustrates the SIENA system model that consists of the overlay network of brokers as well as the client-server and server-server protocols.

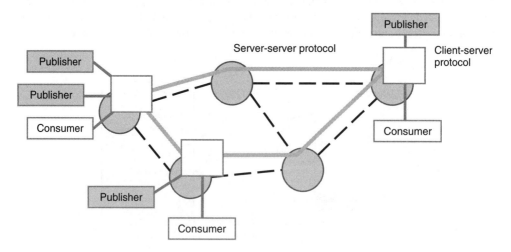

Figure 9.2 Overview of the distributed SIENA system.

IP multicast is another similar mechanism, but it differs from SIENA because multicast groups are not content-based. IP multicast groups are channels, and they partition the IP multicast addressing space into disjoint groups. Thus IP multicast is not very expressive when compared with content-based routing. On the other hand, IP multicast is one implementation choice for delivering datagrams. This requires that there is a mapping from a content-based query and publication to IP multicast groups.

9.3.1 Event Namespace

The event namespace is essentially flat in SIENA meaning that there is no structural correlation between event names. Events in SIENA are attribute-value pairs where every attribute has a name and a value. The types supported are null, string, long, integer, double, and boolean. A SIENA filter consists of three entities:

- attribute name,
- constraint operator, and
- constraint.

Wildcards in the attribute name are not supported and there must be an exact match between the attribute name and the name of a published event. Filtering clauses are supported and a filter may have several of them, joined by AND operators. The filter will pass the event only if each filtering clause or component returns true. Operators supported by SIENA are equal, less than, greater than, greater than or equal to, less than or equal to, string prefix, string suffix, always matches, not equal, and substring.

Patterns in SIENA utilize event attribute values and also combinations of events. A SIENA pattern is generally a filter sequence which will match to a (temporally ordered) notification sequence. SIENA ignores events that arrive in the wrong order, which this may happen because of network latencies.

9.3.2 Routing

As we noted above, events in SIENA are sets of attribute-value pairs, matched with filters. Subscription information, advertisement information, and filters maintained in the event system servers form the basis for routing events to other servers. Subscribers are allowed to specify filters to define and delimit the subscription; also similarly advertisements include filters. SIENA systems knows and evaluates the filters. The general policy is to replicate events downstream and filter them upstream. Therefore replication of events is postponed to as late as possible, reducing the event transmission bandwidth. On the other hand the upstream filtering handles events as close to the sources as possible. This aims to reduce the volume of irrelevant event transmission over the network.

The filter syntax allows building complex filters out of simpler, more general filters that can be evaluated upstream. In fact, a filter must be less general than the one used upstream, in order to be applicable at all. The concept of upstream filtering is used elsewhere in SIENA structure as well.

Event patterns are factored into elementary filters and then delegated to servers using a process which makes an effort to assemble those subpatterns that are acceptable for delegation to other servers. Covering relations are utilized to detect if a filter is covered by a notification, if a notification is covered by a subscription or by an advertisement, or if a subscription is covered by advertisement. Let us take an example: subscription S2 is covered by subscription S1 if S1 is true always when S2 is true. Servers will find the most generic subscription covering a given subscription set and propagate this, hereby minimizing the downstream data structures. There is a computation cost, however, to be paid closer to the subscriber as the subscriptions involve matching and evaluating. Still, SIENA demonstrates that the complexity of the covering relation will be reasonable for a scalable service.

There are two separate notification semantics in SIENA:

1. Subscription-based semantics where subscriptions can be introduced at every event service node. Notifications are routed when they cover a specific subscription.
2. Advertisement-based semantics where routing servers use the information from event publishers in order to route incoming subscriptions. Only subscriptions related to an active advertisement are forwarded.

The poset (a partially ordered set) that we discussed in Chapter 8 is the main concept used by SIENA for realizing the data structures of routing tables. We should note that the scalability of the routing table update processing is restricted in those environments where the updates arrive frequently. The poset is a generic routing table element and it can support various routing configurations. The key routing configurations supported by SIENA are:

- Centralized configuration with a single broker.
- Hierarchical configuration of brokers.
- Peer-to-peer configuration, in which brokers propagate routing updates to neighbouring brokers.
- A hybrid configuration.

The hierarchical configuration is simple, because brokers always propagate subscription updates upwards towards the root and published events upstreams and also downstreams towards matching brokers. This configuration is limited in scalability, because the root node may become a bottleneck. The peer-to-peer configuration is more sophisticated and can avoid the bottleneck, especially when the brokers are configured in a cyclic graph topology. On the other hand, this complicates the routing process and requires that subscriber-specific shortest spanning trees are computed in order to prevent routing loops.

SIENA requires that the clients of the system are able to compute covering relations between filters. When a client subscribes a filter, any filter covered by a new filter from the same client will be removed. Similarly, when a client unsubscribes, any filters that are covered by the removed filter will be removed. The client has to establish any filters that should still be active at the same time with the unsubscription.

9.3.3 Forwarding

The forwarding table is a mapping between filter sets (predicates) and interfaces that are connected to neighbouring nodes. Predicates are defined as disjunctions of filters and every filter is a conjunction of a set of elementary conditions. Elementary conjunctions must return true as this enables a filter (and predicate) to form a mapping to an interface. A filter can be mapped to many outgoing interfaces.

The forwarding algorithm will compute an iteration in the set of event attributes, searching for a partial match from the set of those filters, where an intrinsic filter constraint matches with the given attribute. We examined the algorithm in more detail in Chapter 8. If the partially matched filter is free from current associations with an interface, the algorithm will increase the count of matched constraints for the given filter. The filter will reach a match when the counter value equals the number of filter constraints. The forwarding algorithm will check the matching state of all filters, stopping after processing all attributes of the notification or when all filters are processed. The upper bound of the processing therefore depends on the number of interfaces and on the number of attributes and filters. The optimization of the forwarding algorithm is achieved with binary trees and also using lookup indices for filter attributes. However, the SIENA filter matching algorithm may be extended with further optimization methods [5], for example, using an index-based matching structure and attribute filter selection to prune those predicates that cannot be matched.

9.3.4 Mobility Support

Support for mobility and wireless clients has been added to SIENA as well. This involves a generic protocol of handover type utilizing the existing pub/sub primitives publish and subscribe [6]. A generic protocol enables working on top of various pub/sub systems without requiring any changes to the system API. However, the performance of the mobility support is apt to decrease, because the underlying topology is hidden by the API, making all mobility-specific optimizations difficult to realize.

The generic mobility support service of SIENA is based on a ping/pong protocol that will be examined in Chapter 11. It has been implemented by proxies residing on access routers. The following steps belong to the handover phase:

- The client moves from access point A to access point B sending the new local proxy a move-in request.
- The old proxy is pinged and a response (called a pong) will be received. The pong message guarantees the full propagation of subscriptions from B to A.
- The client downloads buffered events (download request).
- The proxy receives the buffered events.
- The messages are delivered to the client with duplicates removed.

9.3.5 CBCB Routing Scheme

The *combined broadcast and content-based* (CBCB) routing scheme, due to Carzaniga *et al.* [7] is a covering-based routing scheme. CBCB is a *two-layer* routing scheme that combines a content-based routing protocol with a broadcast routing protocol. A broadcast-based protocol is needed to build and maintain forwarding state that allows for sending messages from a node to all other nodes in the network. Broadcasting is used to gather information about the network. There are several well-known ways to implement a broadcasting function, such as minimal spanning trees or reverse path forwarding. After the basic routing and forwarding information is acquired by the broadcasting protocol, the content-based layer in the scheme is used to prune the broadcast trees by utilizing the content-based data.

In the following, we consider a network of nodes such that each node has a content-based address p_n, where n is an identifier unique to each node. The node addresses can be abstracted by aggregating the addresses at the nearest router. The content-based address of a router is a disjunction of its local clients' content-based addresses. In other words, the clients' filters define the subnetwork maintained by the router.

The content-based layer of the CBCB scheme uses a *push-pull mechanism*. The nodes push routing data to their neighbours by using *receiver advertisements* (RAs). RAs is similar to a subscription and should not be confused with an advertisement of available content as discussed previously. In addition, the nodes can pull routing information from the network with the *SR/UR protocol* by sending *sender requests* (SRs) and then receiving *update replies* (URs).

9.3.5.1   Receiver Advertisements

Receiver advertisements are the primary technique of propagating content-based routing data. An RA is sent by a node whenever its content-based address changes or a periodic timer expires. An RA can be represented as the pair (n, p_n), where n is a node identifier and p_n is its content-based address. The RA can also contain additional information. Upon receiving an RA (n, p_n) from interface i, a router handles the message based on the following rules:

- If there is an address p_i associated with interface i in the routing table, the router performs a covering test:
  - if p_n is already covered by p_i, the router drops the RA. This is the *RA ingress filtering rule*.
  - Else the router uses the broadcasting function to find the set of next-hop destinations on the broadcast tree rooted in n. The router updates its routing table by setting the address associated with interface i as $p_i := p_i \vee p_n$.

9.3.5.2   The SR/UR protocol

Simply using only the receiver advertisements may lead to *inflation* of content-based addresses. For example, in the case of a router r that has associated an address p_i with interface i. When the router receives an RA from interface i, it applies the ingress filtering rule to it and drops the RA if its address is already covered by p_i. Now, the reason the router received an address already covered may be due to an unsubscription.[1] Because the RA was dropped, the neighbouring routers still forward all notifications matching p_i to r, even if there is no interest in parts of them. In order to avoid such address inflation and false positives, the routers periodically send out sender requests to update its routing table.

---

[1] The operation differs from SIENA, which propagates unsubscription messages and installs uncovered subscriptions while removing the unsubscribed filters.

An SR contains an identifier of its issuer, an *SR number* and a timeout field. The SR number is used to differentiate between several SRs from the same issuer.

When a router issues an SR, it is broadcast to all routers. A receiving router estimates a new timeout for it and forwards it downstream on the broadcasting tree. If r is a leaf router, it responds with an update reply. A UR consists of the SR issuer identifier, the SR number the UR is reply for, and an address. In the case of a leaf node, the address is its own content-based address. A noinleaf node must wait for URs from all interfaces it forwarded the SR. When all URs have been received, it sends an UR with the address (disjunction of predicates in the received URs) and sends it via the reverse path of the SR.

The SR/UR protocol is based on broadcast and may result in large amounts of control traffic in the network if all routers frequently issue SRs. The system features several improvements over the basic protocol. An intermediate router may be allowed to use an UR to update its own routing table and cache an UR so that it can be used the next time a similar SR is issued. In addition, the use of SRs can be limited to avoid periodic broadcasting.

9.4   Elvin

Elvin is a client-server based event notification service that also routes according to message content [8]. In Elvin clients can establish sessions with servers, subscribing and publishing notifications. A notification of Elvin is a name-value pair like in SIENA, using as basic primitives both 32- and 64-bit integers, a 64-bit double precision floating point, a UTF-8 encoded string and an array of bytes. Subscription is effected using C-like logical expressions. Lukasiewiczs tri-state logic is utilized to evaluate the expressions with an additional value of indefinite (so that we have true, false, indefinite). Elvin binds to languages like C, C++, Java, Python, Smalltalk, Emacs Lisp, and Tcl.

Elvin is also content-based allowing routing decisions based on the whole message, content included. A decoupled security model is typical of Elvin, which does not use the traditional point-to-point model where keys are necessary for authentication of communication between publishers and subscribers. In Elvin the key sets of producers and consumers can overlap allowing for multiparty authorization.

Elvin includes a light multicast-based protocol for service discovery. The clients may use the protocol for discovering the server and dynamically registering if the server has been deployed on the network. Multicast is also used to distribute active router advertisements for the clients.

9.4.1   Clustering

Scalability can be augmented with the use of local clustering. It will also balance the distribution of local loads. Elvin utilizes clustering to implement an address space which is single-subscription based and distributed. Elvin uses a dependable multicast protocol which allows clustered routers to communicate over an IP network. Load over the network is minimized because the routers can force appropriate server changes for clients.

A router in this clustered system is a daemon running on a server. Routers will share part of the the client subscription information with all nodes. The partial shared information allows the router to detect whether a notification has or has not subscribers within the cluster. No unnecessary information is shared. In server-to-server communication the forwarding algorithm first uses a term list to decide the needs. The local router analyzes the messages and builds the appropriate set of destination routers by matching the message with the list of terms. Then it informs the cluster using multicast mechanism and packets with unique identifiers for the matching destination routers. Unfortunately the Elvin forwarding system also may generate unnecessary notifications straining the cluster resources.

The structure of the Elvin cluster is built on a topology where a single master router, which handles the management and maintains the data, has a set of slaved routers. These slaves are listening to the management level traffic within the cluster. They also maintain the information about nodes, server subscription terms, states, URLs open to client connections and router statistics.

A master server has the task of receiving the join-type packets and controlling the cluster. An RPC-style communication between routers and clients uses positive and negative acknowledgments and the semantics applied is of best-effort, at-most-once type. However, if the server drops notifications it must warn the client. If a master server fails, an election protocol is invoked to choose a new one.

9.4.2 Federation

The cluster-level protocol is not the only one needed in Elvin. A separate protocol exists for building a federated system out of the distributed server clusters. This protocol is built using a spanning tree of the clusters, and it supports pull filters in order to constrain the intercluster notifications.

9.4.3 Quench

An operation exists in Elvin which allows the publishes of information to evaluate the expressed subscriptions. The aim of this evaluation is collection information about those events that are no longer needed and can be terminated, that is, *quenched*. The operation is supported by all publishers of events.

This operations is also useful if one wants to examine which kinds of notifications are wanted. As semantic extension of the subscribing mechanism quench is implemented in the Elvin client/server protocol allowing all clients to request notification if there is a change of subscription information in a server. Furthermore, the client is enabled to query information in subscriptions by the attribute names. The client receives the data in the form of an abstract syntax tree. An automatic quench is also supported in the client library.

9.4.4 Mobile Support

Elvin utilizes a proxy-based model in order to support mobile users. This is a very necessary mechanism because Elvin's design principles make it fundamentally nonpersistent, but all notifications, including undelivered ones, must be stored somewhere. The solution is to use a proxy component in the client-server architecture. Servers will see the proxy just as a normal client but the clients see the proxy as server, effectively making the proxy as a service mediator for Elvin.

Elvin originally did not have support for grouping of the client subscriptions but because the proxy handles multiple clients with more than one set of subscriptions, the concept of a session was added which supports grouping. Sessions can span over sets of applications and clients because one holder may have multiple devices with the same information needed for all of them.

The problem of undelivered notifications is solved in Elvin using the TTL mechanism (time-to-live). This defines the storage space management for subscriptions, with the further tool of specifying the maximum for the kept notifications. There are weaknesses in the proxy mechanism, though. Discovering a proxy is not supported, and neither is roaming which could be awkward because the proxies are not transparent and the client must connect and reconnect to a specific proxy knowingly in order to get notifications.

Note that the Elvin proxies should be stateful while the normal Elvin servers are stateless which further complicates the use of Elvin proxy system as a tool of roaming and client migration.

9.4.5 Nondestructive Notification Receipt

A situation which is rapidly becoming normal is the case where a user has several devices that subscribe content. For this the Elvin proxy system includes support for notification receipt that is nondestructive, that is, messages are not automatically removed in the proxy after delivery. Additional management tools are then required controlling subscriptions, because a session may have several clients and Elvin must ensure that a notification delivery to a specific client is not duplicated. Furthermore, several sessions might be alive for a client causing possibly multiple matches and multiple notifications, which should be prevented.

9.5 JEDI

JEDI (Java Event-based Distributed Infrastructure) is a distributed event system architecture the main feature being a set of tree-connected *dispatching servers (DS)* [9]. In the JEDI tree the servers reside on a node connected to exactly one parent DS node (excepting the root node). The number of descendants is zero or greater and the subscriptions (and unsubscriptions) move root wise upwards from DS nodes. The system processes event notifications the same way, that is, notifications are forwarded by a DS node to its single parent upwards. Figure 9.3 illustrates a JEDI network.

When a DS receives an event it will determine if one or several of its descendants have registered an interest. Such events are forwarded down the tree, which is possible because the DS in JEDI will always know the interests of its descendant nodes. This means that event requests and notifications are propagated upwards causing towards the root. The nearer the root a node resides, the greater is the communication and processing overhead from the event traffic and bottlenecks may form which could break the tree into isolated

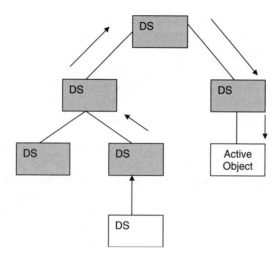

Figure 9.3 Overview of the JEDI system.

regions. This potential segmentation must be dealt with by the JEDI system either by mending it or by creating a new tree with a fresh root node.

The JEDI event system can be summarized as follows:

- An event is an ordered set of strings. The event parameters follow the name of the event.
- The Event Dispatcher can subscribe to an event or alternatively to a pattern of events.
- The event patterns filter events using parameter matching.
- Causal ordering of messages is preserved. If $e1$ is the cause of $e2$, it is delivered first to subscribers. The causal ordering feature enables synchronization between a pair of components using event generation.

Extensions written to JEDI give support mobility of clients and also ad-hoc type configuration [10]. JEDI emphasizes compositionality and reconfigurability supporting mobility with moveOut and moveIn operations. Mobility is supported by the JEDI dispatchers that

- allow clients to disconnect;
- allow clients to move to a new dispatching server;
- allow clients to connect while keeping all notifications.

The system defines temporary storage for notifications. These are controlled by the dispatching servers which also prevent message duplication and guarantee the causal ordering of notifications.

When the client connects to a new dispatching server, this will directly contact the old DS and download the accumulated notifications of the client. The old server notifies its parent (DS) telling it to route subsequent notifications for this client to the new DS. The JEDI dispatching tree routes notifications from publishers to subscribers.

Migration of clients between dispatching servers will cause load changes in the DS system. The basic JEDI system does not address dynamic load balancing; however, load-aware extensions have been considered. This might imply recreating the whole dispatching topology to take care of the load changes. Adapting publish/subscribe mechanism to dynamic environments is accomplished in JEDI by adding a new routing algorithm (spanning tree) to the event routing system. In the algorithm each subscription is handled by a delegate leader, which accepts a specific subscription type functioning as the leader of these subscribers. The leader is responsible of the distribution of the group in the tree; the DS nodes (dispatchers) store the identity of all group leaders.

9.6 PADRES

One of the main problems of developing and executing enterprise applications is the need to decouple strictly enterprise-related parts from the other mechanisms. There are systems that target problems like these aiming to simplify IT development and maintenance. Such a system is PADRES that supports development of enterprise type applications within distributed content-based middleware[2] [11]. PADRES includes fine-tunable subscription language that supports interactions between components and allows event management functions and in-network filtering and processing, while addressing the problems of scalability.

PADRES is an overlay network with broker and client components:

- Brokers route messages (advertisements, subscriptions, publications) using content-based routing algorithms.
- Publishers advertise and publish messages. These are forwarded to the subscribers who have registered interest in receiving such messages.

The key features of PADRES include system administration and monitoring, rule-based routing protocol, rule based matching algorithm, future and historic event correlation, detection of failure, recovery and dynamic load balancing.

Typical user cases supported by PADRES system are:

- target-oriented resource discovery and scheduling,
- distributed transformation, deployment and execution,
- distributed control and monitoring, and
- decentralized secure choreography and orchestration.

9.6.1 Modular Design

PADRES brokers enable message routing in general overlay topologies. This is accomplished with extensions incorporated to the standard content-based routing protocol. This keeps the simple publish and subscribe client interface while not requiring changes to the internal message matching algorithm used by brokers. This allows PADRES to be easily integrated into existing publish/subscribe systems. In The PADRES broker system

---

[2] http://www.msrg.utoronto.ca/projects/padres/index.php.

all brokers are modular and built on two or more queues. One queue for input and one or more output queues representing unique message destinations.

PADRES is temporally bidirectional, allowing subscriptions to data published either in future or in the past. In this PADRES differs from existing content-based pub/sub systems. In the case of future publications, PADRES utilizes the normal pub/sub messaging paradigm but the historic databases are linked to the brokers through a database binding. Publications are stored in publication order in the databases and released (republished) by the brokers when an appropriate request is received.

9.6.2   Load Balancing

The PADRES load balancing solution consists of the following parts:

- detector,
- mediator,
- load estimation tools,
- offload algorithms determining the subscribers to offload.

The process begins with the detector which observes when an overload or load imbalance occurs. The detector then initiates a trigger which will tell the mediator to establish a load balancing session between the offloading broker (the higher load broker which executes the offloading) and the load-accepting broker (which accepts load from the offloading broker). The specific performance metric used for balancing determines which one of the offload algorithms is invoked on the offloading broker. The algorithm will produce the set of subscribers to delegate to the load-accepting broker. The choice is based on estimating the load reduction/increase at each broker utilizing the given load estimation algorithm. Lastly, the mediator will coordinate the migration of subscribers ending the load balancing session of brokers.

9.6.3   Composite Events

There are many benefits of using distributed composite event detection:

- Redundancy in detection will be eliminated, because the detection results are shared.
- If expressions of composite subscriptions issued by clients do overlap, the detection will be executed only once.
- Distributed detection significantly reduces network traffic. This results from the forwarding of a composite subscription in the network as far as possible before splitting it.
- Composite events are detected closer to network data sources.
- After a match only a single notification is sent, and not a set of publications. This reduces the number of publications which must be routed in the distributed system.

PADRES brokers support composite event detection, and they distribute the detection over the network in order to minimize detection costs. We discuss composite events in more detail in Chapter 11.

9.7   REDS

REDS (REconfigurable Dispatching System) is a modular open source distributed pub/sub system developed at Politecnico di Milano[3] [12]. The key feature of REDS is reconfigurability that is supported in two distinct ways:

- Configuration of middleware by selecting the proper mechanisms for message formats, filters, routing strategy.
- Dynamic reconfiguration of the topology and addressing the problems of the overlay network maintenance.

The REDS system is based on the notion of a generic broker that is organized as a set of components that implement well-defined interfaces as shown in Figure 9.4. Developers can parameterize system behaviour and features by implementing the interfaces. Such features include message formats and filters. The REDS transport layer consists of the mechanisms needed to transport content and control messages through the network. The transport system hides the concrete wire protocols that are used to transport data.

The Router encapsulates the routing and forwarding engines This component is responsible for routing messages with the help of three subcomponents. A SubscriptionTable is a routing table that contains subscriptions and it is used to match incoming messages to subscriptions. A RoutingStrategy implements a specific pub/sub routing and forwarding scheme. The default scheme is based on the SIENA model. A ReplyManager that

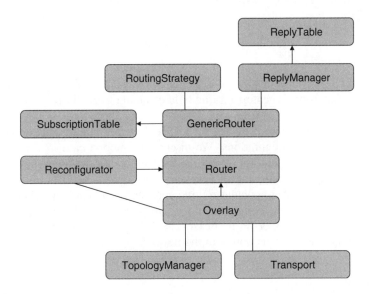

Figure 9.4   Components of the REDS pub/sub system.

---

[3] http://zeus.ws.dei.polimi.it/reds/.

oversees replies to messages. This is an innovative feature of the system and clients can send replies to any messages that they receive.

The Reconfigurator component is responsible for updating the pub/sub topology when reconfiguration of the overlay broker network is detected. The Reconfigurator and the TopologyManager work together in order to establish a new overlay topology. Routing of messages happens on two levels, first there is the overlay layer that can be reconfigured, and then there is the pub/sub routing that happens on a higher level. We discuss topology reconfiguration in Chapter 11 in more detail.

The overlay layer is responsible for providing the basic communications facilities over the transport and network layers. This layer is divided into Transport and TopologyManager components. The former is responsible for communication between two adjacent brokers in the overlay network using TCP or UDP. The latter is responsible for maintaining the overlay network topology and cope with dynamic reconfigurations of the topology. The system has several implementations of the TopologyManager for wired and wireless environments.

9.8   GREEN

GREEN is a pub/sub system that has an extensible filtering model [13]. The filtering models are implemented as plugins and they can be changed easily. GREEN system can be configured to meet application developer requirements. The system supports type based, content based and proximity based filtering.

As an example of XML based filtering, we can consider the following filtering condition [13]:

```
//RoadTraffic/[%type%=$TrafficLight$&%colour%=$Red$]?# \\
DISTANCE# < $100$.
```

This example filter indicates that the subscribing component should be notified if an event of type RoadTraffic, is produced by a traffic light changing to colour Red within a distance of 100 units.

Two basic configuration are provided for GREEN, namely the mobile ad hoc network and wide area network configurations. We discuss the mobile case in Chapter 11 when examining reconfiguration, and briefly outline the wide area configuration here. The configuration consists of the components necessary for content-based pub/sub: publishers, subscribers, SOAP messaging, content-routing core, and an event dispatcher. The system works on top of an existing communications substrate, such as content-based routing overlay, Chord DHT, Scribe, or IP multicast. The default broker overlay plugin is similar to Hermes and uses rendezvous points in the network.

9.9   Rebeca

Rebeca is a distributed event system that supports mobile users and context-aware subscriptions [14, 15]. The system follows the SIENA pub/sub model and is built on an acyclic router configuration and the advertisement semantics. The Rebeca project investigated many extensions to pub/sub routing and forwarding, for example matching algorithms, filter merging, mobility, and security subsystems.

The Rebeca architecture consists of two elements [16]:

- An extendable pub/sub broker core that provides the minimal functionality.
- Plugins that extend the minimal core and that can be inserted at runtime into Rebeca brokers.

The minimal functionality of a broker includes the basic pub/sub API and a pipeline that consists of message handling stages into which components can be inserted (the plugins). For example: in a simple configuration a broker would have an input stage during which a message is deserialized and accepted for processing, a matching and forwarding module that inspects the message and makes a forward decision, and finally an output stage that serializes the message and sends it to the next hop neighbour. This modular structure supports the configuration of the pub/sub system to meet various application requirements, and environments.

The minimal core has been extended with various plugins. Plugins implement, for example, mobility support, reconfiguration of the topology, reliability, scopes, and so on. A scope plugin can be used to restrict the visibility of publications into certain zones defined by a configuration. This is useful when organizational boundaries need to be enforced.

The mobility subsystem supports both logical and physical mobility of clients. The mobility protocol uses an intermediate node between the source and target of mobility, called Junction, for synchronizing the servers. If the brokers keep track of every subscription, the Junction is the first node with a subscription that matches the relocated subscription propagated from the target broker. If covering relations or merging are used this information is lost, and the Junction needs to use content-based flooding to locate the source broker.

9.10   XSIENA and StreamMine

The *XSIENA (eXtended Scalable Internet Event Notification)* system is a scalable and generic event nofication service [17, 18]. This service is based on the SIENA content-based pub/sub system. XSIENA is based on an acyclic overlay network, and utilizes Bloom filters for efficient forwarding of events. We give a brief overview of the system here and return to the routing and forwarding structures of XSIENA in Chapter 8.

XSIENA follows the SIENA model and publishers advertise content with advertisements that are propagated in the network. The subscriptions messages then take the reverse path of the advertisements, and the publication messages the reverse path of subscriptions. The key idea of the system is to represent subscriptions and published events as a set of Bloom filters in order to optimize their processing. This is accomplished with two data structures, namely a poset structure for determining the covering relations between filters that are used in subscriptions and advertisements, and a tree structure for matching events. This mapping of filters and events to Bloom filters is carried out at the edge of the network. The key idea is to utilize fast Bloom filter based matching of events in the core network.

The poset structure is used to derive a set of Bloom filters for each subscription and published event. The Bloom filters are then disseminated with the subscriptions and its original filter, and the intermediate brokers can update their routing tables with this information. A publisher needs then to position a new event to the existing Bloom filter based routing table, and then send the event and its set of Bloom filters for delivery to subscribers. The forwarding system can then match the message to neighbouring brokers and subscribers based on the Bloom filters that position the event with respect to existing subscriptions.

StreamMine is a distributed event processing system intended for large-scale and real-time event stream processing [19] that is based on XSIENA components. One application example for the system is call fraud detection in a telephone system that requires the processing of tens of thousands or even hundreds of thousands of events per second. This kind of data rate requires both parallel processing and distribution support.

The system is based on a content-based pub/sub middleware. The system allows application specific ordering constraints and utilizes a novel solution based on speculative execution of events on multicore systems. Ordering conflicts are determined dynamically and Software Transactional Memory is used to roll back to recover from ordering conflicts. The technique allows the parallel processing of events that should normally be processed sequentially.

The StreamMine system builds on the XSIENA data structures, namely the Bloom filter poset and tree. Thus event processing decisions are made with the efficient Bloom filter representation that summarizes the content of an event.

9.11   Fuego Event Service

The Fuego event service was developed in the Fuego Core project at the Helsinki Institute for Information Technology HIIT. The event service addresses the challenges in the mobile computing environment by providing an asynchronous content-based pub/sub system that supports client mobility [20].

A key component of the system is the event router that connects the publishers and subscribers and mediates event messages between them. Typically, an event router consists of two parts: a set of neighbouring routers and a set of local clients. Both sets are associated with a *routing table* that contains information about which event messages should be forwarded to which neighbouring router or local client.

In order to support event filtering and event delivery, an event router needs to provide an interest-registration service and also have an interface for publishing events. Subscribers define their interests using this interest-registration service.

9.11.1   Fuego Middleware

The event service and event router are part of the Fuego middleware service set. Figure 9.5 presents the Java-based Fuego service architecture APIs and the corresponding runtime system for mobile computing. The architecture realizes a set of generic service elements pertaining to communication and data synchronization. High-level services use the extendable messaging and RPC facilities provided by the messaging service. Jetty and Apache Axis are external Java-based frameworks. Jetty is used to provide support for servlets, and Axis is used to facilitate service deployment and interoperability with existing web services. The proposed architecture allows the deployment of existing Axis web services for wireless users.

The primary data representation format of the system is XML according to the XML Infoset specification. Since XML parsing is a time consuming activity, and XML

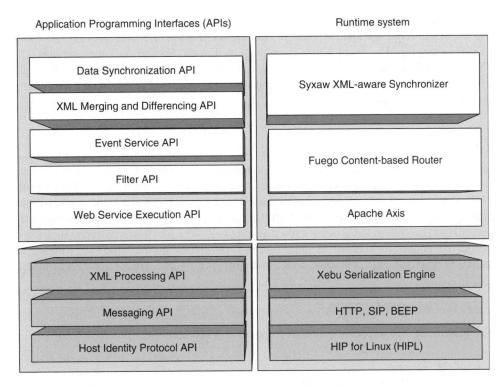

Figure 9.5 Fuego APIs and runtime system.

documents are not very space-efficient, a more efficient XML encoding is used for transmitting most XML content. Use of XML motivated the selection of SOAP [21] as the primary communication protocol in the architecture, transported using optimizations for wireless links [22]. SOAP is used for one-way and request-response messaging for relatively small XML documents and fragments, and HTTP is used for bulk transfers of data.

The Host Identity Protocol implementation for Linux is an optional component of the architecture, which is used for secure mobility and multihoming support. The HIP architecture is currently being standardized by IETF and it defines a new cryptographic namespace between the network and the transport layers.

9.11.2 Event Service

The proposed event service for mobile computing consists of two parts:

- The client-side API is similar in functionality to JMS [23] and offers the basic publish/subscribe functionality and session management.
- The server-side provides an extensible framework for content-based routing with optimizations and mobility support. The generic router implementation allows pluggable routing algorithms and routing table user interfaces.

Events are represented according to the XML Infoset specification. All remote API calls use the SOAP request-response protocol, and the notification of events uses asynchronous SOAP messaging. The event router is implemented as an Apache Axis web service. The client side API implementation uses wireless SOAP by default; however, a lightweight version of the API was created for J2ME end systems that uses HTTP 1.1 and a proprietary binary message format.

9.11.3   Filtering

Filtering is a central core functionality for realizing event-based systems and accurate content-delivery. Filtering may be optimized by using *covering relations* or *filter merging* [24]. The event service supports both through generic data structures for content-based routing. Any objects that implement methods for covering and merging are automatically optimized. The system has a default filtering language, which is based on typed tuples and disjunctive normal form based attribute-filters. Typically, events and filters are represented as lists of typed tuples. In this case, an attribute filter is a 3-tuple *<name, type, constraint>*.

9.11.4   Client-Side API

The client-side API supports expressive operation with three mechanisms:

- multiple sessions,
- expressive pull functionality, and
- fast subscriptions.

The first approach allows clients to create multiple sessions at different access servers for subscriptions with different maximum queue sizes and delivery options. The client may have several sessions at different access servers, for example to support different modes of operation. The second approach is realized by pulling the notifications that match subscription identifiers or arbitrary filters. Thus the client may subscribe different events, which are stored in the session—and when running on a small client only the essential events may be retrieved using the pull operation.

9.11.5   Event Router

The Fuego router is a high-level software toolkit for developing and deploying efficient application-level content-based networks. The distributed network of routers provides the event service.

The router toolkit does not specify any particular routing topology, but rather allows the developers to use the generic API methods available, leverage the efficient data structures for various configurations, and develop various mobility protocols. As an example, the Fuego router toolkit has been used to implement an event channel based configuration.

The server-side system consists of a set of event routers. Each router has two components: a local routing table and a remote routing table. The local routing table stores filters set by local clients and provides queue management for mobile and wireless clients. Disconnected clients may retrieve queued events upon reconnection using push or pull semantics. The remote routing table is responsible for communicating with other routers and forwarding events in the distributed system. In order to support extensibility, the local and remote routing tables and algorithms are separate objects, which may be changed if necessary.

This modularity allows the implementation of various distributed event routing semantics and router topologies. Subscription semantics and advertisement semantics are examples of two different interest propagation mechanisms. The former propagates subscription messages throughout the system and events are routed on the reverse-path of subscriptions. In the latter, advertisements are propagated throughout the system and subscriptions are routed on the reverse-path of advertisements. Supported routing topologies include hierarchical, event channel, and peer-to-peer topologies. The system also has separate user interface modules for the routing tables.

The following features are modular and extensible in the implementation:

- External routing algorithm. A data structure toolkit is provided for creating different routing algorithms, such as *hierarchical*, *peer-to-peer*, and *event-channel* configurations.
- External routing table management interface. An HTML base interface is provided that is customized for different configurations.
- Filtering language. The filtering language may be changed and an example is provided.
- Notification data types. The default data types are fixed, but they may be extended. Custom data types require additional XML serialization code. Custom data types that are not built into the filtering language are ignored in routing.
- Message transport protocols. New messaging protocols can be added in a similar fashion to the HTTP protocol (through a Jetty servlet).

9.11.6   Data Structures for Content-Based Routing

The poset structure is based on the *filters poset* (FP) used in the SIENA system. The FP is a directed acyclic graph structure that stores filter by their covering relations. It is used to optimize routing tables and compute destination interfaces, or the *forwards* set, for messages. New data structures for improving the FP performance, the *poset-derived forest* and variants, were proposed in Fuego that are simpler and more efficient than FP in certain cases, because they store only a subset of the relations.

A set of engineering guidelines for content-based routing systems were presented:

Hierarchical Routing. The forest data structures should be used for hierarchical routing and for finding the noncovered set.

Peer-to-peer Routing. For peer-to-peer routing, the forest structures should be used when there are many local clients. The forest requires more complicated *forwards* set management. Alternatively, the forests should be used to find the noncovered set of the clients and then the FP should be used to manage only external routing information.

Matching. The data structures have similar matching performance and it is not on the same level as more optimized matchers. Matching performance is proportional to the number of filters and also the number of interfaces. The forest should be used to find the non-covered set that is propagated and an additional matcher data structure should be used to quickly match notifications to the local clients. This is a two-phase process: first notifications are matched for the covering set by the poset or forest, and then they are matched by the matching data structure.

Filter Merging. Hierarchical routing systems are easy to extend with filter merging. The merging of local filters is easy to implement for both hierarchical and peer-to-peer routing. On the other hand, merging of external filters for peer-to-peer routing tables is more difficult. We observed significant performance benefits from the merging of interface specific filter sets.

9.11.6.1 Mobility Support

Mobility of event publishers and subscribers is a central topic in any event service and Fuego is no exception [25, 26]. When we consider what a support for true mobility requires, we rapidly notice that the main need is of synchronizing source and target servers and at the same time updating the event routing topology which tends to be arbitrary. The cost of these things in terms of router traffic volume and exchanged infrastructure messages is typically high when filter merging or covering relations are used. Furthermore the use and propagation of covering relations or merged filters will cause loss of subscription source server identities.

Fuego supports mobility by setting up buffering for messages at the access servers. Also custom handover protocols are enabled in the router implementation that satisfy mobility-safety by updating the routing topology. For large networks with typical cost problems Fuego includes a separate simulator to evaluate the handover cost.

One solution which has been successfully utilized for mobility support is setting up rendezvous points to coordinate subscription propagation and mobility management. Rendezvous points at their simplest may be event channels each responsible for a single event type and which are updated using a single message or RPC-operation. Such an event channel is a rendezvous point over a subspace of the content space. A directory service may be used for lookup of channels. This kind of mobility support is easy to manage, scalable and also fault-tolerant because replication is used.

If mobility is to be maintained in an environment where router topology has either acyclic or cyclic graph the solution is more complicated. The rendezvous point will operate in such a way that subscriptions and advertisements always propagate towards it. Reverse paths are used so that subscriptions are sent on the reverse path of advertisements, and notifications use the reverse path of subscriptions. Now the cost of mobility is, in the worst case, limited by the path length to the rendezvous point. Finding a rendezvous-point in order to route messages towards it is not a straightforward task. DHT based data structures, where the event type is hashed to an overlay node identifier, may be used in routing

There are three useful techniques that may be utilized if a mobility-aware pub/sub system is to be created: overlay-based routing, rendezvous points and completeness checking. In overlay-based routing the pub/sub routing is separated from the underlying network routing. The system is therefore enabled to cope with possible errors in network-level, node failures and the content-based flooding problem. The rendezvous points and their inherently better coordination of topology updates make life easier for mobile nodes. Updates will propagate unidirectionally for a single rendezvous point. The completeness checking will accept only fully established subscriptions and advertisements, complete in the topology which enables performing the optimization of covering filters.

9.12   STEAM

The *Scalable Timed Events and Mobility (STEAM)* event system is one of the systems meant to help building and maintaining ad-hoc wireless networks, in this case WLAN type networks that find increasing use for instance in traffic management applications [27].

The STEAM protocol is a wireless application-level broadcast protocol. It uses subject-based filtering at the event producer and content-based filtering at the subscribing client. STEAM has the following main design features:

- The event model used is implicit allowing the entities to subscribe locally to interesting event types without a centralized broker.
- STEAM addresses the problems caused by mobility (the dynamic reconfiguration of the network topology) by using three different filters, based on subject, proximity, and content. A subject filter matches with the event and maps onto a proximity group. A proximity filter works on the geographical attribute of the proximity group.
- The system notifies interested parties by using a group communication service. The interest groups are bound geographically. The identification of nodes utilizes beacons.

In STEAM events are defined to have a name (determining the structure of the event) and a set of typed parameters. When the event type is deployed a proximity filter is established for the type, specifying the space where events occur. When an event is published the publisher will match the subject and proximity, the subscribers then matching the content. We will readily see that the publisher's proximity filter must receive information about the subscriber's location. The methods of getting, maintaining and securing this information are not detailed in the architecture.

In STEAM architecture event publishers will advertise the event types they intend to produce. The publishing activity for a type is bound with a geographical area (the proximity), and the events of this type are published within this area. Note that the proximities do no necessarily correspond the physical area of the communication; they are independent of it and support to multihops is enabled in the routing layer.

STEAM utilizes PGCS (Proximity-based Group Communication Service) where groups are assigned specific geographical areas. The rules prevent nodes not located in the group area from joining the group and active detection of proximities using Proximity Discovery Service (PDS) is provided. If a prospective client has matching subscription profile and if it is located appropriately the associated events will be delivered. Either a subscription or an announcement must match the group before joining is allowed.

STEAM allows for movements of the the clients although the proximities are static. The results from experimental testing (e.g. case of traffic lights) suggest that in ad-hoc environments distributed filtering is advantageous and leads to a significant reduction in the volume of transmitted traffic.

9.13   ECho and JECho

ECho is a high-performance data transport mechanism that is based on event channels [28]. ECho uses channel-based subscriptions, similarly to the CORBA Event Service. ECho's derived event channel mechanism implements filtering by adding an application-supplied derivation function F to all listeners of a particular event channel, and by transferring all events that are generated by the sources and passed through the filters to a derived event channel. This scheme resolves issues in the delivery of unwanted events. ECho is especially optimized for streaming data and data transmission. ECho has been shown to perform better than Jini (distributed Java events), CORBA Event Channels, and XML-based messaging. ECho was developed at Georgia Tech and the source is available for academic research purposes.

JECho is a distributed event system that has been recently extended to support mobility using opportunistic event channels [29]. The central problem is to support a dynamic event delivery topology, which adapts to mobile clients and different mobility patterns. The requirements are addressed primarily using two mechanisms: proactively locating more suitable brokers and using a mobility protocol between brokers, and using a load-balancing system based on a central load-balancing component that monitors brokers in a domain. The mobility protocol is, in principle, similar to most mobility protocols (SIENA, Rebeca, . . .). The filtering model is based on stateful user-defined objects, called modulators, which may transform the event stream. This allows more fine-grained filtering than nonstate-based predicate matching. However, possible security problems are not addressed, and it may be difficult to do optimizations between similar modulators. In addition, client-based filtering is not addressed and it may also be difficult to implement efficiently. For example, a mobile producer should download all relevant modulators from the broker. Furthermore, no session management is provided so all user-specific modulators are relocated.

The system supports load balancing and resource monitoring, which are novel features for mobility-aware event systems. The paper presents simulation results for different scenarios, for example, relocation overhead and mobility patterns. Mobility patterns are examined in a 100-node network using BRITE and the evaluation includes scenarios such as random walk, salesman, pop-up, and fixed. Moreover, end-to-end delay and mobility/communication ratio are measured using a real system with two subnets.

9.14 DHT-Based Systems

Wide-area DHT structures are frequently used as the basic communication substrates for pub/sub systems. We examined well-known DHT solutions in Chapter 3. In this section, we consider pub/sub systems built on top of the DHT structures. The pub/sub systems rely on the DHT for efficient and scalable message routing as well as resiliency and fault-tolerance. The overlay abstracts concerns pertaining to network dynamics and churn from the pub/sub system.

9.14.1 Scribe

Scribe is a topic-based publish-subscribe system that explores the scalability of the notification service in peer-to-peer environments. Scribe is built on top of Pastry, which is a scalable, self-organizing peer-to-peer location and routing system. Scribe provides an application-level multicast. Pastry is based on uniform ID keys that are used as host addresses. The system routes a message to the closest possible key [30]. In an N node network, Pastry can route to any node in less than $log_{2b}N$ hops on average, where b is configuration parameter. Delivery is guaranteed unless $\frac{l}{2}$ nodes with adjacent node IDs fail simultaneously, l is a configuration parameter with a typical value of 16. Node IDs are sequences of digits with base $2b$. A node's routing table contains $log_{2b}N$ rows with $2b - 1$ entries each. The entries in row n refer to the node whose node ID has the same n digits with the current node, but the $n + 1$:th digit is different.

The uniform distribution ensures that the node ID space is even. The algorithm selects one node ID from several possibilities based on a metric, say, cost or latency. Each node also maintains the IP addresses of the nodes that have numerically closest larger and smaller node IDs. A node forwards the message to a node whose ID matches with the key with a prefix that is at least one bit longer than the prefix with the current node. If such a node cannot be found, the message is forwarded to a node whose prefix is numerically closer than the current node's ID. This node is found from the set of larger/smaller IDs.

Scribe provides a best-effort notification delivery on top of Pastry and specifies no particular event delivery order. Moreover, Scribe does not support filtering, buffering or mobility. A publisher of events is a root of a multicast tree. Each topic has a unique topicID, and the Scribe node with a node ID numerically closest to this topicID acts as the rendezvous point for that particular topic. The rendezvous point forms the root of a multicast tree. In other words, the responsibility for a given topic (group of subscribers) is hashed over the set of the servers. When a subscribe message is routed towards the rendezvous point, each intermediate node adds the previous node as to its children table. This information is used in the multicast protocol, which is similar to reverse path forwarding.

Events may be published directly, if the IP address of the rendezvous point is known. However, subscriptions need to be routed within the peer-to-peer topology. Access control can be enforced at the rendezvous point. Pastry can route around faulty nodes by resending the subscription and thus repairing the multicast tree.

Each Scribe multicast group is represented by a Pastry key, called the *groupId*. A multicast tree for a given groupId is created by taking the union of the Pastry routes from each group member to the groupId's root. Content can then be sent using this multicast tree by using reverse path routing from the root towards the leaves of the tree. Figure 9.6 illustrates the subscription and publication operations in Scribe.

The group membership management is decentralized and efficient, because it builds on the existing, proximity-aware Pastry overlay. The introduction of new members to the multicast tree is easy. A new member simply sends a message to the groupId. Thus Scribe can support large numbers of members per group. The groups can also be dynamic.

Pastry's proximity aware routing and Scribe's multicast group management can be combined to support anycast communications. Anycast is useful when performing resource discovery. With Anycast, any node in the overlay can send an anycast message to a Scribe group. The anycast message is routed towards the groupId and forwarded to the nearest member by relying on the local route convergence property.

The Scribe multicast routing state is distributed and maintained in a decentralized fashion. Each node in a tree only maintains its immediate predecessors and successors in the tree. This can be seen as a significant scalability advantage over other overlay multicast schemes, such as Bayeux [31] discussed next. As a result, Scribe does not require excessive signalling traffic in order to gather global state information.

9.14.2   Bayeux and Tapestry

Bayeux is a decentralized source-specific, explicit-join application level multicast system based on the Tapestry unicast routing infrastructure. Bayeux supports load balancing and takes into account locality. The system is based on the prefix routing scheme employed by Tapestry, which is similar to the longest prefix routing in the CIDR IP address allocation architecture. Addresses (Ids) are assumed to be evenly distributed and they are generated using the SHA-1 hashing algorithm. During routing the nth hop has a suffix of at least length n with the destination ID. Receivers are organized into a multicast tree rooted at the source [32, 33].

The senders advertise multicast sessions as Tapestry files on the sessions root node, which is identified by the corresponding hashed session ID. Clients join the multicast group by sending messages to the address identified by the session tuple. Bayeux constructs a distribution tree to deliver data to session members. The distributed multicast tree is set by the response message (TREE message) from the root, which activates the forwarding path to the recipient. For improved fault tolerance the system supports the replication of root nodes in a process called tree partitioning.

The main difference to Scribe is in the way that the multicast tree is constructed. Bayeux sends a subscription message always to the root of the tree. The root node maintains the membership list for the topic-based multicast group. A response message from the root installs state in the intermediate overlay nodes that forward the message. Thus the new node becomes part of the forwarding tree.

Scribe can be seen to have a more scalable approach to the construction of the multicast tree. In Bayeux, the root node has to keep membership information pertaining to all members of the multicast group. Moreover, group membership management introduces overhead since each control message must be sent to the root, and the root then sends the reply back. To prevent the root node from becoming a bottleneck for performance, a

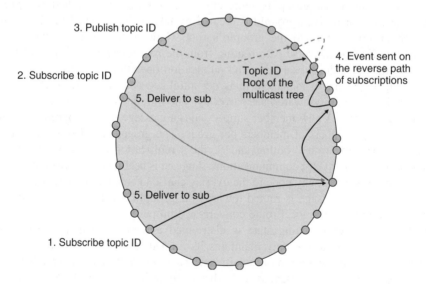

Figure 9.6   Overview of Scribe.

partitioning scheme for the multicast tree has been proposed that shares the load among several root nodes.

9.14.3 Hermes

Hermes [34, 35] is a DHT-based overlay event delivery system. Hermes leverages the features of the underlying overlay system for message routing, scalability, and improved fault-tolerance. The system utilizes special rendezvous point nodes for managing the creation of multicast trees over the network. These nodes coordinate the propagation of subscription and advertisement messages. Each rendezvous point is responsible for at least one event type. The rendezvous points and the types can be chained into distributed type hierarchies. The responsible rendezvous point is located by hashing the event type to the flat namespace of the overlay [34].

In Hermes rendezvous points are established using a special message sent to the rendezvous point. The address of the message is the hash of the type identifier. The message establishes the event type at the rendezvous point. Event type conforms to a schema and the schemas can be enforced. Thus the system supports type-safe subscriptions.

Hermes supports two routing algorithms:

- In type-based routing all messages are sent towards the rendezvous point: subscriptions, advertisements, and notifications. This form of routing does not support filtering, but compares event based on their type. Subscriptions and advertisements are local to a branch of the multicast tree rooted at the RP and they are not forwarded by the RP. This means that notifications have to be always sent to the RP.
- Type/attribute-based routing is similar to type-based routing, but supports filtering with covering relations and, instead of sending all notifications to the RP, notifications are sent on the reverse path of subscriptions. In this case, advertisements are only sent to the RP. Subscriptions are always sent on the reverse path of advertisements. The RP forwards subscription messages to overlapping advertisements. The type/attribute-based routing is more suitable to scenarios, where event traffic is not uniformly distributed, because notifications are not always sent to the RP.

A rendezvous-point-based system functions using the following scenario (Figure 9.7):

- The Publisher
  - The publisher sends an advertisement of an event type (adding a filter if the routing is type/attribute-based) (1).
  - The advertisement is forwarded to the rendezvous point (2).
- The Subscriber
  - The client subscribes to an event of a certain type (using a filter in type/attribute-based routing) (3).
  - The subscription message is forwarded to the rendezvous point, uncovered (type or filter) by intermediate brokers (4–6).
  The Event Publication
  - The publisher publishes an event (7).

- The event message is sent via the multicast tree with its root at the rendezvous point (8–11).
- If the routing is type-based, events conforming to the advertisement from the publisher will be sent along the forward path of the advertisement to the RP. It is the responsibility of the rendezvous point to forward the event message on the reverse path of the subscriptions.
- If the routing is type/attribute-based, the rendezvous point will send the subscriptions on the reverse path of advertisements. The advertisement-conforming events from the publisher are sent on the reverse path of subscriptions.

One notable feature of Hermes is support for connecting RPs into type hierarchies. In *subscription inheritance routing* advertisements are only sent to the RP that maintains the event type. Subscriptions are forwarded by the RP to all RPs with descendant types. In *advertisement inheritance routing* the RP forwards the advertisement recursively to all RPs of all ancestor event types. Also notifications are forwarded to all ancestor event types, because they are sent on the same forward path as advertisements.

As noted above a main feature of Hermes is the connection between rendezvous points and type hierarchies. Only the RP which maintains an event type will get advertisements and will then forward the subscriptions to all rendezvous points with descendant types. This is the subscription inheritance routing. In another model, the advertisement inheritance routing the RP will recursively forward the advertisement to all RPs which are bound to an ancestor event type. Similarly the notifications are forwarded to all ancestor event types. This results from the fact that they use the same forward path as advertisements.

Hermes detects server and RP failures by using heartbeat messages. In case of failures the underlying overlay enables locating a replacement server to handle the responsibilities of a failed node – the routing tables are sent towards the rendezvous point and the overlay provides a new server with associated route. Furthermore, the RP replication is supported by Hermes using synchronization of advertisement and subscription status between replicas which are positioned in the same multicast tree. This allows reduction of message

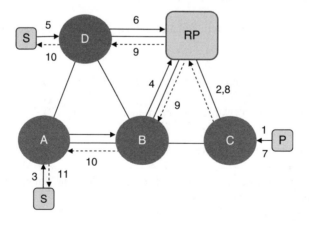

Figure 9.7 Routing in Hermes.

propagation overhead but does not touch directly the issue of load balancing between rendezvous points.

The Hermes advertising model is basically a typical advertisement semantics model. There are differences, though. For example, the messages in type-based routing or advertisements and subscriptions in type/attribute based routing will be sent towards the rendezvous point, making the routing topology essentially constrained by the RP. Furthermore, the path from the subscriber to the RP is used to introduce subscriptions and (in type/attribute-based routing) subscriptions will be propagated using the reverse path of overlapping advertisements. Note also that advertisements may be introduced only on the path between the advertiser and the rendezvous point.

The differences between general advertising model and the Hermes implementation are important because Hermes makes the advertisement a local property belonging to a branch of the multicast tree which has its root in the RP and enabling the use of virtual advertisements modelling. A Hermes rendezvous point will have virtual advertisements for all events of the type managed by itself. Therefore subscriptions are propagated towards it.

9.14.4 Other Systems

9.14.4.1 DADI, Meghdoot, and MEDYM

DADI (Discovery, Analysis and Dissemination of Information) is a research project at Princeton that focuses on discovery, analysis and information dissemination on the Internet. The project investigates the event-based model with emphasis on wide-area pub/sub. The DADI effort includes a number of subprojects that cover the different layers of operation, namely the system, algorithm, and application layers. The system layer encompasses architecture design for pub/sub and subscription-based content delivery. The algorithm layer covers routing and matching algorithms for content-based pub/sub. The application layer pertains to Internet-scale persistent search.

Meghdoot [36] is one of the early examples of a content-based pub/sub system entirely based on a structured overlay infrastructure, namely CAN [37], with rendezvous-based event routing. Meghdoot uses structured subscriptions with either numerical or string attributes. A subscription is mapped to a CAN point, whose coordinates are the bounds of each range constraints. A published event is mapped to a CAN region spanning all the possible subscriptions that can map to the event. A generic architecture for content-based pub/sub independent of the specific infrastructure has also been proposed [38].

Match Early with DYnamic Multicast (MEDYM) [39] partitions the event space into non-overlapping partitions with balanced load. Each server acts as a matcher for one or more partitions. A channelization technique is presented that partitions the event space into a number of multicast groups. A multicast tree is built for each group that spans servers with subscriptions for any event in that group. Multicast can be either performed through IP Multicast if available or with application-level multicast.

9.14.4.2 SUB-2-SUB

The SUB-2-SUB system is a peer-to-peer-based event notification mechanism that clusters subscribers according to their interest similarity [40]. SUB-2-SUB is a content-based system, in which similar filters are clustered into topics. Participants are not organized into a structured overlay. Every topic forms a separate ring structure, which results in the degree of the overlay growing linearly with the number of topics created from filters. An event matching a given topic is first routed towards the correct ring and subsequently efficiently disseminated within the ring itself. Information about subscriptions is periodically exchanged between the nodes using the CYCLONE gossip protocol.

9.14.4.3 TERA

TERA is a topic-based pub/sub system based on a peer-to-peer architecture [41]. Nodes in TERA are organized into two layers, the global overlay network that connects all nodes and topic layer overlays that connect all nodes interested in the same topic. TERA utilizes random walks and access point lookup tables to deliver events to topic layers containing all nodes with matching filters. An event is first delivered to the access point node for the given topic that then sends the event within the topic specific overlay. The peer-to-peer is managed based on gossiping and the view exchange technique discussed in Chapter 3.

9.15 Summary

The pub/sub systems examined this chapter build on top of the network layer basic packet routing capability and extend it with overlay pub/sub routing features. These systems can be classified as basic overlays or DHT-based overlays depending on the layering. The former builds the pub/sub directly on top of the network layer whereas the latter builds the pub/sub system on top of an overlay routing substrate, typically a DHT algorithm. The latter category relies on the scalable lookup operation, resilience, and fault-tolerance features of the underlying overlay substrate. Figure 9.8 summarizes the examined pub/sub systems.

SIENA, JEDI, Rebeca are examples of pub/sub systems that build directly on top of the network layer. They are overlay networks that consist of brokers that maintain content-based routing tables. They also support various optimizations for reducing the routing table maintenance cost, such as covering relations. To our knowledge, covering relations were first introduced in the SIENA project and they support the optimization of event-based communication. Optimizations that aggregate content-based routing state, such as covering relations and merging, are problematic for mobility support because they lose information pertaining to the original subscribers and advertisers. This can be solved by using techniques that partially flood control messages on the reverse path of messages that were used to build the routing table. We have discussed the various issues pertaining to content-based routing in Chapters 7 and 11.

The overlay routing topology should follow the network level placement of routers in order to be efficient. Many DHTs work by hashing data to routers/brokers and using a variant of prefix-routing to find the proper data broker for a given data item. This basic primitive lends itself well for pub/sub, for example by hashing an event type into a rendezvous point. Hermes and Scribe are examples of pub/sub systems implemented on top of an overlay network and are based on the rendezvous point routing model. Hermes is a content-based pub/sub system and Scribe is a topic-based system.

The basic routing algorithms of SIENA and similar systems do not cope well with topology changes. The routing algorithms should be able to adapt to the network environment

System	Subscription model	Infrastructure	Routing
Gryphon, Rebeca, SIENA, JEDI, Elvin, XSIENA	Content	Brokers, overlay	Filtering, filter aggregation, various extensions
Padres	Content	Brokers, overlay	Filtering, dynamic configuration, load balancing
REDS	Subscription, request-reply, pluggable	Pluggable modules: structured or unstructured	Pluggable modules: wide-area, mobile ad hoc network, reconfiguration of topology
GREEN	Various (pluggable components)	Pluggable modules: structured or unstructured	Pluggable modules: wide-area, mobile ad hoc network
Fuego	Content	Various, federated clusters	Rendezvous
STEAM	Subject, proximity	Proximity-based Group Communication Service	Proximity based
ECho and JECho	Channel (ECho) and stateful modulators (JECho)	Brokers, overlay	Filtering, load balancing and mobility (JECho)
Scribe	Topic	DHT (Pastry)	Rendezvous
Hermes	Topic and content	DHT	Rendezvous with filter aggregation
Meghdoot	Content	DHT	Rendezvous
Sub-2-Sub, TERA	Topic	Combination of gossip and infrastructure supported overlay	Filtering, interest-based clustering

Figure 9.8   Summary of the research systems.

and at the same time take the supply and demand of information into account. Several solutions have been presented for the runtime configuration of the pub/sub network, for example the REDS and GREEN systems. In addition to reconfiguration, also probabilistic algorithms have been proposed for improving the robustness of pub/sub systems. In Chapter 3 we examined gossip algorithms for pub/sub that are the basis for a family of broadcast and multicast algorithms with good scalability properties.

References

1. IBM (2002) *Gryphon: Publish/subscribe over Public Networks*. (White paper) http://researchweb.watson. ibm.com/distributedmessaging/gryphon.html.
2. Strom RE, Banavar G, Chandra TD, *et al.* (1998) Gryphon: An information flow based approach to message brokering. *Computing Research Repository (CoRR)*. Available at: http://arxiv.org/corr/home.
3. Bacon J *et al.* (2000) Generic support for distributed applications. *IEEE Computer* **33**(3), 68–76.
4. Carzaniga A, Rosenblum DS and Wolf AL (2001) Design and evaluation of a wide-area event notification service. *ACM Transactions on Computer Systems* **19**(3): 332–83.
5. Carzaniga A and Wolf AL (2003) Forwarding in a content-based network *Proceedings of ACM SIGCOMM 2003*, pp. 163–74, Karlsruhe, Germany.

6. Caporuscio M, Carzaniga A and Wolf AL (2003) Design and evaluation of a support service for mobile, wireless publish/subscribe applications. *IEEE Transactions on Software Engineering* **29**(12): 1059–71.

7. Carzaniga A, Rutherford MJ and Wolf AL (2004) A routing scheme for content-based networking. *Proceedings of IEEE INFOCOM 2004*. IEEE, Hong Kong, China, vol. 2, pp. 918–28.

8. Sutton P, Arkins R and Segall B (2001) Supporting disconnectedness-transparent information delivery for mobile and invisible computing. *CCGRID '01: Proceedings of the 1st International Symposium on Cluster Computing and the Grid*, p. 277. IEEE Computer Society, Washington, DC.

9. Cugola G, Di Nitto E and Fuggetta A (2001) The JEDI event-based infrastructure and its application to the development of the OPSS WFMS. *IEEE Transactions on Software Engineering* **27**(9): 827–50.

10. Cugola G, Di Nitto E and Picco GP (2000) Content-based dispatching in a mobile environment. *Workshop su Sistemi Distribuiti: Algorithmi, Architectture e Linguaggi*.

11. Jacobsen HA, Cheung AKY, Li G, Maniymaran B, Muthusamy V and Kazemzadeh RS (2010) The PADRES Publish/Subscribe System. *Principles and Applications of Distributed Event-Based* Systems, pp. 164–205.

12. Cugola G and Picco GP (2006) REDS: a reconfigurable dispatching system. *Proceedings of the 6th International Workshop on Software Engineering and Middleware*, pp. 9–16. SEM '06. ACM, New York, NY.

13. Sivaharan T, Blair G and Coulson G (2005) GREEN: A configurable and re-configurable publish-subscribe middleware for pervasive computing. *Proceedings of DOA 2005*, pp. 732–49.

14. Fiege L, Gärtner FC, Kasten O and Zeidler A (2003) Supporting mobility in content-based publish/subscribe middleware *Middleware*, pp. 103–22.

15. Zeidler A and Fiege L (2003) Mobility support with rebeca. *Proceedings of the 23rd International Conference on Distributed Computing Systems*, pp. 354, ICDCSW '03. IEEE Computer Society, Washington, DC, pp. 354–61.

16. Parzyjegla H, Graff D, Schrter A, Richling J and Mühl G (2010) Design and implementation of the rebeca publish/subscribe middleware. In *From Active Data Management to Event-Based Systems and More* (ed. Sachs K, Petrov I and Guerrero P) vol. 6462 of *Lecture Notes in Computer Science*. Springer Berlin/Heidelberg. pp. 124–40.

17. Jerzak Z (2009) *XSiena: The Content-Based Publish/Subscribe System*. PhD thesis TU Dresden. Dresden, Germany.

18. Jerzak Z and Fetzer C (2008) Bloom filter based routing for content-based publish/subscribe. *Proceedings of the Second International Conference on Distributed Event-Based Systems*, pp. 71–81, DEBS '08. ACM, New York, NY.

19. Fetzer C, Brito A, Fach R and Jerzak Z (2010) StreamMine. In *Principles and Applications of Distributed Event-Based Systems* (ed. Hinze A and Buchmann AP), IGI Global, pp. 394–410.

20. Tarkoma S, Kangasharju J, Lindholm T and Raatikainen K (2006) Fuego: Experiences with mobile data communication and synchronization. *17th Annual IEEE International Symposium on Personal, Indoor and Mobile Radio Communications (PIMRC)*, pp. 1–5.

21. W3C (2003) *SOAP Version 1.2*. W3C Recommendation.

22. Kangasharju J, Lindholm T and Tarkoma S (n.d.) Requirements and design for XML messaging in the mobile environment *Second International Workshop on Next Generation Networking Middleware*, pp. 29–36.

23. Sun (2002) *Java Message Service Specification 1.1*.

24. Tarkoma S (2008) Dynamic filter merging and mergeability detection for publish/subscribe. *Pervasive and Mobile Computing* **4**(5): 681–96.

25. Tarkoma S and Kangasharju J (2007) On the cost and safety of handoffs in content-based routing systems. *Comput. Netw.* **51**: 1459–82.

26. Tarkoma S, Kangasharju J and Raatikainen K (2003) Client mobility in rendezvous-notify. *Intl. Workshop on Distributed Event-Based Systems (DEBS'03)*, pp. 1–8.

27. Meier R and Cahill V (2003) Exploiting proximity in event-based middleware for collaborative mobile applications. *DAIS*, pp. 285–96.

28. Zhou D, Chen Y, Eisenhauer G and Schwan K (2001) Active brokers and their runtime deployment in the ECho/JECho distributed event systems. *Active Middleware Services*, pp. 67–72.

29. Chen Y, Schwan K and Zhou D (2003) Opportunistic channels: Mobility-aware event delivery. *Middleware*, pp. 182–201.

30. Castro M, Druschel P, Kermarrec AM and Rowstron A (2002) Scribe: A large-scale and decentralized application-level multicast infrastructure. *IEEE Journal on Selected Areas in Communication (JSAC)*, **20**(8): 1489–9.

31. Zhuang S, Zhao B, Joseph A, Katz R and Kubiatowicz J (2001) Bayeux: An architecture for scalable and fault-tolerant wide area data dissemination *The 11th International Workshop on Network and Operating Systems Support for Digital Audio and Video (NOSSDAV'01)*, pp. 11–20.

32. Zhao BY, Kubiatowicz JD and Joseph AD (2002) Tapestry: a fault-tolerant wide-area application infrastructure. *SIGCOMM Comput. Commun. Rev*. **32**(1): 81.

33. Zhuang SQ, Zhao BY, Joseph AD, Katz RH and Kubiatowicz J (2001) Bayeux: An architecture for scalable and fault-tolerant wide-area data dissemination. *Proceedings of the Eleventh International Workshop on Network and Operating System Support for Digital Audio and Video (NOSSDAV 2001)*.

34. Pietzuch PR (2004) *Hermes: A Scalable Event-Based Middleware*. PhD thesis. Computer Laboratory, Queens' College, University of Cambridge.

35. Pietzuch PR and Bacon J (2002) Hermes: A distributed event-based middleware architecture. *ICDCS Workshops*, pp. 611–18.

36. Gupta A, Sahin O, Agrawal D and Abbadi AE (2004) Meghdoot: Content-based publish:subscribe over P2P networks. *Proceedings of the ACM/IFIP/USENIX 5th International Middleware Conference (Middleware'04)*, pp. 254–73.

37. Ratnasamy S, Francis P, Handley M, Karp R and Schenker S (2001) A scalable content-addressable network. *SIGCOMM '01: Proceedings of the 2001 Conference on Applications, Technologies, Architectures and Protocols for Computer Communications*, pp. 161–72. ACM, New York, NY.

38. Baldoni R, Marchetti C, Virgillito A and Vitenberg R (2005) Content-based publish-subscribe over structured overlay networks. *International Conference on Distributed Computing Systems (ICDCS 2005)*, pp. 437–46.

39. Cao F and Singh JP (2005) MEDYM: Match-early with dynamic multicast for content-based publish-subscribe networks. *Proceedings of the ACM/IFIP/USENIX 6th International Middleware Conference (Middleware 2005)*, pp. 292–313.

40. Voulgaris S, Rivire E, Kermarrec AM and Steen MV (006)Sub-2-Sub: Self-organizing content-based publish subscribe for dynamic large scale collaborative networks. In *IPTPS06: the Fifth International Workshop on Peer-to-Peer Systems*.

41. Baldoni R, Beraldi R, Quema V, Querzoni L and Tucci-Piergiovanni S (2007) TERA: topic-based event routing for peer-to-peer architectures. *DEBS '07: Proceedings of the 2007 Inaugural International Conference on Distributed Event-Based Systems*, pp. 2–13. ACM, New York, NY.

10

IR-Style Document Dissemination in DHTs

In this chapter, we present a concrete example of a keyword-based pub/sub system implemented on top of a DHT overlay network. The chapter considers the complexity of the problem and presents an efficient solutions based on rendezvous points on the DHT based network. The chapter illustrates how the already discussed DHT-based solutions can be utilized for keyword based content dissemination.

10.1 Introduction

Differing from content-based pub/sub systems, there exist another kind of content filtering systems, which leverage information retrieval (IR)-based approach to filter publications. In such systems, both publications (equally content or documents) and subscription filters are modelled by a set of keywords. Based on the keywords in publications and filters, the IR-based approaches, such as boolean-based or Vector Space Model (VSM)-based filtering [1], are used to decide whether or not publications successfully match filters.

The keywords-based IR filtering approach is very promising, due to the following reasons. First, keywords have become a de-facto standard for users to find content of interest. Second, using keywords offers users the fine-grained content filtering mechanism, and avoids the coarse-grained content filtering (for example, the topic-based filtering used by RSS feeds applications cannot offer fine-grained filtering and lead to the false positive issue [2]). Finally many real applications have adopted the keyword-based approach, such as Google (and Microsoft Live) Alerts, RSS aggregator [3], Web search-based advertisement display, etc.

Meanwhile, the potential of Peer-to-Peer (P2P) technologies for building distributed applications at a very large scale has been commonly recognized. Existing P2P systems such as Bittorrent and eMule connect millions of machines towards providing Internet-scale content sharing services. Due to the desirable properties of scalability, fault tolerance, and short routing paths offered by DHTs, a significant body of research work has been dedicated to studying IR-style search, filtering and dissemination in DHTs [2, 4, 5].

Publish/Subscribe Systems: Design and Principles, First Edition. Sasu Tarkoma.
© 2012 John Wiley & Sons, Ltd. Published 2012 by John Wiley & Sons, Ltd.

In this chapter, we introduce the data model of the IR-style content dissemination, review an example system in DHTs, namely STAIRS [2, 6], report the recent process and finally discuss the IR-style content dissemination with the traditional content-based pub/sub.

10.2 Data Model and Problem Statement

This section gives the data model that is used by the IR-style content filter and dissemination system, and state the problem definition of IR filtering on DHTs.

10.2.1 Data Model

First in term of publication, there exist various types of content: textual documents, annotated binary content, media, etc. Each of such content could be explicitly or implicitly modelled as pairs of terms and weights. For example, for popular Web documents and weblogs, we can easily represent each of them by a set of terms. In many cases, such terms are associated with weight scores to indicate the importance of such terms. For example, in IR community, the score function such as *term frequency * inverse document frequency* (*tf*idf*) scheme can be used to compute the weight score of the terms in documents. A larger weight score indicates more importance of the associated term.

Other content types (such as binary or media) typically include associated tag keywords and descriptions. Such tags can be viewed as the equivalent of terms for the purpose of content search. Even if the binary content without tags (such as images or videos), there exist algorithm to represent the content by pairs of dimensions and associated weights [7].

Next, in terms of subscription filters, they are similarly represented by a set of terms. Also, each term could be associated with a preference weight, such that users can personally tune the weight of the term.

Given the above publications and subscription filters, there are various ways of the filtering criteria to determine whether or not publication successfully match filters. In the following sections, we respectively present a threshold-based filtering criteria.

Threshold-based Filtering: The boolean-based filtering does help filter out useless documents. However it is too coarse, in particular given a lot of documents containing input keywords appearing subscription filters. For example, when the input keywords appear in a very large number of documents, users could too many documents, which could overwhelm users. Thus, an advanced approach is to measure the relevance score between a document and a filter.

We now assume that each document is represented by a set d consisting of $|d|$ pairs of $\langle t_i, s(t_i, d) \rangle$, where t_i is a term, and $s(t_i, d)$ represents the score or importance of t_i in the document. $s(t_i, d)$ can be pre-computed, for example, using the *tf*idf* scheme.

Each subscription filter is represented by a predefined threshold and a set of $|f|$ terms $\{t_1, .., t_{|f|}\}$. Similar to d, the notation f is used to indicate both a filter and a set of query terms associated with this filter. The predefined threshold, denoted by $T(f)$, can be specified with either a default value T (set by the system) or a personalized value (set by each user).

Then, a document d successfully matches a filter f, provided that

$$S(f, d) = \sum_{i=1}^{|f|} s(t_i, d) \geq T(f).$$
(10.1)

In the above equation, $S(f, d)$ follows the VSM model to compute the relevance score between f and d. If a subscriber does not specify a threshold, the system assumes that the filter uses a *default* threshold T (that works well for general users). The personalized threshold indicates the strength of the desired relevance between input keywords and a document to be received. A higher value of $T(f)$ indicates the expectation of more relevant documents, and vice versa. $S(f, d) = 0.0$ means that there is no term appearing in both f and d.

10.2.2 Problem Statement and Challenges

Given the data models above, we want to design a document filtering and dissemination scheme in a distributed hash table (DHT) based P2P network, satisfying the following requirements:

User Requirements. include (i) no or low false dismissals (i.e. subscribers should correctly receive expected documents satisfying either of two data models); and (ii) a timely manner (i.e. subscribers expect to timely receive fresh documents).

Efficiency Requirement. indicates a low document publishing cost. The publishing cost is measured by the overall number of hops to forward every document to all satisfied subscribers.

With the defined problem, designing a desirable scheme that satisfies both above requirements is challenging. If we relax either of the user requirements, we achieve a useless, though efficient, scheme. For example, if a *high false dismissal ratio* is allowed, we can heuristically forward document d to some randomly chosen nodes. This heuristic scheme consumes a low forwarding cost, but subscribers may not receive useful documents. Or if the *timely requirement* is not emphasized, we can adopt the full-text search scheme [5], where documents are pre-stored in a DHT. After that, searching documents only involves the communication between the query initiator and home nodes of the query terms appearing in filter f. However, this scheme only finds outdated documents, instead of those documents as fresh as possible.

Now, the challenges of designing an *efficient* scheme meanwhile satisfying user requirements are illustrated as follows. First, naively broadcasting document d to all nodes in a DHT incurs the lowest efficiency. Next, an improvement over such a naive solution is to forward document d only to the home node of every distinct term inside d [4, 5, 8]. However, this approach still suffers from a high publishing cost. For example, it has been shown that the bandwidth consumption of full-text document publishing in the DHT is six times of the super-peer system [5]. The high publishing cost depends on two factors: (i) the cost of forwarding document d to one destination home node (called the *unit-publishing cost*), typically equal to $O(\log N)$ hops, where N is the total number of nodes in a DHT; (ii) a large number of distinct terms per document (called the *publishing amount*). For example, in a data set of our experiments, there are thousands of terms per document. In particular, [4, 5, 8] store document d (more precisely, the pointer of a raw document d) to home nodes of all distinct terms inside this document. As a result, the total publishing cost, related to the multiplication of two aforementioned factors, is very high.

10.3 STAIRS: Threshold-Based Document Filtering in DHTs

10.3.1 Overview of DHT-Based P2P Networks

A number of structured P2P routing protocols have been recently proposed, including Chord [9], Pastry [10], Tapestry [11], etc. These P2P systems provide the functionality of a scalable distributed hash table (DHT), by reliably mapping a given object key (e.g. a term with the string type) to a unique live node in the network. For simplicity, we call the node, which is responsible for the object key, the *home node* of this key. These DHT systems have the desirable properties of high scalability, fault tolerance, and efficient routing of queries. Typically, each node is assigned with $O(\log N)$ links, and the average number of routing hops (in short *hop count*) is guaranteed with $O(\log N)$.

10.3.2 Solution Framework

There are three basic operation in the document dissemination system: *filter registration*, *document forwarding*, and *document notification*.

Filter Registration: First, to register a filter f, we propose to register f on the home node of *every* term $t_i \in f$. Thus, f is registered on $|f|$ home nodes. We call this "*full registration*". In Figure 10.1, for the filter f_2 containing three terms B, D, and E, we register f_2 on three home nodes respectively for the terms B, D, and E. Similarly, the filter f_3 is on three home nodes respectively for the terms C, E, and F.

Real keywords-based data sets indicate that end users prefer to use short queries consisting of on average 2–3 terms per query. Thus, the proposed full registration could lead to 2–3 folds of storage redundancy, and will not incur significantly high storage redundancy. Instead, such slightly higher redundancy offers chance to greatly reduce the document publishing cost, which will be given in Section 10.3.3.

Document Forwarding: Second, to publish a document d, we want to forward d on home nodes of some selected terms. If the home nodes register filters matching d, we consider the forwarding is valid. Otherwise, such forwarding only wastes the network bandwidth. Thus, we expect the forwarding is valid without wasting network bandwidth. Furthermore, in order to use the low network bandwidth, we expect that the number of valid forwarding, for example, the number of selected terms, is low. Meanwhile all filters matching d are registered on the home node that d is forwarded to, and no filter falsely misses expected documents.

A naive solution is to simply select all $|d|$ terms from the document d, and forward d on the home nodes of all such $|d|$ document terms. Unfortunately, in real web documents, the

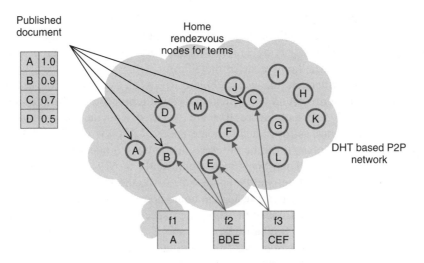

Figure 10.1   Basic STAIRS Framework.

number $|d|$ is typically tens and even thousands, indicating significantly high document publishing cost.

An improvement over the naive solution is to find the terms commonly appearing in d and registered filters, and forward d to the home nodes of such common terms. For example, bloom filters could be used to encode all query terms inside registered filters. By checking whether document terms appear inside the bloom filters, it is possible to find such common terms (though with false positive issue). Though the number of such common terms is surely smaller than $|d|$, the improvement approach is still not the best way to reduce the forwarding cost. For example, in the threshold-based model, each user defines the personalized threshold, and if the defined thresholds are well leveraged, it could be possible to significantly reduce the document publishing cost. We will present the algorithm in Section 10.3.3.

Document Notification: Finally, after a document d arrives at home nodes of selected terms, we can match d with locally registered filters. For the filters that match d, we can notify the associated users of the document d. Here, an important issue in the notification phase is to avoid duplicates, which is shown as follows.

Due to the "full registration", a filter f is registered on the home nodes of the $|f|$ query terms. If we select two of the $|f|$ query terms for document forwarding, the document d can duplicately find the matching filter f on such home nodes. This causes the duplicate notification. For example, in Figure 10.1, f_2 is registered in home nodes of three query terms $\{B, D, E\}$. When d is forwarded to three home nodes of $\{B, D, E\}$, these home nodes could duplicately notify the subscriber specifying f_2 of d.

To avoid the duplicate notification above, we introduce the *significant term* $t_{f,d}$ as follows. Given document d and filter f, among those terms which commonly appear in both f and d, we call the term with the highest value of $s(t_i, d)$ the *significant term* (STerm) of f and d, denoted by $t_{f,d}$. Such a significant term makes sense, because its term score $s(t_{f,d}, d)$ contributes *most significantly* to $S(f, d)$. In case of a tie in the highest score, we can break the tie by some rule such as choosing the term that appears the earliest in the document. With the definition of STerm $t_{f,d}$, we give the following notification rule. Given a qualified filter f satisfying $S(f, d) \geq T$, the subscriber specifying f is notified of document d only by the home node of the STerm $t_{f,d}$.

To illustrate this notification rule, we use Figure 10.1 as an example, where document d contains the pairs of terms $\{A, B, C, D\}$ and associated term scores. Also for simplicity, we assume that all 3 filters of Figure 10.1 are specified with the same threshold 1.0. Consider that document d is forwarded to home nodes of 4 terms $\{A, B, C, D\}$. When d reaches the home node of B, it can be observed that, among 3 query terms in the locally registered filter f_2, term B contributes most significantly to $S(f_2, d)$, and thus it is the significant term $t_{f_2,d}$. Therefore, via the home node of B, the subscriber specifying f_2 is notified of d. Meanwhile, when d reaches the home node of D, by the notification rule, the home node of D will not notify the subscriber specifying f_2 of d, because term D is not a significant term $t_{f_2,d}$. Similarly, terms A and C are significant terms of $t_{f_1,d}$ and $t_{f_3,d}$, respectively. Thus, those subscribers specifying f_1 and f_3 are uniquely notified of d, via home nodes of A and C, respectively.

Based on the notification rule, *no nodes are required to remember the history information of those already notified documents*. Instead, in Sieve [12], proxy nodes of subscribers have to remember the entire history information and then filter out duplicate publications. Clearly, the cost used to maintain the entire history information is nontrivial.

10.3.3 Document Forwarding Algorithm

The document forwarding algorithm proposed in STAIRS mainly leverages the following observation. Consider a filter f with $|f|$ query terms. Due to the full registration, the filter f is registered on $|f|$ home nodes. If d is forwarded to any of the $|f|$ home nodes (instead of forwarding d to all of the $|f|$ home nodes), using a single forwarding is enough to find the filer f. Based on this, we propose the following selection algorithm to select a significantly small number of terms from d.

Default Selection Algorithm: Recall that, only if the condition $S(f, d) \geq T$ holds, document d is forwarded to the subscriber specifying f. Such a condition helps document forwarding. If sorting all terms in d in *descending* order of $s(t_i, d)$, the threshold T can be position in the sorted list, as illustrated by the following example. In Figure 10.2 (a), all term scores are sorted in descending order, the sum of term scores from M to E is 0.68, and the sum from M to D is 1.18. Then, the threshold $T = 1.0$ is positioned between D and E. Since the sum of term scores from M to E is less than $T = 1.0$, we call the terms from M to E *Below Threshold Terms* (BTerms), and the remaining terms in d (i.e. terms from D to A) *Threshold Terms* (TTerms).

Formally, BTerms and TTerms are defined as follows. Given a sorted list of term scores in document d with $s(t_1, d) \geq \ldots \geq s(t_{|d|}, d)$, for any term t_h with $1 \leq h \leq |d|$, if $\sum_{i=h}^{|d|} s(t_i, d) < T$ holds, then any term t_i with $h \leq i \leq |d|$ is a BTerm, and term t_i with $1 \leq i < h$ is a TTerm. With BTerms and TTerms, we can claim the following results with respect to document d and threshold T. (i) The sum of term scores for any subset of BTerms in document d is less than T. (ii) Given document d with $S(f, d) \geq T$, a qualified filter f satisfying $S(f, d) \geq T$ contains at least one TTerm of d.

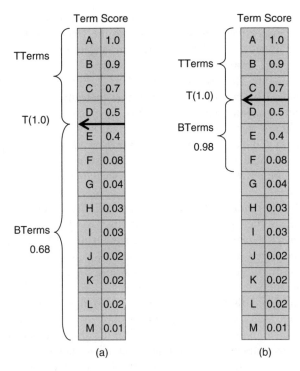

Figure 10.2 Default Forwarding: (a) using a default threshold T = 1.0; (b) improved for short filters with $|f|_s = 3$.

The correctness of the above claims is obvious (the proof can easily be derived by contradiction). For example, in Figure 10.2(a), for the full combination of BTerms from M to E, the sum of term scores, 0.68, is less than the threshold 1.0. Then, for a partial combination for BTerms (e.g. $\{E, F, G\}$), the sum of their term scores is further less than T = 1.0. Next, given a filter $f_2 : \{B, D, E\}$ (in Figure 10.1) satisfying $S(f_2, d) \geq 1.0$, two TTerms $\{B, D\}$ appear in f_2.

Recall that "full registration" guarantees that a qualified filter f satisfying $S(f, d) \geq T$ must be registered in the home node of a TTerm of d. Consequently, when document d is published, we can only require that d be forwarded to home nodes of TTerms in d, without causing false dismissals. We call this approach *partial forwarding*.

Compared with the simple approach that selects all $|d|$ terms, the above partial forwarding reduces the forwarding cost. For example, when document d of Figure 10.2(a) is published, d is only forwarded to home nodes of 4 TTerms $\{A, B, C, D\}$. Instead, full forwarding forwards d to home nodes of 13 terms from A to M.

Optimization for short filters: Real query logs indicate that the major input queries are composed of on average only 2-3 terms per query. This unique property helps further selecting a smaller number of TTerms.

Recall that the above definitions of TTerms and BTerms in implicitly assume that a filter is composed of an *arbitrary* number of terms. Now suppose $|f|_s$ is the largest length among all filters. Given $|f|_s$, we redefine TTerms as follows: in the sorted list of term scores $s(t_i, d)$, the sum of $|f|_s$ consecutive term scores should be less than T. Then, among these $|f|_s$ consecutive term scores, we use the term t_h with the highest one to indicate a term in d is either a TTerm or a BTerm: any term in d having a higher term score than $s(t_h, d)$ is a TTerm; otherwise, it is a BTerm.

For example, in Figure 10.2(b), $|f|_s$ is equal to 3. Since the sum of 3 term scores related to F, E and D is only 0.98, we determine that any term t_i with $s(t_i, d) \leq 0.5$ is a BTerm; and the remaining terms as TTerms. Compared with Figure 10.2(a) where the sum of term scores starts from the bottom line, now we only require the sum of $|f|_s$ consecutive term scores to be less than T. Obviously, with a smaller value of $|f|_s$, we boost BTerms to a higher position in the sorted list, and select a smaller number of TTerms. Thus, forwarding d to home nodes of a smaller number of TTerms consumes less forwarding cost.

Improvement for personalized filters: Besides using the default threshold T, users are allowable to define their personalized thresholds. Using a single default threshold T is not enough to meet the personalization subscription. It could either select too many terms (incurring too much network traffic) or select too few terms (incurring the false negative issue, i.e. some filters f satisfying satisfying $S(f, d) \geq T(f)$ may not receive the expected documents).

Given the personalized thresholds, STAIRS proposed to use a hybrid structure of histogram and bloom filters, named HiBloom, to summarize filters. The histogram (typically an equi-width histogram is used to capture filter thresholds, and bloom filters are used to encode query terms. Suppose the histogram contains b buckets. By dividing the whole range of filter thresholds into b even intervals, the k-th bucket ($1 \leq k \leq b$) is associated with a range of $(lb_k, ub_k]$ where lb_k and ub_k are the lower/upper bound of this bucket. For any filter f with a threshold inside $(lb_k, ub_k]$, query terms in f are encoded by a bloom filter, denoted by bf_k; moreover, among all filters inside such a bucket, we maintain the minimal threshold, denoted by T_k. Therefore, each bucket is uniquely associated with a bloom filter bf_k and a threshold T_k.

By leveraging the HiBloom data structure, a document d can match each bucket as follows. With the bloom filter bf_k, we can first find the common terms between bf_k and d. Among the common terms, the threshold T_k further helps select a smaller number of terms as before. When considering b buckets, all such distinct selected terms are together used to forward d onto the associated home nodes. Due to the fine-grained bloom filters and thresholds, this approach can help reduce the forwarding cost and avoid the false negative issue.

10.4 Recent Progress and Discussion

This section first reports two more recent work of STAIRS, and then compares the IR-Style document filtering with traditional content-based pub/sub.

10.4.1 Recent Progress

Following the technique framework of STAIRS, [13] considered a more advanced filtering model. Though the threshold-based scheme helps define the personalized subscription conditions, it is not easy for users to set reasonable thresholds. Instead, a better way is using a hybrid solution. First, the system implicitly sets a default threshold for each filter, such that the relevance score between a filter and a to-be-disseminated document is no smaller than the default threshold. It guarantees the absolute qualify of disseminated documents. Meanwhile, it uses the top-k filtering model to guarantee the relative qualify of disseminated documents. That is, besides $|f|$ query terms $\{qt_1, \ldots, qt_{|f|}\}$, a user further defines a number of k. Both the query terms and the number k are used to declare the filter f. Suppose the total D published Web documents are published within a period (such as a day or an hour). Then, the top-k filtering model ensures that each user defining a filter f receives the top-k most relevant documents. The top-k documents mean that among the D documents, such top-k documents are those having the largest relevance scores. To disseminate the top-k documents to each user, the main challenge is dynamically changed thresholds associated with the filter f to prune irrelevant documents. Thus, using the HiBoom scheme to summarize such dynamical thresholds is infeasible. Instead, [13] proposed to prune irrelevant documents during the forwarding path towards the home nodes of the default selected terms. Such in-network pruning mechanism need not the synchronization of such dynamical thresholds between the home nodes and the intermediate nodes on the forwarding path.

Besides, [14] considered a general case by assuming that each filter consists of $|f|$ query terms (without a defined threshold or a top-k number). Thus, d successfully *matches* f (and similarly, f matches d) if and only if there is a term t that appears inside both d and f. In other words, this is a *boolean-based content filtering* approach. After the matched documents are found, any filtering model, either the top-k or threshold-based scheme, can be adopted to filter out irrelevant documents. Since neither the term scores nor thresholds are used in this model, [14] alternatively proposed a new approach to select terms. First, [14] identified the problem to minimize the number of selected terms is NP-hard. Next, in order to work practically in DHTs, [14] proposed a cost model approach to merge filters across the nodes of a DHT. Then with a smaller number of forwarding, documents are forwarded to the home nodes registered with merged filters.

Besides the above STAIRS-based work, there is some other IR-based filtering work, such as [15, 16]. For example, [15] heuristically selected the top-k terms with the largest term scores. However, there is no guarantee that the selected terms must be valid and furthermore, some matching filters could be falsely missed. [16] essentially adopted the the attribute-value model, which is the same as the content-based pub/sub. Nevertheless, it allows an attribute, that is, the abstract, to support the keyword filtering.

10.4.2 Discussion

Finally, the comparison between the IR-like filtering (in short IF) model and the filtering model in content-based pub/sub as follows.

First, in the form of content: most of content-based pub/sub focus on *structured* information, which follows predefined schemas; while the IF targets at unstructured web articles (or summary articles like RSS feeds, or preprocessed with if*idf scores).

Second, due to the content of different forms, filters in the content-based pub/sub systems typically consist of selective predicates; while the IF adopted the keywords-based full-text filtering semantics.

Next, for the matching approach, most content-based pub/sub systems explore filter relationships (like covering, overlapping, etc) to build various kinds of filter indexing structures (e.g. poset). The IF solutions typically use the inverted list-based approach. For example, SIFT [17] maintained local inverted lists, and STAIRS implicitly built distributed inverted lists.

Finally, in term of the proposed solution, the content-based pub/sub [18] adopts the sing-registration and full-matching policy, the centralized IF (e.g. SIFT) adopts the full-registration and full-matching policy. That is, a filter f is registered to all posting lists associated with $|f|$ query terms, and to match a documents, SIFT retrieves all posting lists associated with the $|d|$ document terms. Instead, the STAIRS is based on the full-registration and partial-forwarding policy, and the main purpose is to reduce the forwarding cost.

10.5 Summary

This chapter reviewed the keywords-based content filtering and dissemination in DHTs, including the data model, solution framework, and the core term selection algorithm in order to reduce document forwarding cost. In addition, this chapter reports two more recent following work, and compare the IR-style filtering approach with traditional content-based pub/sub. Due to the significantly different data model with traditional content-based pub/sub, the work STAIRS proposed a novel partial-forwarding policy to reduce the forwarding cost.

References

1. Berry MW, Drmac Z and Jessup ER (1999) Matrices, vector spaces, and information retrieval. *SIAM Rev.* **41**(2), 335–62.
2. Rao W, Chen L and Fu A (2011) STAIRS: Towards efficient full-text filtering and dissemination in dht environments. *The VLDB Journal* **20**(6): 793–817. 10.1007/s00778-011-0224-z.
3. Rose I, Murty R, Pietzuch PR, Ledlie J, Roussopoulos M and Welsh M (2007) Cobra: Content-based filtering and aggregation of blogs and rss feeds. *4th Symposium on Networked Systems Design and Implementation (NSDI)*, 11–13 April 2007, Cambridge, MA.
4. Li J, Loo BT, Hellerstein JM, Kaashoek MF, Karger DR and Morris R (2003) On the feasibility of peer-to-peer web indexing and search. *IPTPS*, pp. 207–15.
5. Yang Y, Dunlap R, Rexroad M and Cooper BF (2006) Performance of full text search in structured and unstructured peer-to-peer systems. *INFOCOM*.
6. Rao W, Fu AWC, Chen L and Chen H (2009) Stairs: Towards efficient full-text filtering and dissemination in a dht environment. *ICDE '09. IEEE 25th International Conference on Data Engineering*, pp. 198–209.
7. Liu Z, Parthasarthy S, Ranganathan A and Yang H (2007) Scalable event matching for overlapping subscriptions in pub/sub systems. *DEBS*, pp. 250–61.
8. Nguyen LT, Yee WG and Frieder O (2008) Adaptive distributed indexing for structured peer-to-peer networks. *CIKM*, pp. 1241–50.
9. Stoica I, Morris R, Liben-Nowell D, et al. (2003) Chord: a scalable peer-to-peer lookup protocol for internet applications. *IEEE/ACM Trans. Netw.* **11**(1), 17–32.
10. Rowstron AIT and Druschel P (2001) Pastry: Scalable, decentralized object location, and routing for large-scale peer-to-peer systems. *Middleware*, pp. 329–50.
11. Zhao BY, Kubiatowicz J and Joseph AD (2002) Tapestry: a fault-tolerant wide-area application infrastructure. *Computer Communication Review* **32**(1): 81.
12. Ganguly S, Bhatnagar S, Saxena A, Izmailov R and Banerjee S (2006) A fast content-based data distribution infrastructure. *INFOCOM*.

13. Rao W and Chen L (2011) A distributed full-text top-k document dissemination system in distributed hash tables. *World Wide Web* **14**(5–6): 545–72.
14. Rao W, Vitenberg R and Tarkoma S (2011) Towards optimal keyword-based content dissemination in dht-based p2p networks. *P2P*, pp. 102–11.
15. Tang C and Dwarkadas S (2004) Hybrid global-local indexing for efficient peer-to-peer information retrieval. *NSDI*, pp. 211–24.
16. Tryfonopoulos C, Idreos S and Koubarakis M (2005) Publish/subscribe functionality in IR environments using structured overlay networks. *SIGIR*, pp. 322–9.
17. Yan TW and Garcia-Molina H (1999) The SIFT information dissemination system. *ACM Trans. Database Syst*. **24**(4): 529–65.
18. Fabret F, Jacobsen HA, Llirbat F, Pereira J, Ross KA and Shasha D (2001) Filtering algorithms and implementation for very fast publish/subscribe *SIGMOD Conference*, pp. 115–26.

11

Advanced Topics

In this chapter, we consider advanced features of pub/sub systems. We start with security considerations and examine a number of security solutions for pub/sub. Then we examine topics such as composite subscriptions, filter merging, load balancing, channelization, reconfiguration, mobility support, congestion control, and the evaluation of pub/sub system. Many of the topics pertain to the pub/sub routing topology, its organization and configuration. We observe that the dynamic content delivery environment requires active monitoring and configuration in order to be efficient, available, and fault tolerant.

11.1 Security

This section considers security challenges and solutions for pub/sub systems. We survey well-known solutions and discuss the key security services for pub/sub. The pub/sub communications model is characterized by its one-to-many and many-to-many communications, and this decoupled multicast nature requires solutions for unwanted traffic and the prevention of various attacks on the network and content.

11.1.1 Overview

The problems of security risks emerge from several sources, the most important being the distributed environment, in which subscribers, publishers, and brokers operate. Is is crucial that both the network and the content it carries are protected against attacks. We must ensure the authenticity of data and content and, furthermore, require sufficient levels of confidentiality, integrity, anonymity, availability, and access control and authorization. This has created the necessity for a number of security services for pub/sub. These services are built on top of existing solutions and rely on symmetric and asymmetric cryptography. Pub/sub security thus builds on existing security solutions and adapts those for the multicast environment.

In general, there needs to be a way to establish trust towards the entities of the pub/sub system. Typically, this trust is established with the help of a trusted third party that

Publish/Subscribe Systems: Design and Principles, First Edition. Sasu Tarkoma.
© 2012 John Wiley & Sons, Ltd. Published 2012 by John Wiley & Sons, Ltd.

assigns keys to the entities. Each entity has one or more public keys and the corresponding private keys. The public keys are known to all and a public key can be distributed in a certificate form signed by the trusted third party with its private key. A cryptographic key distribution mechanism is a crucial part of the distributed security solution. Thus a trust chain can be built that links the trusted third party to a key and allows the build-up of trust in the system. Public key cryptography is typically used to authenticate entities and then generate symmetric keys for encryption. Public key cryptography allows digital signatures that are instrumental in authenticating entities and ensuring content integrity.

General pub/sub security has been addressed by many research projects as well as commercial systems, especially requirements, authentication, confidentiality, and payment processing. Security-aware pub/sub systems include Hermes, EventGuard, and Rebeca. The EventGuard system comprises of a set of security guards to secure pub/sub operations, and a resilient pub/sub network. The basic security building blocks are tokens, keys, and signatures. Tokens are used within the pub/sub network to route messages. In this solution, the tokens contain limited filter information in an encrypted format that enables the network to route encrypted messages. The system addresses the mitigation of unsolicited bogus messages originating from subscribers and publishers.

An overview of pub/sub security topics was given in [1]. They propose several techniques for ensuring the availability of the information dissemination network. Prevention of denial-of-service attacks is essential and customized publication control is proposed to mitigate large-scale attacks. In this technique, subscribers can specify which publishers are allowed to send them information. A challenge-response mechanism is proposed, in which the subscriber issues a challenge function, and the publisher has to respond to the challenge. The use of the mechanism in a distributed environment was not elaborated.

Pub/sub broker networks are vulnerable to message dropping attacks. For example, overlays such as Hermes and Scribe may suffer from bogus nodes. The prevention of message dropping attacks has a high cost and only a few systems address them. The EventGuard uses an r-resilient network of brokers.

Secure event types and type-checking was proposed in [2]. Secure event type definitions contain issuer's public key, version information, attributes, delegation certificates, and a digital signature. Secure event types and schemas are important for the prevention of unwanted traffic. Scope-based security was discussed in [3], in which trust networks are created in the broker network using PKI techniques. A proxy-based security and accounting solution was proposed in [4] for untrusted broker networks.

11.1.2 Security Threats

The field of security threats, realized or impended is very extensive and incoherent. This is caused by the wide range of techniques used in malicious attacks but also by the extreme mutability of attack methods. An ever-growing set of attack methods and risks is part of the network the reality and the risks increase when the underlying networks are changed. Paradoxically, the process of the fortifying the networks against security risks is itself a cause of increased risk: any alteration is bound to bring forth new methods of attacks. We can say that the cyclic alternation of security solutions and new malicious attacks is never-ending and also spiralling.

The perpetrators of malicious attacks use a manifold of often ingenious methods to invade and assault networks. From the multitude of schemes we can isolate this list of most dangerous attack scenarios:

- Attacker controls malicious nodes that disguise as legal nodes.
- Attacker overflows the network with data (pollution). This can target content or the routing tables.
- Attacker returns maliciously incorrect data.
- Attacker denies data existence or supplies fraudulent routing info to it.
- Attackers seeks a quorum in k-redundant networks.

Evidently, the methods used to combat the attack are as diversified as the technologies used by the adversaries. Within the overlay and pub/sub networks we can group the general defensive methods into following sectors:

- Securing the basic routing.
- Securing the content-based routing and routing tables.
- Securing the data and content.
- Securing the authentication and access control.

A common feature of the solutions is the need to have fixed points, used to build trust in the system. A logically centralized identity management is generally needed to implement the trust mechanism. Without some form of centralized management a distributed system is impossible to defend against malicious nodes. The most grave danger lies in the possibility that attackers can achieve a quorum – to prevent this we must have a way to control the creation of node identifier, for instance securing the node identifiers through public key cryptography. Furthermore, we need a mechanism to bootstrap trust towards the public keys.

11.1.3   Security Issues in Pub/Sub Networks

The pub/sub overlay environments face their own particular set of security problems. The many-to-many communication model has special needs as to authenticity, integrity, confidentiality and also availability. We have to be able, for instance:

- Publication authenticity: Guaranteeing that only authentic publications are delivered to the subscribers.
- Subscription authentication: Guaranteeing that only the real subscribers to the service will get publications matching their interest.
- Publication and subscription integrity: Preventing unauthorized, perhaps malicious, modifications of pub/sub messages,
- Publication confidentiality: Performing the content-based routing when the publishers do not trust the pub/sub network.
- Subscription confidentiality: Removing the need of the subscribers to reveal their subscriptions to the network.
- Protection: Protecting the pub/sub network from attacks, such as spamming or flooding attacks, both selective and random message dropping attacks, and other *Denial-of-Service (DoS)* attacks.

There are not many systems that provide a true pub/sub security framework. Basic solutions build on transport layer security, such as the TLS protocol, and then sign and encrypt messages based on credentials and certificates obtained from a trusted third party. Messages can be signed and encrypted selectively with specifications such as the W3C's XML Signature [5] and Encryption standards [6]. In the following we examine pub/sub specific security solutions with EventGuard, Quip, and Hermes.

11.1.4   EventGuard

EventGuard aims to provide authentication for publications and confidentiality and integrity for both publications and subscriptions [7]. The system also enables availability and performance metrics while improving scalability and ease of use. The EventGuard system is modular and operates above a content-based pub/sub core. Figure 11.1 illustrates this modular design that consists of two main ingredients:

- A Security Guard suite that can be seamlessly plugged into a content-based pub/sub system.
- A pub/sub network design which is flexible and favours scalable routing and facilitates managing of node failures and possible message dropping DoS attacks.

EventGuard has altogether six guards, that secure the following critical pub/sub operations:

- subscribe,
- advertise,
- publish,
- unsubscribe,
- unadvertise,
- routing.

EventGuard also supports a metaservice that generates tokens and keys, used as identification of the publications (such as a hash function over publication topic). Keys may

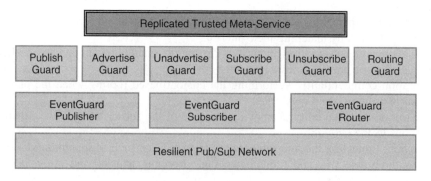

Figure 11.1   Overview of the EventGuard system.

be used for encrypting message contents. Actually, communication with the provided metaservice is mandatory for all pub/sub operations before sending any messages. The encryption, signatures and the creation of tokens [7] use the El-Gamal algorithm.

11.1.5 QUIP

The securing of content distribution in peer-to-peer pub/sub overlay networks requires protocol-level support. QUIP is such a protocol [8], designed to provide encryption and authentication mechanisms for pub/sub systems that are already existing. QUIP has the following security features:

- Protection of content from unauthorized users.
- Protection of payment methods and authentication of publishers.
- Protection of the message integrity.

QUIP ensures that subscribers are able to authenticate the messages received from publishers. Secondly, it allows the publishers to strictly manage who receives their content. QUIP makes use of the public key traitor tracing scheme, with two main advantages:

- QUIP can invalidate the keys of any subscribers without touching any other keys.
- QUIP assigns each subscriber a unique key. This can be used to detect and resolve key leaks.

Thus QUIP incorporates an efficient traitor-tracing scheme with a protocol for secure key management allowing publishers two critical powers:

- Restricting their messages to authorized subscribers.
- Adding and removing subscribers without affecting the keys of the other subscribers.

One field not assessed by QUIP is related to the problem of subscription privacy. QUIP assumes that a single trusted authority exists. This authority (the key server) is fully responsible for keys and payments handling.

The key software in a QUIP network is the QUIP client which is mandatory for each participant. Anyone in the pub/sub network willing to use QUIP must download the client in advance. The client assigns the participant with a unique random ID and also discloses the public key of the key server. A QUIP participant gets a certificate linking their public key to their identifier from the key server at the initialization phase. For instance if a publisher wants to publish a protected object, it will contact the key server receiving a content key in return. This is used to encrypt the content. A subscriber who wants to read this publication also contacts the key server and obtains the content key. The system allows for requiring a payment in any of these phases.

11.1.6 Hermes

In the pub/sub architecture access control features are intimately bound to assigning privileges to all parties in the pub/sub system and then enforcing access to resources based on the privileges. Hermes [9] is a distributed event-based middleware pub/sub architecture that features a role-based access control model. We have examined Hermes previously in Chapter 9.

The central concept in Hermes is the event type. Hermes supports several features derived from object-oriented language model, for instance type hierarchies and supertype subscriptions. Atop of the overlay routing network sits a scalable pub/sub routing algorithm. Hermes accordingly avoids global broadcasts since rendezvous nodes are created and used. Hermes comes fully integrated with fault-tolerance mechanisms in the routing algorithm. These are able to overcome breakdowns of the middleware thus yielding a robust and scalable system.

The main goal of Hermes lies in creating an architecture where security management and control are provided by the pub/sub middleware. Access control has been made fully transparent for publishers and subscribers. Event owners can decide the access policies to the event that they publish. The ownership is asserted with a X.509 certificate. Roles are assigned for users with privileges assigned for each role.

Privileges are not assigned directly to users but through their roles. This method has two main benefits:

- Role-bound privileges are easier to manage than those directly assigned to users.
- Control of policies is strictly decoupled from the specific software under protection.

Need for authentication process affects both publishers and subscribers. In fact all requests directed to brokers are delivered under their specific credentials. Brokers examine the credentials and may then accept, partially accept or reject the request. For the expression of the policies there exists a a specific policy language provided by the OASIS security framework. *Role-Based Access Control (RBAC)* is a frequently employed technique for scalable security administration that decouples principals and privileges with roles. Privileges are not directly assigned to principals, the users of the system, but to roles that are then associated with the principals. Thus the technique separates administration of the users and their association with roles from the actual privileges that users use to access services. Thus users can leave and join different roles in an organization and the service policies do not have to be modified, because they are tied to roles.

The OASIS policy specifies for each method the role credentials and their constraints that authorize invocation of the method. The system supports fine-grained privilege management, for example setting privileges to particular message types and specifying security requirements on the attribute level. These security requirements are maintained when pub/sub operations, namely advertise, publish, and subscribe, are invoked.

Hermes uses predicates as a base for the access control and the involved decisions. The system utilizes generic predicates operated as black-boxes. Decisions made by a predicate may be based on the message size and other message properties. Pub/sub restriction predicates exist, as well. In the latter case the pub/sub system is fully cognizant of the predicates, and the decision can be made based on the content of the event and its position in the event type hierarchy. Take, for example, a subscriber attempting to subscribe to (for this subscriber) an unauthorized event. Hermes then tries to detect if there exist access privileges for any subtypes of this event, effectively transforming the original access (subscription) request into a different subscription scope. This approach is impossible if the brokers are not trusted to use the access control policies.

The critical web-of-trust is built in Hermes utilizing certificate chains. In such a chain an event owner signs the trust broker certificates, and the brokers sign the certificates of their adjacent brokers, effectively facilitating a chain of trust. If there exists a trusted root certificate which the publishers and subscribers can show for the event owners, the trust chain delivers verification to the local brokers for processing the targeted event.

11.1.7 Encrypting Attributes

Following the secure pub/sub model proposed in conjunction with the Hermes overlay pub/sub system [9, 10] the covering and overlapping algorithms may be extended to support encrypted attributes. A router must be able to process encrypted filters and notifications to ensure privacy and security of the distributed system. The integration of two security areas need to be considered:

- Encrypting notifications and attributes in notifications.
- Encrypting filters and attribute filters.

The secure pub/sub model is based on key-classes defined using symmetric keys. The routers share keys using a cryptographic key distribution mechanism. In this model, each subscriber and producer selects one or more key-classes for the subscription or notification from the set of key-classes it is authorized to use. Key-classes are used to encrypt attributes in notifications and attribute filters in subscriptions. An encrypted attribute can be matched with an attribute filter only when they have the same key-class. Notifications and filters with non-overlapping key-classes are uncomparable. If several key-classes are used for an attribute or attribute filter, it must be disjunctively encrypted multiple times using the key-classes. This means that there will be several copies of the attribute or attribute filter – one for each key-class [9].

Key groups of brokers are used to manage access to encryption keys and also to issue and reissue encryption keys to brokers. The component responsible for verification is the key group manager. It will certify the authorization of a broker to join a given key group. Therefore the key group manager should be trusted by the event type owner in that access control will be correctly applied. This can be accomplished for example, if the key group manager is a member of a trusted party, perhaps the domain of the type owner.

Authorizing the membership of key groups is managed using capability structures, essentially the same type as for managing pub/sub requests. The methods of access control are different for clients and brokers but the system still supports consistent capability representation

11.1.8 Privacy

There is an inherent trade-off between the privacy of the subscriber and the accuracy of the result set. The former can be described in terms of how well the exposed information uniquely characterized the subscriber, and the latter how well the returned data items match the subscribers real interests. Privacy contrasts content-based pub/sub, because a broker needs to know the content of an event and the subscriber interests in order to perform the forwarding decision. On the other hand, distributed content-based pub/sub offers certain level of anonymity through the decoupled nature and filter aggregation techniques; however, these are not very strong in maintaining privacy requirements. Thus additional solutions are needed.

Subscriber privacy is enhanced by guaranteeing that a subscriber cannot be distinguished from a set of subscribers when the network delivers the interests and matching content. This delivery can happen in the form of broadcast within a certain area, or delivered using unicast or multicast across multiple brokers. Physical broadcast can be implemented in such a way that specific recipient identifiers are omitted; however, given the knowledge that only a single entity is interested in the data is sufficient to pinpoint the subscriber. Publisher privacy and anonymity requires that it is not possible to identify the publisher based on the published events and network traces.

A general way to maintain anonymity in the network is to route a message through a network that hides the identity of the sender. The Mist system handles the problem of routing a message though a network while keeping the location of the sender hidden from intermediate systems [11]. The system consists of a number of routers, known as Mist routers, ordered in a hierarchical structure. The anonymity degree metric has been proposed for evaluating route selection strategies that maximize the degree of anonymity of a system. Mist is a general network service and it does not address pub/sub functions such as filtering and multicast. The EventGuard system discussed above supports encrypted subscriptions and event messages. These are necessary features for a privacy enhanced distributed pub/sub.

A general way to anonymize tuples is to generalize an attribute, for example a number to a range. In this case, larger the range, the more information loss is introduced by the anonymization process. This approach can be used for both published events and subscriptions in a pub/sub system. For the case of subscriptions, the more general a subscription is made, the more unwanted traffic, false positives, will be generated. Both the information loss due to generalization and the false positives can be measured. The techniques proposed for k-anonymity and l-diversity are suitable for anonymizing a published event or a stream of events for a single user or for multiple users [12]. Their application for pub/sub is a new research topic in this area. Location privacy has become an active research topic in recent years. The system model typically includes a set of clients and a centralized server that brokers points of interest to the clients. The k-location anonymity technique uses a cloaked region to represent the client location and this region needs to contain at least $k - 1$ other client locations.

11.2 Composite Subscriptions

Composite subscriptions present new challenges for distributed pub/sub systems. Detection of complex event sequences would be handled best with a special event language for expressing concurrent event patterns. Such a one is the Core Composite Event Language of the CEA system [13]. This compiled language produces automata that support regular expression-type patterns. The system transforms the specification language to finite state machines. Formal semantics are given for the interval time model. The problem of translating nondeterministic automata to deterministic is not discussed other than to mention that the current implementation uses nondeterministic automata with a list implementation.

The key benefits of the this approach are the distribution of the detection task, an automata-based detection engine, and the use of an interval time model to detect the causality of events [14]. A Lamport Logical Scalar Clock gives a causal ordering if such exists (but not a strict causal ordering). CEA is limited to considering the basic situation and does not examine alternative paths nor the overlay network's dynamic load characteristics.

The Hermes composite event service [9] builds on the CEA system. It includes composite event detection automata as well. Hermes automata are finite state type with additional support for temporal relationships and concurrent events. This is used for analyzing event streams. The utilization of extended finite state automata has many benefits in composite event detection, listed below:

- The computational model of finite state automata is widely used and theoretically very well understood. Its implementation is simple.
- The expressive power of such automata is restricted, leading to limited and predictable usage of resources. This essentially augments the safe distribution of detectors in the event-based middleware.
- Regular expression languages typically avoid the risks of redundancy and incompleteness as they include operators specifically designed for detection of patterns. This is important when one defines a new composite event language.
- Complex expressions of a regular language are easily decomposable when distributed detection is utilized.

Composite event subscriptions use a core composite event language to specify the set of composite events that an event client is interested in. Let us take a little closer look on the features required from such a language.

Obviously the central features of any such language are the operators. Many are not dissimilar from the corresponding operators in regular languages, like concatenation, alteration and iteration but also special operators are needed for instance with features belonging to detection automata and specifically related to the construction of composite event detection automata from subautomata. The composite event service should enable the distributed detection of composite events in an (event-based) middleware, typically accomplished using a decomposition of composite event expressions into subexpressions. These will be registered by separate distributed detectors in the system. The use of decomposition of composite event expressions into subexpressions is important for several reasons:

- Popular subexpressions can be reused among event subscribers. This saves network bandwidth and also computational strain to the system.
- The communication volume will be reduced. This results from the positioning of subexpression detectors close to primitive event publishers.
- Replication of subexpressions is possible for load-balancing and increased availability.
- Computationally expensive expressions can be decomposed. This protects detectors from overload.

Two important decisions must be made when developing a language like this. They are

1. The optimal strategy for the decomposition of composite event expressions
2. The optimal placement of composite event detectors in the system.

These decisions might include internal conflicts. For instance, minimizing network bandwidth might force to maximize the reuse of subexpression detectors. This conflicts with the possible targeting of decreasing notification latency which would require replication of detectors around the network thereby causing increased bandwidth consumption. To solve this problem we must take the static and dynamic features of the application and the network into account and reach a trade-off solution.

Another solution is provided by the Padres system that includes an expressive subscription language where a composite subscription is a Boolean function of the atomic subscriptions [15]. In Padres one may specify constraints on the content of the publication. Furthermore, publications can be correlated in a distributed environment and the detection of event patterns will be handled by the broker network in a distributed manner.

The main features of a topology-based composite subscription routing are the following:

- The subscription routing requires an acyclic overlay.
- It will not take into account any dynamic network conditions.
- Composite subscriptions are routed as units towards the potential publishers. The forwarding continues until the subscription reaches a suitable broker. The data sources needed to satisfy the composite subscription will be located in different directions (in the overlay network).
- A broker (the join point broker) splits the composite subscription and routes individual partial subscriptions separately to potential publishers.
- The matching publications are routed back to the join point broker, where the composite event detection is executed.

11.3   Filter Merging

We examined various pub/sub routing strategies in Chapter 7 including the filter covering and merging techniques for optimizing filter processing. Both covering and merging aim at reducing the number of filters propagated in the network. Filter covering determines whether not a filter is already contained in an active filter that has been sent to the neighbours. If so it can be dropped. This is easily visualized by imagining filters as subspaces and if a subspace is contained in another subspace, all the points (events) that the contained subspace selects are also selected by the containing subspace.

Filter merging extends this to filters that do not have a covering relation, but can be fused together by applying filter merging rules [16]. A filter merging algorithm takes two or more filters as input and outputs a merged filter if merging can be done. The result of the merging is said to be perfect when it is equivalent to the input filters. In other words, the merger of the filters does not result in any false positives or false negatives. Otherwise the merger is called imperfect and it can result in false positives. False negatives are not allowed in order to meet the safety requirement discussed in Chapter 7.

Merging introduces some complexity into the pub/sub engine; however, it can result in the reduction of the routing table sizes and processing costs of the pub/sub routers. Merging, as well as covering, is a generic optimization technique and it can be applied to subscriptions and advertisements.

Filter merging may be applied in different places in the event router. The two typical use cases are: the merging of filters from local clients, and the merging of the external routing table. The former is simpler, because it can be done locally and the resulting mergers can be directly given to the remote routing table for processing. The latter is

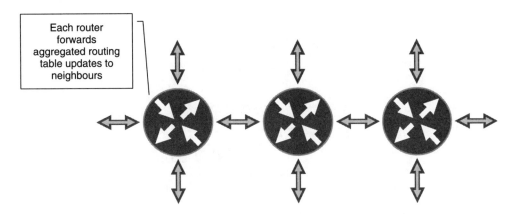

Figure 11.2   Merging in multihop environment.

more complicated, because updates need to be sent to neighbours. Merging requires that the mergers that are sent to the neighbouring nodes are tracked and when a merger is removed due to unsubscription, the resulting active filters need to be re-merged and installed without resulting in false negatives.

Figure 11.2 illustrates filter merging in a distributed environment with many pub/sub routers. Each router maintains a routing table that is used to forward events to interested neighboring routers. The routers propagate subscription and unsubscription messages. Each router also aggregates the subscriptions by analyzing the routing table and merging mergeable subscriptions. Each router forwards aggregated routing table updates to neighbours. When a router receives an update from its neighbour, the update replaces existing subscriptions covered by the update. When an existing merger is modified, for example when unsubscription happens to a part of the merger, a new update is prepared that replaces the previous update and installs the currently active subscriptions at the neighbour. The key requirement is that the neighbours have an accurate view to the active subscriptions so that published events are not dropped.

The optimal merging of filters and queries with constraints has been shown to be NP-complete [17]. Therefore, in practice a heuristic scheme is needed that balances the processing cost with accuracy. The first filter-merging-based routing mechanism was proposed in the Rebeca distributed event system [18]. This mechanism merges conjunctive filters using perfect merging rules that are predicate-specific. Routing with merging was evaluated mainly using the routing table size and forwarding overhead as the key metrics in a distributed environment. The merging mechanism was applied in the context of a stock application for simple predicates.

Later merging was applied to the general content-based routing scenario and integrated with the poset and forest structures [16]. The root sets of these structures are a good basis for merging; however, bookkeeping is required to be able to support distributed operation with many neighbouring brokers. The basic idea is to determine the noncovered set of filters and then apply a scanning or counter based merging algorithm to find mergeable candidates for the merging process. Upon successful merging, the merger is installed to the applicable neighbouring routers. When a merger is removed, the still active constituent

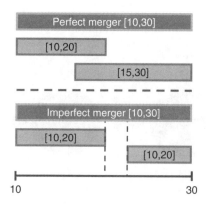

Figure 11.3 Comparison of perfect and imperfect merging.

filters are activated at the same time as the merger is removed from neighbours. The update process needs to be atomic in order to prevent lost messages.

Figure 11.3 illustrates the differences between perfect and imperfect mergers. Two range filters that completely cover a range can be merged into a single filter that covers the range. This merger does not result in false positives and hence it is called a perfect merger. An imperfect merger, on the other hand, does cover the ranges of its constituent filters, but may be larger than the union of the constituent filters. Thus an imperfect merger can result in false positives. The key metrics in assessing the efficiency of these mergers are the merging cost, size of merger, and the accuracy of the merger.

The structural equivalence of filters is necessary for both perfect and imperfect merging. Typically, filters with differing number of tuples are nonmergeable. This depends on the merging rules and the filtering language. The covering algorithm for filters removes those filters that have differing number of tuples and are covered by other filters so the merging process is simplified by focusing on filters with the same number of tuples. The distinctive attribute filter for perfect merging, if one exists, is also found during the mergeability test. Both perfect and imperfect merging mechanisms use the same attribute filter merging mechanism.

Figure 11.4 illustrates the benefits of filter merging for typed-tuple-based filters. The example shows a simple content-based routing core with one input port and one output port. Incoming filters are tested for covering and mergeability. The set of input elements consists of four filters (F1, F2, F3, and F4). We observe that there are no covering relations in this set, so without filter merging support they would need to be propagated by the routing core to the router behind the output interface. Since there are no covering relations, the routing core will not drop any of the filters. In the SIENA model, filters that are covered by a filter from the same input port can be dropped in order to minimize state information and unnecessary signalling between routers.

Assuming that only filters with the same structure are merged, the filters F1 and F2, and later F3 and F4 can be merged. F1 and F2 can be combined by combining the two ranges. The figure presents the state at the router behind the output interface.

Now, if the filter F3 is deleted, the routing core determines the set of uncovered filters, and their possible mergers, and reinstalls necessary state by sending an update message to

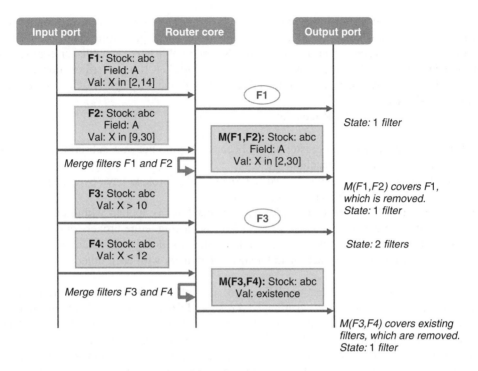

Figure 11.4   Example of merging.

the output port containing all information regarding the state change. The update message triggers the routing table modification process.

Merging can be implemented as a pluggable module for content-based routers and integrated with the routing table data structure. In the general case, such a module requires access to the first level (hierarchical) or the two first levels (SIENA peer-to-peer) of the structure. Access to the two levels is required in the peer-to-peer case, because of the way the destination set is determined, namely a subscription is never sent to the source of the subscription. As noted above merging can also be utilized in many places inside the content-based router, for example for client subscriptions or for the external routing table.

11.4   Load Balancing

Load balancing is an important middleware feature. A pub/sub load balancer is responsible for distributing load in such a way that hotspots are alleviated and that the parallel processing capability of the system is maximized. There are many ways to realize load balancing in a distributed pub/sub system. Offloading subscribers and publishers from one router to other routers is one particular technique for achieving load balancing in a distributed system. Moreover it is possible to offload the maintenance of a content-based routing table to clients and other routers [19].

The Padres system is built on top of a publish/subscribe system called *Padres Efficient Event Routing (PEER)* [20–22]. The PEER system consists of two different kinds of brokers, namely the cluster-heads and the edge brokers. The former have many connection with other brokers, whereas the edge brokers only have one neighbouring connection. Edge brokers serve subscribers and the cluster heads serve publishers. A publisher connecting to an edge broker will be redirected to a cluster head broker. If a subscriber connects to a cluster head broker, it will forward the subscriber to a randomly chosen edge broker that is not overloaded.

The key features of PEER include:

- Little processing delay for cluster heads, because they do not save subscribers and only forward publications to matching clusters.
- Load balancing is only applied to edge brokers. Edge brokers have overhead, because they need to match publications to subscribers.
- PEER has two levels of load balancing. The first level is applied within clusters, and the second one between clusters. In local load balancing, edge brokers only exchange information with edge brokers in the same cluster. In global load balancing, the edge brokers from different clusters load balance subscribers with each other.

PEER utilizes a load detection framework in order to detect overloaded brokers. In order to be able to detect overloaded brokers, PEER exchanges load information between brokers. If a broker finds a broker that is more loaded than others, the load balancing system is activated. The load level messages are disseminated both within the cluster and between clusters. Global load level messages include summary information about the situation including utilization ratios and matching delays of edge brokers.

A load estimation algorithm is used to estimate the input and output load requirements of subscriptions. This allows the system to estimate the load when offloading subscriptions to brokers that can accept additional load. The algorithm samples the publication rate of filters. The offload algorithm then determines the subscriptions to offload to brokers that can accept additional load.

The offload algorithm consists of three modules:

- Input offload algorithm that aims to decrease the input utilization ratio of the offloading broker.
- Match offload algorithm that aims to balance the processing delay across brokers.
- Output offload algorithm that aims to balance the output bandwidth used by the two brokers. The aim is to offload subscriptions that are similar to the ones at the offloaded broker. This minimizes additional traffic to that broker.

In addition to subscription offloading, system performance can be improved by active publisher placement [21]. The publisher placement algorithm relocates publishers in such a way that the end-to-end delivery delay and system load are minimized. An effective publisher placement strategy places publishers close to clusters that have a high number of subscribers.

The publisher placement algorithm consists of three steps:

- Gathering of publication delivery statistics.
- Identifying the destination target broker for a publisher.
- Migrating the publisher to this target broker and establishing the publisher there.

The placement algorithm only activates the inactive advertisement path between the old broker and the new broker. This minimizes reconfiguration cost of the overlay network. In order to make a publisher migration decision, the system needs to know the approximate number of subscribers in a certain part of the network. A distributed trace algorithm is used for this that gathers subscriber information from the network. Downstream brokers notify upstream brokers about subscribers for a particular publication. The actual migration phase requires several steps. First, the publishing is temporarily paused, then a migration advertisement is issued towards the target broker, and finally the publisher is established at the target broker and the publication activity can continue.

11.5 Content-Based Channelization

The *Event Space Partitioning (ESP)* approach presented in [23] divides the event space into regions that are distributed over the servers. Each region is responsible for a subspace of the event space. Subscriptions are then sent to the overlapping regions. A notification is only sent to one region containing it, because it will have all the subscriptions for the event.

Following the ESP model, one way to formulate the content-based channelization problem is to map the multidimensional event space ES onto n event channels [24]. Each channel is hosted by a server or broker. The workload of a channel in the event space is given by a function that may be time-dependent. The problem is to find a set of n subregions $\lambda_j \subseteq ES$, $1 \leq j \leq n$, such that $ES = \cup_{j=1}^{n} \lambda_j$. Ideally the subregions are disjoint, but it may be that they are also overlapping. In this case the overlapping area should be minimized. In order to maximize efficiency, similar subscriptions should be placed to the same physical host if that is possible under the given constraints. Let $W(\lambda_j)$ give the workload contained by the cells of λ_j. The maximum parallel efficiency will be achieved when all the $W(\lambda_j)$ are equal.

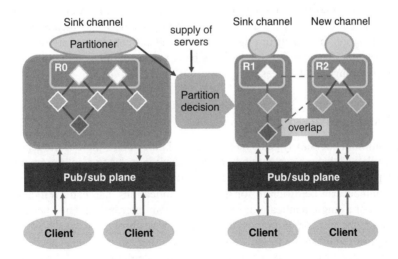

Figure 11.5 Overview of content-based channelization.

Different forms of load balancing are needed in order to ensure scalability:

- Load balancing of content to event channels.
- Splitting and combining channels when necessary.
- Relocation of a single event channel across different computers. This is accomplished by establishing a copy of the channel at the new server, updating the relevant lookup properties, and ensuring that there are no false negatives.

With this dynamic event channel scheme, a notification is forwarded at most once to an event channel for publication and subsequent forwarding to receivers. Subscriptions may be forwarded multiple times. The core of the this system is based on the concept of the event space, ES, which is updated when a channel is split. The event channels are automatically expanded and contracted by the system using covering and merging to cluster similar subscriptions together. Two key operations for load balancing event channels are: the *split* and *join* operations. Figure 11.5 illustrates the channelization technique.

The split operation involves the following phases:

1. Start split procedure when the channel becomes overloaded (a threshold is reached).
2. Find root set (minimal cover).
3. Split the root set into two parts while minimizing overlap and taking into account other constraints.
4. Establish new channel and transfer subscription state to this new server.
5. Independently merge the root sets of the old channel and new channel. Merging is optional and used to reduce the size and complexity of how the content managed by a channel is described.
6. Advertise the updated content space of the old channel, and the new content space.

The join operation involves the following phases:

1. Start the join operation when activity in the channel is below a given threshold.
2. Unadvertise the content space associated with the channel. This means that any pointers that are stored by clients or other servers to the current channel are removed. Traffic that this channel was responsible for falls back to the sink channel (and other channels with overlapping event spaces).

11.6 Reconfiguration

Reconfiguration in pub/sub systems has two crucial parts. First, there is the reconfiguration of the pub/sub middleware system and protocol stack in order to meet the requirements of the application and the environment. Such reconfiguration is necessary, because the operating environment can be different for different applications, and can even change during the execution of a single application, for example from Internet to ad hoc network connectivity. Second, the network topology in which pub/sub brokers forward event messages can change thus requiring the modification of the routing tables. This is needed in order to route around failed brokers and network nodes. The reconfiguration of the content routing graph and the routing table maintained by brokers needs to be efficient and maintain the safety and liveness properties of the system. The reconfiguration needs to be done in such a way that false negatives are avoided and the number of false positives is minimized [25].

We consider first solutions for the middleware component reconfiguration, and then for the reconfiguration of the routing topology.

11.6.1   Middleware Component Reconfiguration

GREEN [26] and REDS [27] are pub/sub systems that feature reconfiguration capabilities. Both systems are based on a component architecture, where each component is responsible for a specific functionality (matching, event routing or overlay management). The systems are specifications of a component framework enabling dynamic pluggable components. This allows such systems to cope with various deployment environments, for example fixed and mobile environments, and structured and unstructured overlay networks. We briefly consider how these systems support the mobile ad hoc environment through modular and pluggable components.

The REDS system has a pluggable topology manager component that supports wireless and ad hoc environments [28]. The wireless ad hoc protocol is designed for building content-based routing systems on top of the ad hoc network. The protocol is a modification of the MAODV protocol designed for multicast communications in ad hoc networks that is a modification of the AODV protocol. AODV is a well-known basic reactive routing protocol for these networks.

The GREEN system has a specific configuration for mobile ad hoc networks that addresses the following requirements:

- Inter-vehicle event delivery.
- End-to-end QoS monitoring.
- Content, proximity, and composite events.

The configuration is realized as a set of component plugins. The key components are the publish and subscribe components that are based on XML events, and the proximity plugin for taking the proximity into account in the subscriptions. Subscribers can utilize rule-based composite events using the provided composite event interface. QoS monitoring requirements for events and event channels can be also specified by the clients.

The event broker plugin needs to address the challenges of the mobile ad hoc network environment. The characteristics are different from the traditional fixed network environment, because the network topology is highly dynamic and the communications are prone to disconnections. This plugin implements a fully distributed broker overlay, in which the mobile nodes perform partial broker functionality. The plugin supports publisher and subscriber side event filtering. In this model, publisher define the event type and the scope for event delivery. Subscriptions are not propagated in this model, but rather they are evaluated at the subscriber side.

11.6.2   Topology Reconfiguration with Failures and Mobile Brokers

Generic topology reconfiguration involves a process that monitors the operating conditions and then when changes are detected triggers a topology update in order to find a more suitable configuration. This kind of reconfiguration is needed, for example, when links between brokers can fail, or when the brokers can move from one location to another.

Each node typically executes a reconfiguration process concurrently. The process is local, but the modified resource is global (the topology).

In the general case, the reconfiguration process is divided into phases:

- First, a node gathers information about the current network topology.
- Then, the node determines whether or not reconfiguration is required.
- Finally, the node reconfigures the topology.

Concurrency control is needed to prevent simultaneous changes to the topology by simultaneously executed reconfiguration processes. Depending on the solution this may require locking of those parts of the network that are undergoing reconfiguration.

Reconfiguration can be used to cluster subscribers and position publishers close to subscribers. A content-aware reconfiguration process adds a link to the graph that has the highest subscriber similarity. The link with the lowest similarity is removed. Figure 11.6 provides a simple example of this kind of reconfiguration based on link weights. If a link with a greater similarity measure than the current is not located, the process terminates. A content-based reconfiguration protocol with concurrency control has been proposed for SIENA [29].

The strawman protocol was one of the early proposals for pub/sub topology reconfiguration. The idea is to implement the reconfiguration based on the pub/sub primitives (the API) [25]. This approach implements the link removal operation needed for the reconfiguration with the unsubscribe operation. When a failed link is detected, the brokers remove subscriptions sent to the link and received from the link. In other words, when a broker detects a failed link, it assumes that the neighbour behind the link has unsubscribed all the events. This effectively stops information delivery across the broken link. When the new link is added, the new broker issues subscriptions and thus the routing tables are eventually updated to the correct state and information flows again.

Figure 11.7 illustrates this process. In subfigure A) the system is routing and forward events properly. In B) one of the links breaks and the reconfiguration process starts and identifies the new link. Now, the topology needs to be updated by the protocol. The

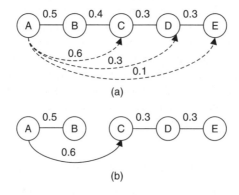

Figure 11.6   Example reconfiguration with interest weights.

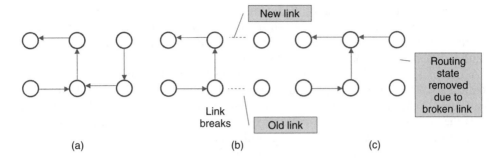

Figure 11.7   Example of the strawman protocol.

subfigure C) illustrates the topology after the reconfiguration, in which information flows through the new link and the old link has been removed.

Now, if the routing state is immediately updated by issuing unsubscriptions from the brokers that detect the broken link, the content delivery stops and results in false negatives. Therefore, the strawman approach was extended with delayed unsubscriptions that do not immediately stop the content delivery. Once a new link is added, the brokers connected by the link exchange routing table and thus mend the topology. The delayed strategy has been shown in simulation studies to reduce the overhead of the strawman algorithm up to 50%.

The delayed unsubscription technique can be improved based on the observation that the reconfiguration is restricted to the brokers on the path between the endpoints of the old new links. A protocol variant starts at the old link and then updates the routing tables along the reconfiguration path. This variant is more efficient than the previous ones, but requires that the path remains the same during the reconfiguration. Another variant protocol exchanges information between the brokers of the old and new links. The broker on the old link sends the subscription tables to the broker on the new link. The broker on the new link then determines what subscriptions need to be active on the new link.

The topology reconfiguration process is not very different from the reconfiguration necessary for mobile subscribers and publishers. The key is the path or subgraph that needs to be updated in order for information to flow to the new link. The notion of mobility-safety can be applied also for reconfiguration protocols in order to ensure that false negatives do not occur during the reconfiguration process.

11.6.3   Self-Organizing Pub/Sub with Clustering

Several reconfiguration solutions have been proposed that take the interest similarity of brokers into account and try to cluster the brokers in such a way that the notification delay and other involved costs are minimized. The problem of finding an optimal topology given a distribution of subscriptions and notifications has been proven to be NP-hard [30]. Therefore heuristic techniques have been proposed for adapting the broker overlay structure to the subscription and notification distributions, the processing costs, and the network topology.

The *maximum associativity tree* algorithm clusters brokers based on their interest profiles. The distributed algorithm takes the interests into account and builds an associativity metric based on the intersection of the interests. The brokers are then clustered in such a way that brokers with high mutual associativity are placed in the same cluster. Thus the algorithm attempts to minimize latency in notification delivery by reducing the average number of hops that the notifications travel [31].

A related system reduced the distance between brokers that receive many identical notifications. This is realized by keeping a cache of processed notifications and exchanging summaries of the cache between brokers to compute the interest overlap. This technique takes the communication and processing costs into account. The resulting algorithm is shown to be NP-hard and a heuristic algorithm is then developed to adapt the routing configuration to the current supply and demand of content [30].

11.7   Mobility Support

Mobility of subscribers and publishers is an important requirement for a pub/sub system. The location of a component may change for a number of reasons, for example physical relocation, load balancing, fault-tolerance, and security. Figure 11.8 illustrates the many faces of pub/sub mobility. The figure presents the three key mobility scenarios in the distributed environment that consists of access brokers and event domains. The domains are connected together either through the same pub/sub routing system or through a federation protocol that connects systems with different technologies and policies. The first

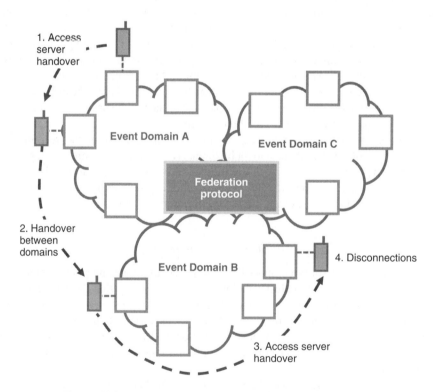

Figure 11.8   Different forms of mobility in a pub/sub system.

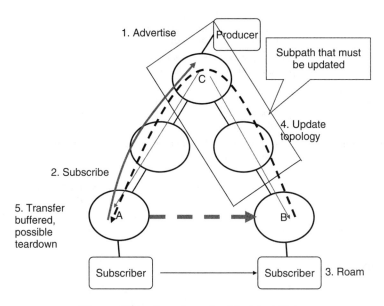

Figure 11.9 Overview of pub/sub mobility.

and the third case pertains to the mobility of the device that is subscribing and publishing events. The device roams from one access broker to another in the same administrative domain and this requires that the routing topology is updated to take the new location into account. In the second case, the device roams from one domain to another. In this case, the routing topology needs to be updated. This involves the federation protocol that may involve security, billing, and other issues. The fourth case involves disconnections, in which the device is connected to the same access broker, but cannot receive events all the time due to problems with the connections. The access broker has to buffer messages for the device. Broker mobility is not considered here and it can be supported through the reconfiguration protocols discussed in this chapter.

Figure 11.9 presents the generic mobility solution for pub/sub. The publisher has advertised content (1) and subscriptions have been connected with the publisher (2). Then a client system relocates from broker A to broker B (3). Now, part of the topology needs to be updated in order to ensure that content flows to the broker B. After the topology has been updated, the old subscription can be removed if there are no other subscribers at A (5).

It is evident that pub/sub mobility support requires buffering of messages for the subscriber, and that the routing topology is updated to reflect the changed locations of subscribers and publishers. This topology update should be as efficient as possible and it should not result in false negatives. Indeed, the mobility-safety property presented in Chapter 7 formalizes this requirement. One easy solution would be to route all messages through a designated home broker; however, this kind of routing is not efficient [32].

Pub/sub mobility protocols are similar in many ways to the more traditional network mobility protocols. Figure 11.10 compares four different solutions for mobility support, namely generic pub/sub, rendezvous pub/sub, i3, and SIP. The key characteristics

	Generic pub/sub	Rendezvous pub/sub	i3	SIP
Target	Subscription or advertisement	Subscription or advertisement	Any object	Session
Mechanism	Generic ping/pong synchronization	Rendezvous node	Rendezvous node	Home registrar
Update process	One or more nonfixed points	One or more fixed points	One fixed point	One fixed point
Buffering	Yes	Yes	No (additional service)	Yes with stateful proxy
Location privacy	Yes	Yes	Yes	No
Authentication	-	-	-	Yes

Figure 11.10   Comparison of different mobility support systems.

are the number of indirection points, mechanism how mobility is supported, the target of the update, buffering of packets and messages, location privacy, and authentication. For example, the SIP framework supports the mobility of sessions through the home registrar, whereas pub/sub configures the routing topology.

The JEDI system is an early example of a pub/sub system with mobility support with explicit move-in and move-out commands used by clients to relocate between brokers. The JEDI broker network is based on a tree-topology that has a potential weak point at the root of the tree. Elvin is another early example that supported disconnected operation with a centralized proxy that buffers messages for the clients. Elvin introduced patterns such as quench and the nondestructive notification receipt. The former pattern allows publishers to ask the network whether or not there are active subscribers for a given publication. This can be used to prevent unnecessary publications and thus save bandwidth. The latter allows subscribers to selectively retrieve messages and leave some messages at the server for later retrieval. This technique can support multiple devices and data sharing across the devices with the help of the broker. The Fuego system introduced a handoff protocol for event channels [33] and a family of subscriber and publisher handoff protocols that maintain mobility-safety. Next, we will investigate the general pub/sub mobility approach proposed for SIENA and a selection of more optimized handoff protocols that have been considered in the Rebeca, Fuego, and Padres projects.

11.7.1   Generic Pub/Sub Mobility

The SIENA event system was extended with generic mobility support that uses existing pub/sub primitives [34, 35]. The motivation of the generic protocol was that it works over various pub/sub systems and requires no changes to the basic API. This protocol

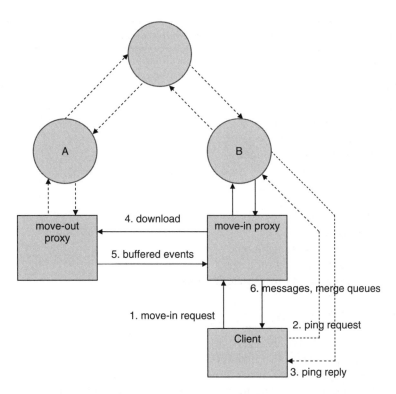

Figure 11.11 Move-in function in mobility support service.

was formally verified to maintain mobility-safety. On the other hand, the performance of the mobility support decreases, because mobility-specific optimizations are difficult to realize when the underlying topology is hidden by the API. Indeed, it has been shown that this generic solution has a high cost in terms of message exchanges and it may not be mobility-safe in all topologies [36].

The SIENA generic mobility support service, the *ping/pong protocol*, is implemented by proxy objects that reside on access routers. Figure 11.11 presents an overview of the process with six steps. In the first step, the client arrives to access point *B* from *A* and sends the *move-in* request to the new local proxy. in the second step, a ping request is sent and a response will be eventually received (3) from the old proxy. The response can also be called a *pong*. The pong message ensures that subscriptions are fully propagated from *B* to *A*. In the step phase, the client sends a *download* request for buffered events and in the fifth step the buffered events are sent to the proxy. In the final step, the client receives the messages and duplicates are removed.

We can observe that the handoff protocol proceeds in four distinct phases: first, the target subscribes to the relocated events, then the target and source synchronize by sending *ping* and *pong* events in order to ensure that the subscriptions have taken effect, the source unsubscribes, and finally any buffered events are relocated. In addition, there may be further costs triggered by changes in the subscription tables of the intermediate routers. Algorithm 11.1 gives an overview of this protocol.

Algorithm 11.1 The generic *move-out sub* and *move-in-sub*

Data: *Op* denotes the operation for the router.
begin
    switch *Op* **do**
        case *move-out-sub*
            Before handoff start:
            Advertise *pong* message.
            Subscribe *ping* message.
            Disconnect client.

            After handoff start:
            Wait for *ping*.
            Publish *pong*.
            Wait for transfer request.
            Send any buffered messages.
            Unsubscribe and reset state if necessary.
        end
        case *move-in-sub*
            Advertise *ping* message.
            Subscribe *pong* message.
            Activate client subscriptions.
            Publish *ping* every epoch *T* until *pong* is received.
            Start out-of-band state transfer by sending a message to *source*.
        end
      end
end

For subscription semantics the basic challenge with the ping/pong synchronization protocol is that the signalling messages are not guaranteed to be subscribed by other brokers on the network and hence the subscription messages will be introduced at every broker on the network. With advertisement semantics, the protocol works in a similar fashion, but the ping and pong messages need to be advertised, which also requires flooding the network.

If the client relocates faster than the ping message is propagated, it has to wait until the target receives the ping subscription. This requires that the pub/sub API allows to query the subscription status of the broker. The SIENA mobility solution does not require specific API support, because the ping messages are continuously resent [35], but this kind of behaviour further burdens the network.

Publisher mobility for subscription semantics does not require additional handoff functionality. Publisher mobility for advertisement semantics needs to implement a handoff protocol following the above model and it has a similar cost structure.

11.7.2 Graph Based Mobility with Optimizations

Content-based routing using subscription and advertisement semantics becomes challenging when the topology needs to be reconfigured with the introduction of mobile subscribers and publishers. In advertisement-based pub/sub networks a successful activation of a subscription may require that an advertisement is first propagated throughout the network, and then a connecting subscription is propagated on the reverse path. Furthermore, optimizations such as *covering* and *filter merging* [18, 37] lose information pertaining to the original interfaces that sent the messages.

Assuming that the pub/sub topology is an acyclic graph, the general graph-based mobility solution is to active only the inactive, also called incomplete, path between the old and new locations. This way the high cost of the above generic ping/pong mobility solution can be avoided. Unfortunately, this path is not known beforehand in content-based routing system and it needs to be discovered. This discovery is performed by sending an update message at the new location towards the old location. The update message will meet old subscriptions at some point due to the acyclic nature of the graph. The path from this meeting point to the new location needs to be updated so that subscriptions flow to the new location. After this the path from the meeting point to the old location can be removed if there are no other subscribers on that path. One major challenge for this kind of protocol is the discovery of the inactive path and the meeting point when covering and merging optimizations are used. These optimizations aggregate routing state and lose the origin interface identifiers that sent subscriptions. In addition, it is known where the destination is located in the topology, IP address cannot be directly utilized in choosing the next hop destination for the update message. Therefore the solution is to flood the update message towards covering interfaces. This selectively flooding has a high cost.

Three useful techniques have been proposed for improving mobility support for the acyclic graph topology [36]: overlay-based routing, rendezvous points, and completeness checking. Overlay addresses prevent the content-based flooding problem, because the destination overlay address can be used to prune unnecessary destinations. The overlay allows the system to cope with network-level routing errors and node failures. Rendezvous points simplify mobility by allowing better coordination of topology updates. There is only one direction where to propagate updates for a single rendezvous point. Completeness checking ensures that subscriptions and advertisements are fully established (complete) in the topology. This is needed to perform the covering optimization and avoid handoff altogether at the new location if the subscriptions are already covered.

The rendezvous based handoff proceeds as follows [36]: 1. The client ensures that the subscription is complete to the RP before mobility. 2. An update message is sent with a subscription towards the RP at the destination server. 3. The update reaches the RP, a message is sent to a that triggers the session transfer (4). Algorithm 11.2 outlines this process.

In general, utilization of rendezvous points will provide increased performance compared with semantics based on basic subscription or advertisement. In mobile solutions rendezvous architecture is utilized to guarantee unimpeded and full propagation of subscriptions and advertisements in the distributed environment. For this goal the a procedure called completeness checking is used which essentially ensures that false negatives do not arise from the topology updates. A similar process also functions in covering optimization aimed at detecting when the topology update should be omitted. The reason behind bypassing an update would be the observation that the subscriptions or advertisements have already been registered at the destination router after their relocation.

Algorithm 11.2 The rendezvous-point-based *move-out-sub* and *move-in-sub*

Data: *Op* denotes the operation for the router.
begin
    switch *Op* **do**
        case *move-out-sub*
            Before handoff start:
            Ensure completeness to RP.

            After handoff start:
            Wait for an update.
            Send any buffered messages.
            Unsubscribe and reset state if necessary.
        end
        case *process-at-RP-or-intermediate*
            if *update message is received and subscription complete* **then**
                Send update to inform that path is complete. Drop message.
            end
        end
        case *move-in-sub*
            if *subscription active* **then**
                if *waiting for completeness update* **then**
                    wait.
                end
                if *subscription is complete to RP* **then**
                    Start out-of-band transfer.
                end
            else
                Send update message to RP and activate subscription.
            end
        end
        end
    end
end

Publisher mobility differs from subscriber mobility, because state is not transferred. Rather, the path between the new *a* and old *b* routers must first be tested that the new advertisement issued by *b* has reached *a*. Then the path must be tested again to ensure that any subscription from *a* has reached *b*. After this, the publisher can be certain that any published messages are not missed by the subscribers. Publisher mobility is easy when the topology is complete. In this case, the covering optimization may also be performed. However, in the typical case when completeness cannot be assumed the path must be tested in both directions.

11.8   Congestion Control

Congestion is a major challenge for pub/sub systems, in which subscribers and brokers can become overwhelmed by the incoming traffic. The phenomenon is very similar to the case of network congestion; however, it differs in that there is no notion of end-to-end communications between the publishers and subscribers. The pub/sub network abstracts the identities of the communicating components. Thus we have the two different levels of solutions: the network level and the pub/sub network level.

TCP's congestion control solutions, such as the Slow Start, Congestion Avoidance, and the *Additive Increase Multiplicative Decrease (AIMD)* principle, are the baseline for implementing congestion control on the network level and for the Internet. Indeed, the brokers typically use TCP to communicate and TCP provides congestion control and flow control between two brokers. Separate solutions are needed for the case of congestion control in the pub/sub network across several brokers. In addition to mitigating congestion, the pub/sub congestion control solution needs to take into account retransmissions due to lost event messages. Retransmissions contribute to congestion and this needs to be taken into account in the solution.

Figure 11.12 illustrates the four congestion control techniques for pub/sub systems:

- Simple backpressure, in which upstream senders need to block until the downstream service can catch up. This technique has been applied in the Cobra RSS dissemination system that implements it a 1 MB buffer for each upstream service [38].
- Rate limited subscriptions allow subscribers to specify notification rate limits that cannot be exceeded. For example, a subscriber can specify that only ten messages per second are allowed. Rate limits can be easily combined with the general content-based filtering model with covering relations and filter merging. This is accomplished by treating the rate limit as one of the attribute filters in subscriptions. The rate limits can also be included in advertisement messages sent by publishers.
- Congestion notification implements an explicit congestion signalling mechanism by which the subscribers and the pub/sub network can inform the publisher that the sending rate should be slowed down.
- Rerouting is a process by which the network routes around a congested broker or a subnetwork. Rerouting requires support for pub/sub network reconfiguration.

In all cases, rate measurement and congestion detection are important. Notification rate can be measured by simply counting the number of events arriving at the current node. Congestion is detected locally by monitoring the queue sizes for incoming messages. When the queue size grows beyond a set limit, congestion has occurred. Congestion outside the subscriber and broker are detected with explicit congestion signalling techniques.

11.8.1   Rate-Control Using Posets

Rate-control policies aim to improve the scalability of content-based systems. Rate-control rules define how many notifications per second or time unit should be forwarded to a

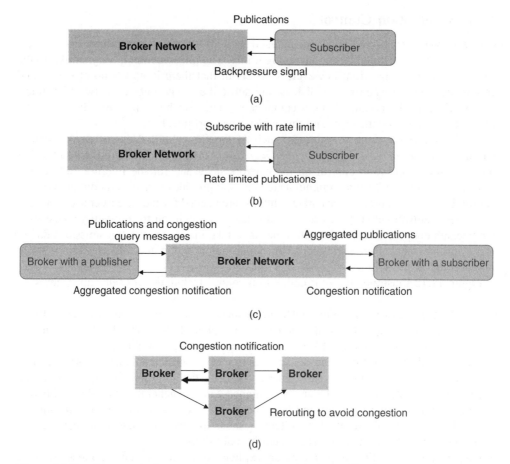

Figure 11.12 Four congestion control techniques for pub/sub. (a) Simple backpressure; (b) Rate limited subscriptions; (c) Congestion notification; (d) Rerouting.

subscribing interface. Some rate-control rules are set by system designers and policies, and some rate-control rules are set by applications.

Rate control rules are represented using attribute filters and thus parts of filters [39]. The rate control rules support the covering relation and are also mergeable. The covering relation is a simple inequality, where a bigger rate-value covers smaller rate-values. For example: (15 events/s) $\sqsupseteq$ (7 events/s).

Therefore, the rate-control extension may be used with the forest or poset-based matchers of Chapter 8 with some modifications. For each filter in the data structure, the notification rate per time interval needs to be tracked. If the rate limit value has been exceeded, it is not necessary to check the filters rooted at that node, because they have also exceeded their limits. This provides a convenient way to prevent unnecessary messaging between brokers. To balance the forest properly during insertion, a covering subgraph is selected with the closest rate-control filter.

Broker and subscription replication complicates event rate determination, because the rate limiters are distributed. Given that there are k brokers with replicated subscriptions and total event publication rate of x events per second. Assuming that the notifications are distributed uniformly to replicated hosts the rate is $\frac{x}{k}$ per broker.

The time-window for which the rate is monitored is important, and this information is typically required at least for the root set of the poset or forest and in some cases for each subscription. Monitoring of the rate for the root-set requires that the covering subscriptions of the root set are updated (their counters are increased) during matching. When the monitored time-period restarts, the counters are reset and the old values may be stored into a history. The current rate value may not be very interesting for a load balancer, but rather knowledge of the recent behaviour of the rate is important and may be used to extrapolate future behaviour, for example, using moving averages or other statistical methods.

11.8.2   Explicit Signalling

A congestion control mechanism for pub/sub systems was developed in conjunction with the Hermes system by Peter Pietzuch [9]. This solution is based on six requirements for such a system: support for bursty traffic, queue size management, recovery control for recovering lost event messages with NACK messages, robustness for preventing malicious clients from harming the system, and publishers are rate limited in a fair manner.

The developed congestion control system has two distinct subsystems:

- A publisher driven congestion control algorithm that controls the publication rate. The rate is adjusted based on a congestion metric. The metric is the rate of publication messages at brokers that host subscribers. Publishers send explicit congestion status requests to downstream subscribers, which then send back congestion status updates. The status updates are aggregated by the intermediate brokers to prevent the flooding of the publisher.
- A subscriber driven congestion control algorithm that manages the rate at which a broker hosting subscribers requests lost events with NACK messages. This algorithm regulates the rate of NACK messages during recovery so that it is appropriate given the available system resources.

11.8.3   Rerouting to Avoid Congestion

Content rerouting can be used to avoid congestion by choosing content delivery paths that avoid the congested parts of the network. The reconfiguration protocols examined in this chapter as well as the mobility protocols are basic mechanisms for implementing content rerouting. Example solutions are:

- The baseline solution is to build several multicast trees that are totally or partly node disjoint. Then it is easy to switch from one multicast tree to another when congestion is detected. The creation and maintenance of these multicast trees requires its own mechanism.
- The multicast trees can be built with a DHT system and rendezvous points that are roots of multicast trees. Replicating rendezvous points and when the primary rendezvous point becomes congested, move traffic to a replica. The multicast tree rooted at the

replica rendezvous point should not have too much overlay with the primary multicast tree to avoid congestion.

- Clustering subscribers and publishers at the edge of the network to avoid hotspots in the network.

11.9   Evaluation of Pub/Sub Systems

The evaluation of pub/sub systems is important to ensure that they work properly and achieve the necessary performance, reliability, security, and other nonfunctional requirements. We briefly consider different evaluation strategies for these systems. Evaluation strategies range from formal methods to software engineering and empirical measurements:

- Formal verification of system properties, such as safety and liveness.
- Analytical models of the system that estimate various performance and scalability metrics.
- Simulation of the system with a discrete event simulator such as ns3 and Omnet++. Simulation can be realized on many abstraction levels. Typically detail, such as network level packet routing, needs to be abstracted in order to simulate large-scale overlay networks.
- Prototype implementation in a physical test network. The network can be a cluster or it can be a global testbed such as PlanetLab.

In addition to the evaluation strategy, the parameters of the evaluation and the analyzed metrics need to be defined. One challenge for pub/sub is that there are many possible applications and very different environments in which they can be executed, for example a chat application can run on a desktop PC and a mobile phone. The network topology can range from a small-scale cluster to a wide-area environment with thousands of routers. For example: for JMS server performance the following important parameters have been defined for cluster-based experiments [40]:

- Expressiveness: Number of filters, filter installation time, filter types, number of topics, complex filters and their structure.
- Server utilization: I/O access utilization, local memory, CPU utilization, rendundacy and resilience, backup-capacity.
- Message throughput: Size of message, number of filters, number of clients, message mode (persistent, durable, transacted).
- Network utilization: Number of network connections (TCP connections and JMS sessions), delay and loss properties of the network, capacity of subscribers, replication grade.

The environment thus needs to be taken into account as well as the workload for the distributed system.

The workload of the system typically includes issues such as:

- Network configuration, such as the network topology, initial broker topology, and links between brokers and their properties.
- Broker configuration, such as maximum queue lengths, broker policies, etc.
- Arrival and departure of subscribers.

- Arrival and departure of publishers.
- Event publication specifications and rates of publishers (distributions).
- Subscription specifications and subscription durations for subscribers (distributions).
- Mobility related parameters (how often mobility occurs, duration of disconnection, where subscriber/publisher is roaming to, etc.).
- In security related experiments the security model, the adversary model, etc.

Typically several different evaluation strategies are needed in order to understand how a component or a system behaves under different scenarios. For example: a new pub/sub routing algorithm can be formally analyzed to prove that it does not result in false negatives even in the presence of reconfigurations (under certain assumptions). The assumptions should not be overly limiting in order for the algorithm to be applicable in real-life settings. This analysis does not yet give a picture of the performance of the algorithm, only about its correctness with certain assumptions. Therefore experiments are needed that evaluate how to algorithm works in a dynamic environment with realistic parameters. These experiments are typically conducted with simulation studies and experiments based on prototype implementations.

An experiment is a set of single measurements where typically one parameter is changed and the others are left unchanged. Each measurement is replicated several times in order to increase confidence in the resulting mean. The results of the replications can then be used to calculate the standard deviation and the confidence level for the measurement. The number of replications need to be set in such a way that the standard deviation of the resulting mean stabilizes.

The next step is then to simulate the algorithm in medium and large scale networks to demonstrate that it is not only correct but also efficient. The topology should represent the realistic operating environment for the algorithm (e.g. ad hoc network, Internet, . . .). The simulation results can then be analyzed, typically against a benchmark algorithm, to assess the performance and feasible parameter space for the system. Analytical models can also be very useful in analyzing systems; however, it may be difficult to obtain closed form solutions for systems with many components. One challenge is the number of parameters that can grow with complex systems. Statistical techniques can be utilized to assess the impact of different parameter combinations, such as the factorial design experiment technique.

The analytical and simulation results can then be compared to empirical measurements made with a prototype system and deployment. This evaluation method requires a software, and in some cases hardware, implementation of the system. The implementation is then configured for the experiment, and the experiment is conducted either in a controlled or uncontrolled environment. A controlled environment is instrumented for the experiment and the necessary system properties and metrics can be measured and then evaluated. An uncontrolled environment, such as people downloading and running the software, can support more large-scale experiments, but it is subject to more measurement errors and unknowns.

Experimental measurements are necessary for instrumenting analytical and simulation models. Thus it is common to conduct micro-benchmarks with simulations and prototypes in order to understand the feasible parameter space and fine-tune the parameters for larger studies. Thus one or more evaluation strategies are needed to assess the characteristics of a solution.

11.10   Summary

Figure 11.13 presents a summary of solutions discussed in this chapter. We have addressed important topics such as security, composite events, reconfiguration, load balancing, congestion control, and mobility support. Many of the topics pertain to the pub/sub routing system and aim to improve its efficiency, reliability, and availability to name some important properties.

Security solutions are needed for protecting the different parts of the pub/sub system: subscribers, publishers, and brokers. A key component is the distributed key distribution mechanism as well as the guards that utilize the keys. Sophisticated solutions are needed to be able to secure routing tables and the forwarding process. EventGuard is a comprehensive security suite that offers also content-based security solutions. Other well-known technique is role-based security that offers indirection between users and privileges through roles.

Composite events are necessary for identifying interesting sequences of events. A composite event language is needed that supports expressive operations, such as temporal conditions between events. A basic composite event system is centralized, but this solution is limited in scalability. Hence different distribution strategies have been investigated that divide a composite event query into smaller parts that are then distributed in the network. The aim of the distribution is to place the composite query parts close to the publishers of the events in order to minimize overhead. Example systems include CEA, Rapide, and Padres.

Reconfiguration is used to adapt the middleware system to new operating conditions. We have observed that pub/sub needs to be adapted for different environments, for example ad hoc networks and the Internet. Application requirements also vary and they may require changes to the pub/sub middleware as well. Therefore middleware level reconfiguration

Solution	Examples	Description
Composite events	Rapide, CEA, Padres	Event correlation with a composite event language.
Security	CEA, Hermes, EventGuard	Security solutions for the pub/sub system: access control, various guards, securing routing and forwarding.
Reconfiguration	REDS and GREEN	An algorithm that is able to configure a pub/sub system's topology at runtime based on a trigger condition. The aim is to find an improved routing configuration with lower cost in terms of the given cost metric
Load balancing	Padres, Fuego	This technique aims at a uniform load distribution across the distributed pub/sub system or part of it. This addresses the problem of a single broker or router becoming overloaded with subscription or event messages.
Congestion control	Hermes, Padres	Congestion control solutions for preventing publishers from overwhelming the network and the subscribers.
Handoff	SIENA, Rebeca, Fuego, Padres	A handoff protocol is used to transfer one or more subscribers or publisher from a source broker to a destination broker. This requires that the routing topology is updated. Subscribers and publishers each need their own protocol variant.

Figure 11.13   Summary of solutions.

is necessary to support multiple operating environments. This is achieved with a modular core architecture that supports various pluggable components, such as routing modules, merging and other optimizing modules, and various transport modules.

In addition to middleware reconfiguration, pub/sub routing topology needs to be configured when changes are detected. For example, a link may fail requiring that a new link is actives that replaces it. Another example is broker mobility, in which brokers can move from one part of the physical network to another. Thus a protocol is needed that actives new links and tears down old links. The protocol needs to be designed and implemented in such a way that it does not lose messages and that it can operate in a concurrent environment.

Subscriber and publisher mobility can be supported in many ways. They can be supported by buffering events at designated routers and then allowing the subscribers to retrieve them later. This effectively supports disconnected operation, but it does not allow subscribers to roam between routers. Several pub/sub mobility protocols have been proposed for supporting roaming subscribers and publishers. We examine the generic ping/pong protocol proposed for SIENA, and then a acyclic graph based protocols and solutions. The mobility protocols update the routing topology and are similar in spirit to the topology reconfiguration protocols.

Congestion control is an important part of a pub/sub system, because it prevents publishers from overwhelming the network and the subscribers with content. We examined different solutions for this problem including backpressure and explicit signalling based schemes.

References

1. Wang C, Carzaniga A, Evans D and Wolf A (2002) Security issues and requirements for internetscale publish-subscribe systems. *Proceedings of the 35th Annual Hawaii International Conference on System Sciences (HICSS'02), Volume 9*, pp. 303–11. HICSS '02. IEEE Computer Society, Washington, DC.
2. Bacon J, Eyers D, Moody K and Pesonen L (2005) Securing publish/subscribe for multi-domain systems. *Proceedings of the ACM/IFIP/USENIX 2005 International Conference on Middleware*, pp. 1–20 Middleware '05. Springer-Verlag New York, Inc., New York, NY.
3. Fiege L, Zeidler A, Buchmann A, Kilian-Kehr R and Mühl G (2004) Security aspects in publish/subscribe systems. In *Third Intl. Workshop on Distributed Event-based Systems (DEBS04*. IEEE.
4. Khurana H (2005) Scalable security and accounting services for content-based publish/subscribe systems. *Proceedings of the 2005 ACM Symposium on Applied Computing*, pp. 801–7, SAC '05. ACM, New York, NY.
5. Hirsch F, Solo D, Reagle J, Eastlake D and Roessler T (2008) *XML Signature Syntax and Processing*, 2nd edn. W3C recommendation, W3C. http://www.w3.org/TR/2008/REC-xmldsig-core-20080610/.
6. Reagle J and Eastlake D (2009) XML encryption syntax and processing version 1.1. W3C working draft, W3C. http://www.w3.org/TR/2009/WD-xmlenc-core1-20090730/.
7. Srivatsa M and Liu L (2005) Securing publish-subscribe overlay services with EventGuard. *CCS '05: Proceedings of the 12th ACM Conference on Computer and Communications Security*, pp. 289–98. ACM, New York, NY.
8. Corman AB, Schachte P and Teague V (2007) QUIP: a protocol for securing content in peer-to-peer publish/subscribe overlay networks. *ACSC '07: Proceedings of the Thirtieth Australasian Conference on Computer Science*, pp. 35–40. Australian Computer Society, Inc., Darlinghurst, Australia.
9. Pietzuch PR (2004) *Hermes: A Scalable Event-Based Middleware*. PhD thesis. Computer Laboratory, Queens' College, University of Cambridge.
10. Huang Y and Garcia-Molina H (2001) Publish/subscribe in a mobile enviroment. *Proceedings of the 2nd ACM International Workshop on Data Engineering for Wireless and Mobile Access*, pp. 27–34. ACM Press.

11. Al-Muhtadi J, Campbell R, Kapadia A, Mickunas MD and Yi S (2002) Routing through the mist: Privacy preserving communication in ubiquitous computing environments. *ICDCS '02: Proceedings of the 22nd International Conference on Distributed Computing Systems (ICDCS'02)*, p. 74–. IEEE Computer Society, Washington, DC.

12. Gedik B and Liu L (2008) Protecting location privacy with personalized k-anonymity: Architecture and algorithms. *IEEE Trans. Mob. Comput*. **7**(1): 1–18.

13. Pietzuch PR, Shand B and Bacon J (2003) A framework for event composition in distributed systems. *Proceedings of the 4th International Conference on Middleware*, pp. 62–82.

14. Pietzuch PR, Shand B and Bacon J (2004) Composite event detection as a generic middleware extension. *IEEE Network* **18**(1): 44–55.

15. Jacobsen HA, Cheung AKY, Li G, Maniymaran B, Muthusamy V and Kazemzadeh RS (2010) The PADRES publish/subscribe system. *Principles and Applications of Distributed Event-Based* Systems, pp. 164–205.

16. Tarkoma S (2008) Fast track article: Dynamic filter merging and mergeability detection for publish/subscribe. *Pervasive Mob. Comput*. **4**: 681–96.

17. Crespo A, Buyukkokten O and Garcia-Molina H (2003) Query merging: Improving query subscription processing in a multicast environment. *IEEE Trans. Knowl. Data Eng*. **15**(1): 174–91.

18. Mühl G (2002) *Large-Scale Content-Based Publish/Subscribe Systems*. PhD thesis. Darmstadt University of Technology.

19. Salo J (2010) Offloading content routing cost from routers. Master's thesis, University of Helsinki.

20. Cheung AKY and Jacobsen HA (2006) Dynamic load balancing in distributed content-based publish/subscribe *Middleware*, pp. 141–61.

21. Cheung AKY and Jacobsen HA (2010) Load balancing content-based publish/subscribe systems. *ACM Trans. Comput. Syst*. **28**(4): 9.

22. Cheung AKY and Jacobsen HA (2010) Publisher placement algorithms in content-based publish/subscribe. *ICDCS*, pp. 653–64.

23. Wang YM, Qiu L, Achlioptas D, Das G, Larson P and Wang HJ (2002) Subscription partitioning and routing in content-based publish/subscribe networks. *16th International Symposium on DIStributed Computing (DISC'02)*, October 2002, Toulouse, France.

24. Tarkoma S (2008) Dynamic content-based channels: meeting in the middle. *Proceedings of the Second International Conference on Distributed Event-Based Systems*, pp. 47–58. DEBS '08. ACM, New York, NY.

25. Picco GP, Cugola G and Murphy AL (2003) Efficient content-based event dispatching in the presence of topological reconfiguration. *Proceedings of the 23rd International Conference on Distributed Computing Systems*, pp. 234–44. ICDCS '03. IEEE Computer Society, Washington, DC, USA.

26. Sivaharan T, Blair G and Coulson G (2005) GREEN: A configurable and re-configurable publish-subscribe middleware for pervasive computing. *Proceedings of DOA 2005*, pp. 732–49.

27. Cugola G and Picco GP (2006) REDS: a reconfigurable dispatching system. *Proceedings of the 6th International Workshop on Software Engineering and Middleware*, pp. 9–16. SEM '06. ACM, New York, NY.

28. Mottola L, Cugola G and Picco GP (2008) A self-repairing tree topology enabling content-based routing in mobile ad hoc networks. *IEEE Transactions on Mobile Computing* **7**: 946–60.

29. Baldoni R, Beraldi R, Querzoni L and Virgillito A 2007 Efficient publish/subscribe through a self-organizing broker overlay and its application to siena. *Comput. J*. **50**: 444–59.

30. Jaeger MA, Parzyjegla H, Mühl G and Herrmann K (2007) Self-organizing broker topologies for publish/subscribe systems. *Proceedings of the 2007 ACM Symposium on Applied Computing*, pp. 543–50. SAC '07. ACM, New York, NY.

31. Baldoni R, Beraldi R, Querzoni L and Virgillito A (2004) A self-organizing crash-resilient topology management system for content-based publish/subscribe. *3rd International Workshop on Distributed Event-Based Systems (DEBS'04)*, Edinburgh, Scotland.

32. Burcea I, Jacobsen HA, de Lara E, Muthusamy V and Petrovic M (2004) Disconnected operation in publish/subscribe middleware. *Mobile Data Management*, pp. 39–50.

33. Tarkoma S, Kangasharju J and Raatikainen K (2003) Client mobility in rendezvous-notify. *Intl.Workshop on Distributed Event-Based Systems (DEBS'03)*.

34. Caporuscio M, Carzaniga A and Wolf A (2002) An experience in evaluating publish/subscribe services in a wireless network. *WOSP '02: Proceedings of the 3rd International Workshop on Software and Performance*, pp. 128–33. ACM Press, New York, NY.

35. Caporuscio M, Carzaniga A and Wolf AL (2003) Design and evaluation of a support service for mobile, wireless publish/subscribe applications. *IEEE Transactions on Software Engineering* **29**(12): 1059–71.

36. Tarkoma S and Kangasharju J (2007) On the cost and safety of handoffs in content-based routing systems. *Computer Networks* **51**(6): 1459–82.

37. Tarkoma S and Kangasharju J (2005) Filter merging for efficient information dissemination. *CoopIs LNCS 3760)*, pp. 274–91. Springer.

38. Rose I, Murty R, Pietzuch P, Ledlie J, Roussopoulos M and Welsh M (2007) Cobra: content-based filtering and aggregation of blogs and rss feeds. *Proceedings of the 4th USENIX Conference on Networked Systems Design and Implementation*, p. 3–. NSDI'07. USENIX Association, Berkeley, CA.

39. Mühl G, Ulbrich A, Herrmann K and Weis T (2004) Disseminating information to mobile clients using publish-subscribe. *IEEE Internet Computing* **8**: 46–53.

40. Henjes R (2010) *Performance Evaluation of Publish/Subscribe Middleware Architectures*. PhD thesis. University of Würzburg.

12

Applications

This chapter considers applications of pub/sub systems and technology. We consider the role of pub/sub as an enabler of a cloud computing platform, SOA and a generic XML-broker for enterprise applications, Facebook Messages and Chat, Pubsubhubbub, Complex Event Processing, online advertisement and games. From the mobile environment, we examine the Apple push notification service and the Internet of Things environment. The patterns and solutions used by the applications are discussed.

12.1 Cloud Computing

Internet services have become ubiquitous, and the recent trend towards building massive datacenters is part of the current network evolution. The evolution pertains not only to being able to support millions and even billions of users, but also to how data connections and network routes are configured and maintained across these hubs of data processing.

From one viewpoint, Internet computing is returning to the old utility computing paradigm of time-sharing systems. This time, however, the latest development has a twist – peer-to-peer networking offers the possibility of offloading computation and storage across the Internet. This draws a landscape in which distributed applications and services collaborate with each other in order to realize the best possible service while minimizing metrics such as monetary cost, energy consumption, or carbon footprint. In order for this cloud-based service ecosystem to work, easy access protocols, data representation formats, well-defined interfaces, and efficient data distribution and dissemination techniques are needed. Indeed, some solutions have already been deployed, such as those used by Google and Amazon in their service platforms.

Although cloud services have emerged, the landscape is still fragmented and significant research and development is needed for scalable solutions – especially across the Internet, and for mobile devices. Taking the policies and security requirements of multiple stakeholders into account is a significant challenge. Moreover, given that billions of mobile phone uses will be part of the Internet, the partitioning of service logic between devices and peers, and datacenters becomes important.

Publish/Subscribe Systems: Design and Principles, First Edition. Sasu Tarkoma.
© 2012 John Wiley & Sons, Ltd. Published 2012 by John Wiley & Sons, Ltd.

The services of Cloud computing can be divided into three categories:

1. Software-as-a-Service (SaaS), in which a vendor supplies the hardware infrastructure, the software product and interacts with the user using a portal.
2. Platform-as-a-Service (PaaS), in which a set of software and development tools are hosted by a provider on the provider's infrastructure, for example, Google's App-Engine.
3. Infrastructure-as-a-Service (IaaS), which involves virtual server instances with unique IP addresses and blocks of on-demand storage, for example, Amazon's Web services infrastructure.

All of the three above categories require signalling and messaging between the virtual machines, services, and applications. Pub/sub is a frequently employed communication solutions also in this environment, and it is extensively used to connect cloud platform components. In the following, we examine state of the art cloud pub/sub solutions and give examples of their usage.

12.1.1   Pub/Sub for Cloud

Publish/Subscribe based cloud applications have increased in importance lately. This is a rational consequence of their integrative power as tools in business to business communication. Cloud hosted intermediary services enable efficient decoupling of publishers and subscribers from each other: they may reside in different networks, need not to be publicly addressable or in fact not be able to talk directly to each other. Because of the inherent cloud infrastructure features all intermediary services are able to correctly scale according to the number of publishers and subscribers. The cloud also has the automatic ability to act as firewall for communication brokering. Note that an explicit permission is needed for the publishers and subscribers to connect the intermediary service, and for sending or receiving messages. Workflow patterns like this are useful because they can:

- Relay events among applications in distributed computers.
- Update business system data.
- Move data between data storage nodes.

12.1.2   The Windows Azure AppFabric Service Bus

An interesting cloud application is the AppFabric Service Bus running in Microsoft datacenters; it belongs to the Microsoft PaaS cloud strategy Azure. The system is based on a suite of middleware services called Azure AppFabric as well as development tools in the form of the Azure SDK and Visual Studio Toolset. The Azure AppFabric Service Bus is responsible for connecting various components together and it exposes Web services through a public endpoint in an Azure datacenter. AppFabric brokers relay messages in the cloud to services running behind firewalls or NAT devices. All endpoints are secured by AppFabric which utilizes a claim based security model (in this case the Access Control service). Other features present are for instance federated authentication for listening and sending in the cloud and a specific naming system for the endpoints within the

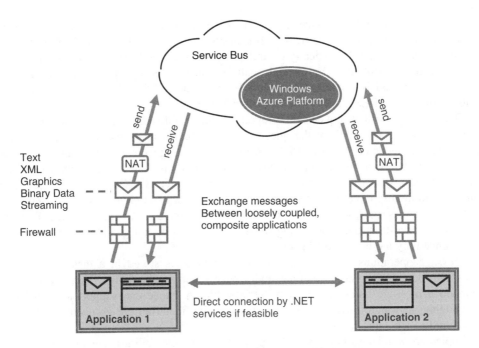

Figure 12.1 Overview of messaging with Azure.

cloud. Figure 12.1 presents the general communications model of the system. Furthermore, AppFabric allows several mutual communication options and supports a service registry for integrated applications.

Integration of applications with the cloud services was originally supported by AppFabric with a relay service which had dual operability: either using a message buffer in the cloud accessible through a REST API or utilizing the *Windows Communications Framework (WCF)* model with special channels for messaging to the relay service. The latter model makes the communication with the relay binding essentially transparent for applications as WCF manages all messaging details at channel level. The buffering is no problem as the message buffer is temporary and the messages disappear when they expire or become consumed.

New features in the AppFabric include the availability at the service bus level of durable messaging that can be either reliable message queuing or durable publish/subscribe messaging. These two differ in the number of separate parties consuming a message published in the service bus. Furthermore the pricing model used by AppFabric is based on the number of used connections.

The main features of AppFabric are:

- The decoupling generated by the service bus is strong and based on service namespaces. A single service namespace creates isolation for a set of endpoints and multiple service namespaces can be used by an AppFabric customer allowing two applications (associated with a single account) listening on separate namespace addresses.
- The communication options include the REST API, the.NET APIs and the WCF bindings. These make AppFabric easy to use from applications.

The AppFabric v2.0 Service Bus supports both queue and topic based pub/sub communication scenarios. The Service Bus allows multiple publishers and subscribers to communicate through the logically centralized system, and subscribers can use filters for customized subscriptions. The pub/sub features can be used by developers with several programming models, for example the.NET API, a new WCF binding for Service Bus Messaging, and a REST API for Queues and Topics.

Queues are persistent sequenced buffers into which publishers send messages. The messages are stored at the tail of the queue and the subscribers receive messages from the head. Thus a queue can have multiple receivers competing to receive messages. A queue has a cursor that points to the current message. The cursor is shared by the subscribers, and moves when a subscriber receives a message. The Service Bus Queues system provides methods to modify the default behaviour of a Queue, for example to modify the temporal visibility of messages, and deferral of message processing. A queue supports either at-most-once or at-least-once messaging semantics. Queues do not guarantee message order and hence they are best effort FIFO.

Message operations rely on various system properties stored in a message, such as message identifier, session identifier, and arbitrary name-value pairs. Each message can contain a message body that must be serializable. A queue can support session based messaging, in which messages with the same session identifier are grouped together and delivered to the same receiver.

Topics are similar to Queues and support the same capabilities, but in addition they support multiple heads that are referred to as subscriptions. Each subscriber receives distinct copies of the messages. Each subscription has its own cursor that is moved when a message is received by the associated subscriber. A subscription can support message filtering, in which a subset of messages in the queue are delivered to the subscriber. After a topic is created, it is necessary to add subscriptions to it. Each subscription has a unique identifier and can have an associated filter and an action. Several different filter formats are supported, for example simple matching based on the correlation identifier and an SQL-92 style syntax that can be used to create a filter using the name-value pair properties of the message. Once a topic has been created it is necessary to add one or more subscriptions to it since, although senders send messages to the topic, subscribers receive messages from a subscription not the topic. The Topic.AddSubscription() method is used to add a subscription to a topic. Each subscription is identified by a unique name. When the subscription is added, a RuleDescription can be used to specify a filter and action for the subscription.

Both Queues and Topics are addressed by using the Service Bus namespace. The namespace relies on URIs, and topics/queues and subscriptions are identifier with the familiar hierarchical notation. For example, *sb://example.service.com/News/Subscriptions/Tech* consists of the namespace *example*, a topic named *News*, and a subscription names *Tech*.

The Service Bus supports three forms of messaging, direct messaging, relay messaging, and brokered messaging. In direct messaging, the publisher and subscriber communicate directly, whereas in relay messaging they communicate through a relay. Both direct and relay messaging are subscriber driven, and the subscribers determine the message consumption rate. If a subscriber cannot process messages fast enough, the publisher has to slow down the sending rate. The brokered messaging contrasts these two forms of messaging, because it relies on a high performance intermediary that stores the messages.

The subscriber can then retrieve messages from the message queue maintained by the intermediary. With brokered messaging, the publisher and subscriber can scale independently of each other.

12.1.3 Amazon Simple Queue Service (SQS)

Amazon Simple Queue Service (Amazon SQS) is a distributed message queue service created by Amazon.com in 2006 as part of their cloud platform. SQS supports message queue operations and provides a Web service interface for developers. The main goal of the SQS service is to offer a scalable hosted message queue that then can be used to connect various service components. This is illustrated by Figure 12.2. This allows developers to easily move data between different application components and keep the components synchronized. SQS can be used together with the virtualization solution Amazon EC2 (Elastic Computer Cloud) in order to realize an automated workflow system. SQS differs from JMS and other message queue solutions, because it is a hosted service maintained by Amazon that sells the service with a pay-per-use business model.

SQS messages can contain any kind of data, and the message bodies are limited to 64KB. If larger messages need to be sent, the developers have to divide the larger data into smaller segments for transmission with multiple messages.

The SQS service offers the at-least-once semantics for message delivery. The message are stored on multiple servers in order to achieve reliability and availability. The system does not guarantee message ordering, and this feature needs to be implemented by the application based on information contained in the messages.

SQS can be used in many ways, for example:

- Combine SQS with other Amazon cloud services in order to connect application components.
- Coordinate distributed processes by placing task description messages to SQS queues.
- Store Web application notifications in SQS queues, and allow Web browser-based applications to retrieve the messages.

As an example, we can consider a video transformation website that uses the Amazon services for data processing. The website allows users to submit videos for processing. The videos are stored in the Amazon S3 storage service. Once a video has been stored, a message is sored in an SQS queue denoting incoming messages for the transformation service. The message contains details pertaining to the requested operations, namely the transformation operation. The transformation engine runs in an EC2 instance and reads the message from the incoming queue, and then starts the transformation process. The modified video file is stored in the Amazon S3 and the pointer is sent in a completion message through the SQS message queue. The Web site can then read the completion message from the queue and then inform the user that the video is ready.

12.1.4 PubNub

PubNub is a relatively new push service hosted in the cloud.[1] It is currently run in the Amazon EC2 infrastructure, and the system provides a set of APIs for various programming languages

---

[1] http://www.pubnub.com/.

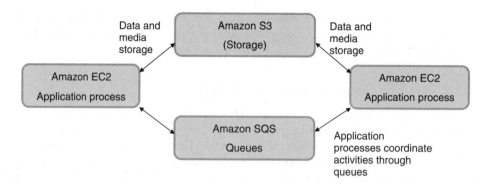

Figure 12.2   Amazon hosted applications communicating with SQS.

for pushing or receiving messages. PubNub pushes data to the different subscribers using a long-lived client-initiated connection implemented with HTTP. Subscribers listen for messages on an specific channel, and then receive messages when they are published on the channel. A message is forwarded to all subscribers on the channel.

12.2   SOA and XML Brokering

Service Oriented Architecture (SOA) is a software architecture where applications are designed after business processes and implemented as interoperable services. SOA consists of a wide variety of services that can be used to support, manage, and coordinate business processes. The aim is a loose coupling of components and services that supports interoperability in many different forms. SOA separates functions into services that are available over a network. The loosely coupled nature in together with interoperable service definitions supports the combination and reuse of the services in the creation and execution of business applications.

SOA is heavily based on message-oriented middleware and pub/sub. SOA components communicate by sending and receiving messages with the help of a message broker or bus component, the Enterprise Service Bus (ESB). The message passing primitive and the broker connect the different component together and realize the loosely coupled nature of the system. The basic message oriented interactions are the basis of more complex interactions.

SOA system typically employ design patterns that have been found to be efficient and work well for this environment. We discussed patterns in Chapter 4, and the Enterprise Integration Patterns have been specifically designed for SOA and the integration of enterprise systems. Several guiding principles can be identified for the development and usage of SOA:

- Services creation, monitoring, composition, and provisioning.
- Reuse, interoperability, componentization, modularity, and composability.
- Standards based components and messaging.
- Loose coupling of the components.

Various architectural features influence the design and implementation of these systems. These features include abstraction of internal service logic, autonomy of encapsulated

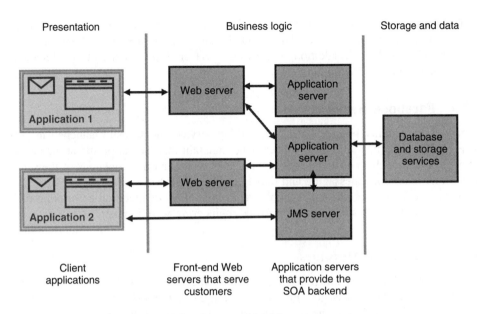

Figure 12.3   An example SOA environment.

service logic, contracts that define agreements between services, encapsulation that can be used to encapsulate services to conform to SOA interfaces, and service discovery that is needed for locating services.

Figure 12.3 illustrates a SOA architecture with JMS as the ESB connecting application servers and application components. The system follows the three tier model with the presentation, business logic, and storage and data tiers. This layered structure aims to improve flexibility and offer separation of concerns. The presentation tier is responsible for executing the client-side part of the application and provide the user interface. The business logic tier is responsible for executing the server side part of the application and for providing resources for the user interface. Web servers are typically located in the frontend part of this tier and they used for generating the user interface. The backend part of the tier hosts application servers that implement the business logic. The storage and data tier is responsible for persistent storage of content. The business logic and storage tiers typically reside in the datacenter.

SOA and ESB rely on standardized solutions in order to achieve service interoperability in distributed environments. Commonly used communications standards include XML as the message format, SOAP and REST as messaging protocols, XPath and XQuery for query processing, and JMS for pub/sub. Given the important role of Web services and XML in SOA, the XML message broker is a key component that provides the ESB functionality. The main functions of XML brokers include the following:

- Filtering matches messages to queries representing interest of the subscribers.
- Transformation restructures matched messages according to receiver-specific requirements.
- Routing involves the delivery of the messages to the receiver.

The XFilter and YFilter that were examined in Chapter 8 are examples of efficient XPath and XQuery matchers [1]. These and similar matchers are the building blocks for efficient XML brokers. In addition, SOAP and REST are typically used to implement the messaging in an interoperable manner.

12.3   Facebook Services

Facebook is a highly popular social networking service and website launched in 2004. The site has more than 800 million users who maintain their social profile at the site and utilize services such as chat and messaging. Many of the internal functions of the service rely on asynchronous messaging and thus efficient and scalable messaging system is one of the core components of the Facebook architecture. The architectural description of the messaging system is based on the Facebook engineering weblog posts.[2]

12.3.1   Facebook Messages

The Facebook Messages integrated many communication channels of the site, namely email, Inbox, Instant Messages, Facebook messages, and SMS. In order to meet with scalability and availability, the messaging architecture was designed from scratch. The messaging system needs to handle over 350 million users sending over 15 billion person-to-person messages per month. The chat service has over 300 million users who send over 120 billion messages per month.[3] The three key requirements for the messaging service were:

- Scalability to support millions of users with existing message history.
- Real-time messaging and operations.
- High availability.

Two general data patterns were observed when the system usage was monitored: a small set of temporal data is volatile and a large set of data is rarely accessed. These observations were then taken into account when choosing the system components for various tasks.

The internal organization of the service is hierarchical (illustrated in Figure 12.4):

- Application servers that are responsible for handling queries and writes to the system. These servers interact with the other services in order to achieve their goals. A server has a layered organization with four key layers: the API for the set and get operations that clients call, the logic for distributing responsibilities, the business application logic, and the data access. The application business logic is responsible for all user data and processing for complex product operations. This part has a write-through cache and interaction modules for Web servers to respect user privacy and system policies. The data access layer is a generic interfaces to the user's metadata. The layer stores the user's metadata and consists of a time sequenced log. This log is used to backup, restore, retrieve, and regenerate user data. The layer also stores snapshots of serialized user objects.
- Cells that are clusters of application servers responsible for a subset of users. A cell consists of an application server cluster, a number of ZooKeeper machines that control

---

[2] http://www.facebook.com/Engineering?sk = notes.
[3] https://www.facebook.com/note.php?note_id = 454991608919.

the cluster, and a metadata store. The Apache ZooKeeper[4] is an Open Source software that implements a centralized service for configuration information, naming, distributed synchronization, and group services. The ZooKeeper is based on the consistent hashing technique discussed in Chapter 3. This technique makes it easy to insert and remove servers from the cluster. The metadata store is responsible for storing the user profiles and it is interfaced through the data access layer.

The structuring of the system into cells results in several advantages. The cells support incremental scalability and help to limit failure scenarios, the system becomes easier to upgrade and deploy, and cells can be hosted at different datacenters.

A separate message store is responsible for the persistent storage of messages. The Facebook Messaging uses the Apache HBase for storing the messages. HBase has good scalability and performed properties for the observed workload and it supports replication and fault-tolerance. Message attachments are stored in Haystack, and the user discovery is done using the Apache ZooKeeper.

The Messages system features several design principles and patterns: it is layered and API driven, the distribution model is hierarchical with cells, cells providing incremental scalability and fault containment, application logic is separated to application servers, and applications can be kept stateless.

12.3.2   Facebook Chat and Messenger

The Facebook Chat is a chat application that integrates with the Facebook backend system.[5] Figure 12.5 presents the general architecture of the Chat application. The application runs inside the Web browser and connects with the server-side system. Instant messages

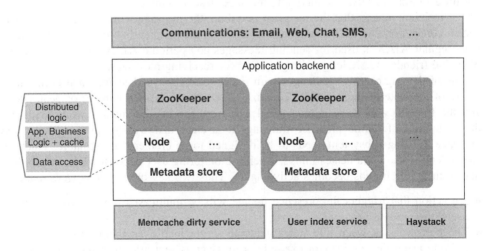

Figure 12.4   Facebook Messages architecture.

[4] http://zookeeper.apache.org/.
[5] http://www.facebook.com/note.php?note_id = 14218138919.

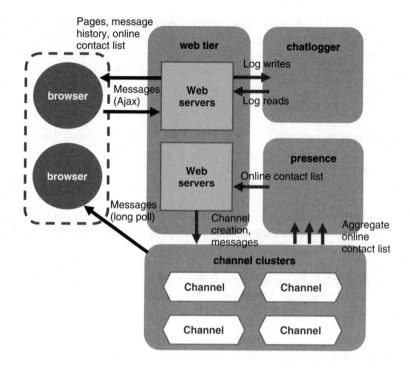

Figure 12.5   Facebook Chat architecture.

are sent to one of the Web servers that then authenticates the user and processes the message. Messages are stored on partitioned channel clusters each cluster being responsible for a subset of users. The incoming message is sent to the channel cluster responsible of the recipient. Given that the recipient is online, the message will be sent using a persistent connection that the backend system has with the recipient's browser. On the client side, regular AJAX is used for sending messages and periodic AJAX for polling the list of online friends. AJAX long-polling is used for receiving messages.

One of the engineering challenges with Facebook Chat was keeping each online user aware of the state of their friends. The size of the friend-list, the number of peak users, and the frequency of users joining and leaving are the key parameters when assessing the scalability of online presence tracking. It is clear that with millions of users the presence tracking subsystem needs to be designed to be efficient and scalable. Tracking user idleness also complicates things, when a monitoring user needs to be notified when one of their idle friends becomes active again. Thus two key observations were made:

- An action that triggers the presence update will be multicast from one user to many users.
- The transitions between idle states generate server load even though the user is not actively using the service (for example, a message is sent after user has been idle for one minute).

The time delivery of messages was another challenge. The solution is based on the client-initiated connection pattern presented in Chapter 4, in which a persistent connection is established with a server component. The connection returns when there is data for the client.

The Facebook Chat is based on a custom Web server written in Erlang that maintains the connections and keeps a log of the chat messages. The subsystems are clustered and partitioned for reliability. Erlang was chosen, because it is a functional concurrency oriented programming language with lightweight user processes and good message passing, distribution, and fault-tolerance properties. The channel servers are responsible for queuing user's messages and sending them to the Web browser with HTTP. The channel servers are also written in Erlang.

The presence updates are sent to the Web server that forwards them to the channel cluster of the user. Each cluster maintains a list of the status of its users and this is periodically updated to the presence servers. The status updates are small and thus it is possible for each presence server to store the presence of all users. Clients then poll a Web server to obtain a list of the online contacts. Web servers obtain this information from one of the presence servers.

The Facebook messenger is a social application available for many different devices that allows people to keep up with their friends, follow their status, and chat. The application provides a chat and presence system that integrates with Facebook. The application relies on the server-side Facebook Messages system.[6] In order to maintain the persistent connection efficiently, the Messenger uses the MQTT protocol examined in Chapter 5. The MQTT is used to establish a connection with the server and routing messages through the chat system.

12.4   PubSubHubbub

Often there arises need of a topic-based publish/subscribe protocol simple enough for general Internet usage. PubSubHubbub is such a protocol. It is an open protocol for distributed communication on the Internet that

- Transforms RSS and Atom feeds into real-time streams.
- Has a single API which supports web-scale, low-latency messaging.
- Has three participating entities: Publisher, Subscriber and Hub.

We could use the term ecosystem to describe PubSubHubbub: an ecosystem of publishers, subscribers, and hubs. Essentially PubSubHubbub will extend the Atom and RSS protocols for utilization with data feeds providing notifications of change updates almost instantly. This works well with the typical case where a client polls the feed server at arbitrary intervals. The PubSubHubbub liberates the client from polling whole feeds because Atom/RSS update notifications are pushed by PubSubHubbub.

The operation of a typical PubSubHubbub system is described in the following:

- Publishers will expose their published content as feeds with hub references included.
- Initially an interested subscriber will pull the Atom or RSS feed for the appropriate topic as usual by requesting it from a server providing such feeds.

---

[6] https://www.facebook.com/notes/facebook-engineering/building-facebook-messenger/10150259350998920.

- The feed (for instance Atom file) is examined by the subscriber. If a hub is referenced or described, the subscriber will register with the hub and subscribe to the feed with the Topic URL. Thus it can avoid repeated polling of the URL.
- The subscriber runs a server that can be directly contacted by the hub when any of the registered topics are updated. The subscriber is thus capable of receiving direct notifications if a subscribed topic has been updated.
- Publishers will notify the referenced hubs (pinging them) after publishing or updating an event.
- A hub can then fetch the updated feed and multicast the content to all registered subscribers.

Figure 12.6 presents the key interactions of the Pubsubhubbub protocol. The publisher includes the hub URL in the RSS/Atom feeds, and keeps the hub updates when content changes by sending a ping message to the hub. The hub can then retrieve the content from the publisher, and send it to subscribers who have registered for the content.

Figure 12.7 presents the light ping technique, in which the hub relays the ping to the subscriber. The subscriber can then upon receiving the ping request the content directly from the publisher. This approach gives the content request decision to the subscriber instead of the hub.

PubSubHubbub protocol is essentially decentralized without a central agency or company. Anyone can run an open hub, publish, update, and subscribe content. For example the hub can be a community hub for free usage or a proprietary publisher hub. One of the most important benefits of the simple and pragmatic PubSubHubbub protocol works via webhook callbacks set by subscribers for the topics of interest. Because PubSubHubbub pings contain the full Atom or RSS body with only the incremental feed content, the publisher load and also the end-to-end latency are reduced. The whole Internet-scale

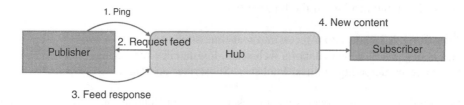

Figure 12.6   Overview of the Pubsubhubbub system.

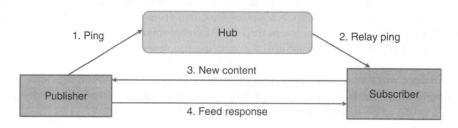

Figure 12.7   Pubsubhubbub with light pings.

and bandwidth saving implementation of publisher and subscriber software is simple and consistent for the pub, sub and hub components. Many publishers have enabled Pub-SubHubbub including Google, LiveJournal, MySpace, and TwitterFeed. Google provides several products with PubSubHubbub support: Buzz, FeedBurner, Blogger, Reader and Google Alerts.

12.5   Complex Event Processing (CEP)

Complex Event Processing (CEP) was pioneered by professor David Luckham at Standford University[7] and it has been studied since the 1990s [2]. CEP is a generic technology that deals with the detection and processing of composite events.

CEP systems provide a declarative way of detecting complex events in real-time. From the data stream point of view, they transforms set of input streams into one or more output streams. A CEP solution typically provides a declarative language for specifying the interesting event patterns. Many CEP solutions are based on an extension SQL that includes time, causality, and pattern matching features.

The Rapide language is a key example of CEP. The main idea of Rapide is to use asynchronous events and their causal relations to model both static and dynamic architectures. In this context, an architecture consists of interfaces, connections, and constraints. When the architecture specification is executed all causal relations are stored and checked against the constraints.

The key requirements for the Rapide system were component abstraction, communication abstraction, communication integrity, dynamicity, causality and time, hierarchical refinement, and relativity. The interdependencies of Rapide components are modelled using partially ordered sets. A pattern language is defined for detecting composite events. An event of a particular action is a tuple of information with a unique identifier, a timestamp, and dependency information. The system supports placeholders and universal quantification over types. Patterns are used in interfaces to define behaviors and in architectures to define connections.

CEP has many applications that include the following:

- Business process monitoring,
- Network monitoring,
- Security event correlation and intrusion detection,
- Risk management,
- Fraud detection,
- Algorithmic trading,
- Monitoring service level agreements in call centers and datacenters.

Esper[8] is an Open Source Java-based CEP engine widely in applications ranging from the stock market to aerospace. Esper is based on the *Esper Processing Language (EPS)* that is an SQL-based continuous query language using insert into, select, from, where, group-by, having, order-by, limit and distinct clauses. Figure 12.8 illustrates the key

---

[7] http://pavg.stanford.edu/rapide.
[8] http://www.espertech.com.

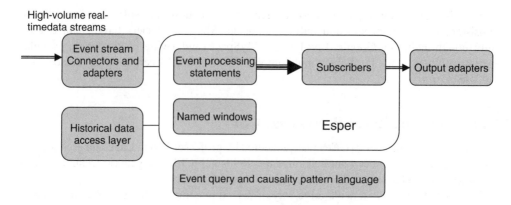

Figure 12.8   Overview of the Esper system.

components the Esper system. The connectors and adapters manage incoming high-volume data streams. the streams are then processed by the core Esper system and matched against the event processing statements created with the event query and causality pattern language. There is also a historical data access layer that stores events pertaining to windows. The output data streams and events are then forwarded to subscribers. Outgoing data can be processed by output adapters.

As a simple EPS query we can consider the following statement that computes the per-second rate using Esper internal timestamps by averaging 7 seconds of market events:

```
select rate(7) from MarketDataEvent output snapshot every 1 sec
```

Data streams can also be easily joined together with a very similar construction to the join in SQL. In order to combine two data streams, the join keys need to be identified. The joined stream can then be processed as a single stream. For example, given a two streams of fraud warnings and account withdrawals, they can be combined based on the account number over a given period of time. This allows the detection of withdrawals from bank accounts suspected of fraud.

Esper has been designed to take take performance issues into account. The key considerations are:

- Complex detection rules, such as event correlation, sliding windows (time, length, sorted, . . .), joining event streams, various event stream operations, combining windows with intersection and union semantics, triggers based on absence of events, and so on.
- High throughput for applications that need to process tens of thousands or even hundreds of thousands of events per second.
- Low latency for applications that need to react in real-time to events.

As an example of CEP, we can take *algorithmic trading*, which has been becoming increasingly popular in the last decade [3]. Algorithmic trading is motivated by the need to minimize transactions costs in order execution as well as being able to react as fast as possible to market dynamics. Algorithms can also hide the buying or selling action by splitting it into smaller pieces.

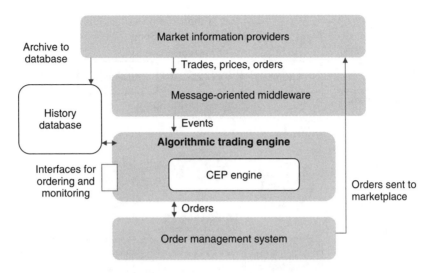

Figure 12.9 Algorithmic trading.

Figure 12.9 illustrates an example algorithmic trading system. A message-oriented middleware connects the different components of the system together. Typically the XML broker would be used to realize the messaging system. The market information providers send information about the trades, prices, and order in the form of events to the algorithmic trading engine. These are also archived to a history database as well. The CEP engine is the heart of the system, which is responsible for implementing the desired behavior by monitoring the events and then triggering actions, typically orders for the order management subsystem.

12.6 Online Advertisement

Online advertising is an emerging application domain for pub/sub systems. Figure 12.10 presents the key components of this application area. The content broker, for example the XML broker mentioned above, is responsible for matching between the advertising campaigns and the user profiles. The advertisers define the advertising campaigns that target certain user groups. Similarly, the users have associated user profiles that define the user group and other relevant information, for example the location of a user. The broker can then based on the campaign information and user profiles target advertisements to the users. This application requires also additional mechanisms than pub/sub, for example databases for storing content as well as privacy enhancing technologies in order to protect the consumers from exposing privacy sensitive information.

The matching algorithms discussed in Chapter 8 can be used to match advertisements to users in online advertising. A Boolean expression can represent an advertisers targeting requirement, and attributes in a user profile represent the user visiting an online page. The same model applies also for other kinds of digital advertising, for example context-aware advertising in mobile settings.

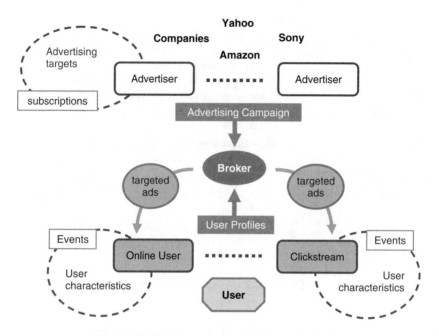

Figure 12.10   Content advertising with pub/sub.

This kind of advertising system works as follows:

- The advertising systems send subscriptions to the pub/sub system. The subscriptions represent their targeting requirements. For example, a subscription might specify a specific part of a Web site (URL), a specific time, a specific age group, gender, visiting pattern, and so on.
- Each time a user visits a Web page, the pub/sub system receives an event that specifies the page in question and the characteristics of the user. The pub/sub system then matches the event against the subscriptions and notifies matching subscribers. The pub/sub system may be able to correlate events in time in order to match queries pertaining to visiting patterns.
- The advertising system is notified that there is a matching event for a specific subscription, and the advertiser can generate a specific advertisement for the user. The ad is typically provided by a specific advertising server that then needs to be notified that a certain ad should be displayed. The advertising decision should be done fast so that the user will see the ad when the page load ends.

The pub/sub content matching system needs to support expressive filtering, Boolean expressions, in order to support ad targeting. Composite event detection and event aggregation may also be needed in order to target to users who regularly visit certain pages. The matching process needs to be fast in order to show the ad at the same time as the Web page.

12.7 Online Multiplayer Games

Pub/sub is a building block for online multiplayer games. The asynchronous and real-time nature of pub/sub lends itself well to the requirements of this application domain. In a multiplayer game, users share a single instance of the game world. Each participant of the game has a limited view to the game world, and interacts with the world and other users through this limited view. A game server or set of servers are responsible for managing the game world state and then informing the participants of relevant changes in the portion of game world visible to them.

Based on the above description, we can observe that pub/sub is a very good candidate for supporting the distributed game world management. In pub/sub terminology, game participants subscribe to receive events pertaining to a specific part of the game world. When interesting things occur in their view, they are then notified and update their own local state accordingly.

Figure 12.11 illustrates how pub/sub can be used to implement a multiplayer online game. A player has subscribed a rectangle area of the virtual game world. A publication is typically a point in the content space. In this case, if a publications coordinates are contained within the subscribed region, the client game engine of the player is notified, and the event is then taken into account when rendering the game world.

The mercury pub/sub system is a notable example of a pub/sub architecture for Internet games [4]. This system supports a query language and distributed matching based on DHT technology.

12.8 Apple Push Notification Service (APNS)

The *Apple Push Notification Service (APNS)* is a service that enables information push to mobile devices. APNS was developed by Apple and released with iOS 3.0 in 2009. The push notification system is part of the iOS Xcode development environmente. The service is based on the client-initiated long-lived connection pattern presented in Chapter 4. The client-side system establishes a long-lived connection to dedicated push servers operated by Apple. Service and application developers can then register with Apple and send push notifications through the APNS system to mobile devices.

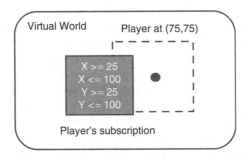

Figure 12.11   Example of a game world with a subscription.

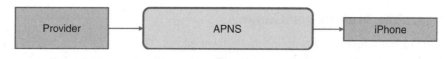

Figure 12.12   Apple APNS service.

APNS usage involves the following steps:

1. Service or application developer connects to the APNS system using a unique SSL certificate. The certificate is obtained from Apple with the developer identifier and application identifier.
2. The APNS is used to send one or more messages to mobile devices. The push operation is application and device specific and a unique deviceToken is needed for each destination device.
3. The service or application disconnects from APNS.

Each push message is addressed to a specific device by using the unique deviceToken that is generated by the APNS system for an application. This token is then given by the application to a server. The server-side system can then send push messages to the application with the token and a developer push certificate. There is a feedback service provided by Apple that can be polled for a list of invalid device tokens.

Figure 12.12 illustrates the push scenario, in which the server-side component of the applications sends a message to the mobile device using the APNS system.

APNS system has some restrictions and limitations. The payload is limited to 256 bytes and thus it is designed for short alert messages and notifications. The system does not provide status feedback regarding the delivery of a message. Only the last sent message will be queued for delivery overwriting any previous queued message for the same application and destination. The messages are only delivered through cellular data and WLAN.

12.9   Internet of Things

The Internet of Things is an emerging environment, in which everyday things and devices discover each other and exchange information in order to implement various higher level services. Example applications domains include the delivery and processing of sensor information in smart homes, factories, structural monitoring, and other contexts, logistics, security monitoring, and health-care. Figure 12.13 presents an overview of the environment, with two *wireless sensor networks (WSN)* consisting of distributed sensors and actuators, and Internet-based applications that interact with the WSNs through a broker. In this book, we have covered several messaging protocols that are suitable for supporting Internet of Things and WSN applications, namely the DSS and MQTT protocols in Chapter 5 and the CoAP presented in Chapter 6. These protocols support basic message based communications and topic-based pub/sub.

Pub/sub is a good candidate technology for the Internet of Things, because it supports loose coupling of the entities as well as data and content centric information dissemination.

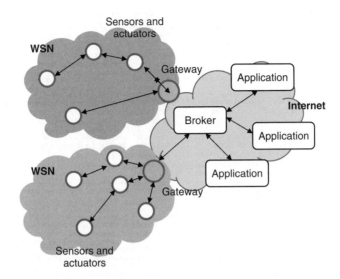

Figure 12.13 Overview of the WSN environment.

We have discussed several research systems that are suitable for this environment, namely the REDS, GREEN, and STEAM systems examined in Chapter 9. These systems have been designed for the *Mobile Ad Hoc Network (MANET)* environment, in which network nodes communicate with short range wireless communication links and they are mobile.

This means that the pub/sub system needs to cope with the wireless communications environment and support node mobility. Thus many of the traditional solutions, such as reverse path routing, cannot be applied in this environment. The general solution has been to design modular pub/sub routers that allow pluggable routing and communications components. This modular pub/sub engine can then be adapted for the various environments, for example the Internet of Things environment.

In addition to the wireless and mobile characteristics, also the limited resources of the sensor and actuator devices need to be considered. The pub/sub system running on these small devices needs to be simple and efficient. The constrained nature of sensors typically results in a multi-tier design, in which more powerful clusterheads or gateways process and route sensor data. Gateways are also used to connect the different WSN deployments to the Internet. The two basic protocol, CoAP and MQTT, are both designed for the small and limited devices and they support the gateway paradigm.

12.10 Summary

This chapter considered applications of pub/sub systems and technology. Pub/sub is a frequently used technology for connecting components in a loosely coupled system. We examined the role of pub/sub as an enabler of a cloud computing platform, SOA and a generic XML-broker for enterprise applications, Facebook Messages and Chat, Pub-subhubbub, Complex Event Processing, online advertisement and games, and the mobile

Application	Solutions and Patterns	Description
Cloud computing	Message queuing, topic-based communication, namespaces, replication and load balancing, filter-based message retrieval from queues, pay-per-use	Pub/sub is used to relay vents between applications and system components. Typically, REST and SOAP-based interfaces for developers with a token-based authentication model with pay-per use. Examples include Azure Service Bus and Amazon SQS.
SOA	ESB, XML routing and transformations, SOAP	ESB connects the application servers and application components together. Typical protocols include SOAP, REST, and JMS.
Facebook Messages	Hierarchical and cluster-based design, stateless applications, concurrency improved with Erlang	Asynchronous messaging for messaging within the Facebook framework. Requirements for high availability, scalability, and performance. Features available for developers through the Facebook API.
Facebook Chat and Messenger	Multi-tier and cluster-based design, client-initiated connection, concurrency improved with Erlang, periodical presence updates and aggregation in the backend	AJAX long-polling for the Web application, MQTT for the mobile application.
CEP	Detection of composite events through a rule-based stateful filterin gsystem, CEP language	The Rapide language and the Esper CEP engine.
Pubsubhubbub	Decentralization through hubs, proxy, client-initiated connection	Supports timely delivery and aggregation of RSS and Atom feeds through the API offered by the hubs.
Online advertisement	Event routing and forwarding, content-based routing, real-time notification	The system needs to support high dimensionality in the filtering engine. Boolean expressions offer a rich basis for an advertisement engine.
Online games	Event routing and forwarding, content-based routing, real-time notification	Pub/sub is used for updating distributed game world state and to send asynchronous updates to game engine instances.
Apple Push Notification	Client-initiated connection, security token	The system pushes small notification messages to smartphones running iOS with a long-lived connection. A security model ensures that authorized services can send notifications.
Internet of Things	Lightweight pub/sub solutions, basic messaging and topic-based routing, content-based routing, reconfiguration	DSS is used in many embedded and wireless enviroments. Basic protocols include MQTT and CoAP.

Figure 12.14   Summary of example applications.

environment. Figure 12.14 summarizes key observations pertaining to the applications examined in this chapter.

Pub/sub can be used in various application domains and settings. The design and implementation differ based on the requirements, for example scalability, expressiveness, availability, and so on. The basic patterns discussed in Chapter 4; however, are the same. A logically centralized service typically is used to handle and deliver events, and accept subscriptions. In this chapter, we examined different designs for the service, for example the Facebook design takes into account the scalability and availability requirements, and the Pubsubhubbub design aims for a decentralized solution for the Internet while minimizing notification overhead and delay. Mobile solutions, on the other hand, are based on lightweight designs and emphasize reconfiguration and adaptation to the operating environment.

References

1. Wu E, Diao Y and Rizvi S (2006) High-performance complex event processing over streams. *Proceedings of the 2006 ACM SIGMOD International Conference on Management of Data*, pp. 407–18. SIGMOD '06. ACM, New York, NY.
2. Luckham DC (2002) *The Power of Events: An Introduction to Complex Event Processing in Distributed Enterprise Systems*. Addison-Wesley, Boston, Massachusetts.

3. Lindström J (2010) Algorithmic trading and complex event processing. Master's thesis, Aalto University.
4. Bharambe AR, Rao S and Seshan S (2002) Mercury: a scalable publish-subscribe system for internet games *Proceedings of the 1st Workshop on Network and System Support for Games*, pp. 3–9. NetGames '02. ACM, New York, NY.

13

Clean-Slate Datacentric Pub/Sub Networking

In this chapter, we consider recent advances in introducing pub/sub features in the protocol architecture at the network layer in order to realize efficient one-to-one and one-to-many communications. We consider two recent architecture proposals, namely the *Content-centric Networking (CCN)*, and the *Publish/Subscribe Internet Routing Paradigm (PSIRP)* and PURSUIT architectures.

13.1 Datacentric Communication Model

TCP/IP protocol suite has scaled well with the incredible growth of the current Internet but it has some inherent limitations for security and scalability for the increasing online video consumption and file transfers. Because anybody can address packets to IP addresses naming destination hosts, unwanted traffic remains a problem. In addition, most of the current traffic is generated by applications that mostly want to share named data and their implementation becomes unnecessarily complex on top of the current network stack optimized for message passing. As the current model is centered around named hosts serving the content, the data becomes location-dependent and the network may not always use the closest cached copy of the data. This limits the availability of the popular content, because it needs to be served by a specific container. Also the authenticity of the data is tied to the container of the data instead of the data itself, which makes the architecture inflexible and less secure because of the additional step in the trust chain.

There has been recently a growing research interest in redesigning the Internet from scratch starting from the network layer to address the above problems. A common approach taken by multiple projects is *datacentric* model of communication. In this chapter two prominent projects and their architectures, *Content-Centric Networking* (CCN) and PURSUIT, are briefly introduced. The datacentric model can be seen as special type of topic-based pub/sub, where each topic is a name for an immutable data item or content chunk potentially made available to the network. Subscribers can express

Publish/Subscribe Systems: Design and Principles, First Edition. Sasu Tarkoma.
© 2012 John Wiley & Sons, Ltd. Published 2012 by John Wiley & Sons, Ltd.

their interest in a name to the network, which then returns the associated data item as it becomes available.

Another way to look at datacentric networking is to think it as the inversion of control between the sender and the receiver compared to traditional message passing of the current Internet: Instead of using IP addresses, TCP ports and protocol-specific data to identify remote continuations, which senders can trigger by sending a message, the receiver expresses its interest in an identified data item, which the network can then return when it becomes available. This way the network can take an advantage of multicast and caching to improve the utilization of link resources and minimize latency. The receiver-driven model has the additional advantage that it makes it possible to develop strong countermeasures against unwanted traffic such as distributed denial-of-service (DDoS) attacks with payload packets.

In general, datacentric communication primitive provided by the network simplifies applications that need to share the same content between multiple nodes while message passing is a good match, when there is a stateful remote object that needs to be manipulated by multiple entities. There exists a deep symmetry between the continuations and data values explained in [1], which is reflected by these two communication paradigms. The term *Content-centric* is also used to describe this communication pattern when the emphasis is on the efficient content delivery while *information-centric* can be used interchangeably to point out that the data items can link to other named data and that the data has structure.

13.1.1 Naming of Data

Naming plays an important role in a datacentric network architecture as it pervades all aspects of the system. Typically, naming and routing designs are coupled and many of the approaches can be divided into two categories based on whether they are optimized for an additional name resolution phase before the actual payload communication or routing is directly based on the names. For example, the names may contain restrictions such as hierarchical structure or embedded location information, which make it possible to optimize routing based on them. There are also many other dimensions in which different identifier structures and semantics can be compared such as the lifetime of the identifiers, (im)mutability of objects, trust model, human-readability etc. In the following subsections we go through the naming in CCN and PURSUIT/PSIRP architectures.

13.1.1.1 CCN

CCN [2] uses opaque, binary objects that have an explicitly specified number of ordered components to name the content chunks in a hierarchical way shown in the Figure 13.1. These names allow subscriptions that match simple sets of publications and at the same time allow naming of data chunks that do not yet exist. These *active names* are, for example, needed for dynamically generated content that is widely used in addition to static web pages. Hierarchical name prefixes can also be used for context-dependent names such as */ThisComputer/freeMemory*. This has the advantage of making some naming schemes simple as no complex enumeration or probing strategies are needed runtime [2]. The names, that can be human-readable, are only meaningful to some higher layers in the stack, but the network does not need to understand them. It is possible to encrypt the names for privacy, but in general, the datacentric model is more natural at securing

Human readable hierarchical name:

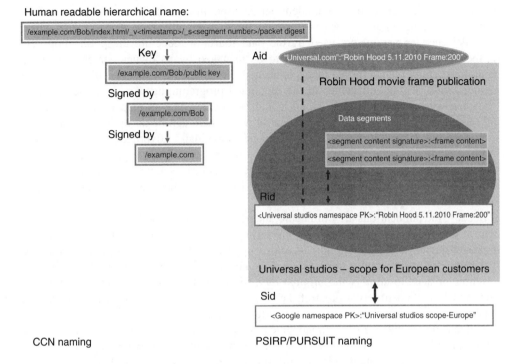

Figure 13.1 CCN and PSIRP/PURSUIT names side-by-side.

the integrity of the data instead of confidentiality as there is no notion of set of receivers defined by the publisher. Actually, it is preferable that the publishers are unaware of the receivers in order to be scalable.

CCN names are also straightforwardly compatible with hierarchical URLs. Individual publications are always immutable, but the name can contain a version number of mutable data and a segment number for large chunks of content. SHA256 digest of the name can be used by links to unambiguously identify any piece of content. For dynamic data, however, the full name is required to reach the data source as it may encode information needed to generate the requested content. Hierarchical names can be ordered in lexicographic order, which makes it simple to ask for the most recent version of the data item. For example, a request for "/example.com/video.avi" with annotation *RightmostChild* matches "/example.com/video.avi/version_2/segment_0" instead of "../version_1/..". However, the original data source must be reached in this model in order to be sure that no newer publications matching the requirement exist.

13.1.1.2 PSIRP/PURSUIT

PURSUIT continues the work of PSIRP project and our explanation here is based on the line of work explained in [3, 5], which is only one possible realization of the abstract PURSUIT functional model. All identifiers in this architecture have the similar *self-certifying*, two-tier, DONA-like structure [6]. An identifier is a (P, L) pair where P is the public-key of the *namespace* owner of the identifier and L is a variable length label of free-form binary data.

Only fixed length hash of the identifier is used in-network to identify a publication, but variable length names are needed for dynamically generated content, where the label encodes parameters used by the data source to generate the publication data on-the-fly. Each namespace is assumed to be managed by its owner, which is why no centralized entity managing the labels is needed. Variable length labels could be used for memorization of function values in the functional programming paradigm such as label "weather(Helsinki,19.11.2011)", which could be cumbersome to map to a solution that only uses fixed-length identifiers in a general way. There can be an unlimited number of such potential publications and they cannot be individually advertised beforehand. The self-certification is achieved by the namespace owner authorizing a publisher to publish a certain set of labels in the namespace. The actual content segments are then signed by the publisher and the certificate chain starting from the namespace owner public key is included in every payload packet.

Because the publications are immutable, the network is always able to use a valid cached copy if one is found locally without considering the freshness of the data. In addition to improved latency, this assumption increases opportunities for parallel computation and scalability for many applications as there is no need for the global connection to be available or need to contact the original data source, which might become a bottleneck. The model is compatible with that of the functional programming, where names are also bound to immutable values as well as function values.

Mutable documents can be implemented with version information embedded in the labels and potentially leaving the stream of older versions still available. This type of "explicit plumbing" of stable identity information in the labels should be recommended generally as the publications can then be handled independently and all assumptions about the semantics of the data is easily visible. Other metadata not directly associated with the identity of the publication should be stored as independent metadata publications.

As a concrete example of how streaming could be mapped to this model, a simple game application could be considered, where each user publishes his/her avatar's coordinate as a stream of versions. The Rids could be of the form (<gameserver namespace public key>, "Avatar X coordinate version Y"). This naming scheme leaves us with the problem of determining the latest coordinate. That can be easily solved with an extra publication that has the current time in its identifier's label part. The contents of this publication then has the most recent version of the coordinates that was available at that time. This same scheme can be used to implement streams as versioned data. Because subscribers can subscribe multiple versions preemptively before they have been published, no additional latency is introduced by this mapping. Also message sinks from many to one can be easily implemented by including the identity of the subscriber and the current time into an Rid and subscribing this Rid. This way the subscription is relayed all the way to the data source that can use the information in the identifier to subscribe to a stream of publications published by the subscriber.

13.1.2 Content Security

Both of the technologies we cover here identify the data items in a *location independent* way instead of addressing hosts or containers of data using their location information. From the point of view of the subscriber, this shifts the emphasis from the question *where* to *what* as anybody can provide the needed data if they have it. This change in viewpoint also incorporates the approach of content-based security: the data itself is secured for integrity and trust is between the subscriber and the original producer of the content. Similar to subscribers, caches and intermediate nodes can also validate the data locally from any source on per-packet basis. Availability of popular data is improved as multiple data sources can naturally be used to serve the data to the network.

Both in CCN and PURSUIT, all content chunks can be authenticated with public-key signatures by storing a signature that binds the name and the content in each data packet. The integrity check is possible to do independently for each packet as the PURSUIT Rids contain the public key of the namespace the Rid belongs to. We call such identifiers *self-certifying*. In CCN, on the other hand, signed data packets carry enough information to be *publicly authenticatable* by containing enough information to allow the retrieval of the public key necessary to verify it. In CCN, the trust model is contextual as there does not exist a single entity that everyone trusts for every application. Keys are just CCN data and content can be used to certify other content. This can be used in conjunction with the hierarchical names so that, for example, key named "example.com" is used to sign another publication named "example.com/Bob" that contains the key that is used to sign content published by Bob. This follows the trust model of SDSI/SPKI, where keys are mapped to identities using locally-controlled namespaces without an external trusted 3rd party.

Both CCN and PURSUIT also support *secure references* that can refer to another data item by the cryptographic digest of its contents instead of only a name. This is possible when the referenced content exists before the creation of the reference and is as secure as the cryptographic hash function used. When this more secure approach cannot be used, there remains always the problem of leaked secret keys that will render the self-certified identifiers insecure. The traditional key management problems, especially the mapping of real life identities to public keys used on the network layer must also be solved on the application layer. There is no single solution that fits all application needs and therefore it is not meaningful to try to solve this problem in a general way. PURSUIT has taken the philosophy that the Rids are only network level identities of data and may not have long life-time. Some application level naming scheme is needed for such persistent identifiers and another mechanism for mapping the application level identifiers to Rids. CCN provides a partial solution to the key management problem by using the hierarchical names, where the trust for top-level contexts can be based on different mechanism, such as manual configuration of trusted keys, and the trust chain to individual data items is then checked hierarchically.

PURSUIT introduces the additional concept of *scope* and every publish operation also determines the scope in which the publication takes place. The scoping of information is orthogonal to the information structures formed by naming schemes and data items referring to other data items. Instead, the scope determines the *global distribution strategy* of the publications inside it. This has the additional security benefit that the subscriber can separate its trust on the distribution strategy from the trust on the content. For example, inside untrusted scopes, some 3rd party data sources may falsely advertise data that they do not intend to serve in order to cause the data to be unavailable to subscribers. CCN cannot solve this problem by simply using data sources with the credentials from the original publisher as this restricts the use of 3rd party data sources opportunistically.

13.2   CCN

CCN aims to preserve the simplicity of TCP/IP and the design choices that made it successful. In CCN, the network "waist" forwards simple content chunks and routes interest
information in the network. The familiar idea is that this very simple core functionality
can be run on top of different types of distribution strategies and link technologies and
anything can be run on top of CCN. In the intradomain case, CCN should work with
small modifications to an IP-based network. Just instead of routing IP packets, the network nodes should be able to use the same algorithms for routing Interest packets and
forward Data packets in the reverse direction in a simple manner.

13.2.1   CCN Node Operation

CCN has two packet types, *Interest* and *Data*, shown in Figure 13.2. Each node broadcasts
matching Interest packets for the content chunks it wishes to receive over its all available
connections. Underneath the network level API, there is a strategy layer that guides the
operation of dissemination of Interest packets. Data flows in the opposite direction to
Interest packets and the pending interest is thought to be consumed by matching content.
Data packet satisfies a pending interest if the name carried in the original Interest packet
is a matching prefix of the name in the data packet. This type of aggregation of names is
similar to DNS or IP hierarchies and has proved to be scalable for routing and forwarding
state in Internet scale while also allowing fast lookups.

The schematic structure of a CCN node is close to an IP router in its operation: a
packet arrives on an interface of the node and the following action is determined by a
longest-prefix matching from a lookup table. Each node is assumed to implement at least
the following three abstract data structures:

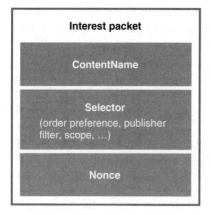

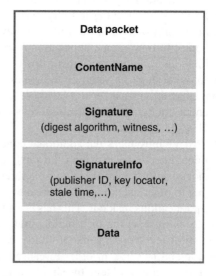

Figure 13.2   CCN packet formats [2].

1. Forwarding Information Base (FIB),
2. Content Store, and
3. Pending Interest Table (PIT).

The FIB is used to route Interest packets towards one or many potential data sources. The potential multicasting of interest information allows parallel queries in contrast to the IP FIB, which is similar to its operation otherwise. Nodes can use different link-specific strategies for the routing of Interest packets. For example, laptops in a local wireless network could simply broadcast their interests to each other and thus continue communication even when the connectivity to the global network is severed. The eventual goal of the CCN project is to specify an abstract machine, that can be used to run programs that implements the *strategy layer* for routing Interest packets. In the current research system reported in [2], the interests are by default sent to all interfaces having the attribute *BroadcastCapable* and after that, in case there is no response, try all other available interfaces in sequence. This strategy allows the query from local sources before falling back to the default global routing method. CCN does not have any separate rendezvous functionality or phase before actual payload communication to find the data such as in PURSUIT.

In order to maximize the multicast potential, Content store keeps received data packets stored in a buffer after they have been served to outgoing interfaces as long as possible using LRU or LFY cache replacement strategy. Basically all nodes can participate in the opportunistic caching of popular data in the network.

PIT data structure stores the pending interests from different inbound interfaces and is used to forward the data downstream back towards subscribers. It can be thought that the routed Interest packets first leave a trace of "bread crumbs" stored as PIT entries that the data then follows in reverse direction and consumes the stored crumbs as it traverses them. Multiple, exactly matching subscriptions are stored only once and only the new interface is added to a list store in the PIT when the same interest is received via multiple interface, forming a multicast structure by consuming resources only once for multiple subscriptions. Each router may also perform validation of the data before storing it in the Content store. This policy eliminates many types of cache poisoning attacks and unwanted traffic.

All data stored in PIT is considered as "soft state" meaning that it can be timed out and erased in case the memory runs out. This adheres to the "fate sharing" design principle [7] on the basis that the subscriber is eventually the only entity responsible for making sure that it receives the data it is interested in. If the Content Store already has a matching Data packet for the newly arrived Interest packet, it is immediately served downstream without further routing the Interest packet. In case both the FIB does not have an entry for the *ContentName* inside Interest packet and the requested data is not stored in the Content Store, the Interest packet is simply discarded as the router does not have a rule to handle it.

13.2.2 CCN Transport Model

The underlying packet network is assumed to be unreliable, for example, because of the intermittent connectivity of a mobile device: both Interest and Data packets can be lost or corrupted in transit. Therefore, subscribers need to perform a retransmission of Interest packets after a certain time-out. Senders are stateless in CCN and all flow management is done by the receivers. The particular strategy may depend on the receiver. For example, how many duplicates of the interest is created is determined by the used strategy and previous response times can be used as a heuristic for determining where to route the following requests for best results etc. Similarly the intermediate links can have their own strategies and the operations in CCN are always local, hop-by-hop decisions. To prevent duplication of looped Interest packets (Data packets cannot loop in CCN), they contain a random *nonce* value so that duplicate packets can be discarded when one is received.

Because one Interest packet retrieves at most one Data packet, it is said that *flow balance* is maintained in the network: basically the receiver can use the number of pending Interests to control how many packets can at most be in transit at one moment similar to TCP window size advertisements. As each packet is subscribed independently, no other mechanism such as the TCP SACK mechanism is required to keep the flow of packets efficient in case of a packet drop.

In CCN, all communication is thought to be local and there is no functionality for end-to-end congestion control in addition to the hop-by-hop flow balance. It is assumed that the routers on the path have enough buffer memory to store all data packets in fly, which guarantees eventual progress in case of congestion. It is claimed that the use of LRU policy instead of FIFO queues decouples the hop-by-hop feedback control loops and damps oscillations [2] thus making end-to-end congestion control unnecessary. On the other hand, delays may still grow without bound and network resources are not optimized also opening vectors for Interest packet DDoS attacks. Sequencing is handled by the hierarchical structure of the names as the name can contain segment numbering; however, the specifics of this are an application level concern.

13.2.3 Interest Routing

For intradomain routing, both IS-IS and OSPF support general type label value (TLV) schemes that are compatible for distributing CCN content prefixes advertised by the data sources. One advantage of CCN naming is that because of it resemblance of IP prefix routing, a router implementing the CCN forwarding model can be attached to an existing network using IS-IS or OSPF without modifying the network. However, in case of multiple advertisements of the same prefix, in CCN the Interest packets need to be sent to all of the announcers instead of a single one in IP. This is because the semantics for CCN advertisement is "some of the content with this prefix can be reached via me".

CCN does not yet offer any solution for scalable interdomain routing. Basically, BGP, used in the current Internet, supports equivalent extension to the IGP TLV mechanism that could be used by domains to advertise their offered content prefixes, but this is mostly scalable only because IP addresses are used for aggregable locations, thus considerably reducing the size of the BGP routing tables. As multihoming and mobility are becoming more and more popular, it is an open question whether the location-independent, potentially human-readable names of CCN can scale to be used for interdomain routing. For example, the number of active DNS domain names is two orders of magnitude larger than the size of current global BGP table and requiring aggregation limits the possible uses of CCN names.

13.3   PSIRP/PURSUIT

In PURSUIT, an immutable association can be created between a *rendezvous identifier* (Rid) and a data value by a publisher and this association is called a publication. A *data source* may then publish the publication inside a set *scopes*, that determine the distribution policies such as access control, network resources, routing algorithm, reachability, and QoS for the publications inside it and may support transport abstraction specific policies such as replication and persistence for datacentric communication. Scopes are identified with names that have similar structure as Rids and these identifiers are called *scope identifiers* (Sid).

It is assumed that the network is composed of domains, that encapsulate resources such as links, storage space, processing power in routers, and information. The concept of domain is here a very general one, and can refer to abstractions of any granularity, such as software components, individual nodes, or ASes. Scopes and domains are, in some abstract sense, the conceptual inverse of each other: Domains introduce resources that can be used to implement the communication event whereas scopes eliminate possible resources based on some requirements for them.

An *upgraph* of a node is the set of potential resources, that can be represented as a network map of domains and their connectivity, to which a node has an access to based on its contracts between its service providers. A language called advanced network description language (ANDL) has been developed in the PURSUIT project for the communication of reachability information in control plane publications in an abstract and general way.

The links in the upgraph maps represent resources abstractly and can be limited to carrying only various transport abstractions or protocols over them. Each transport protocol implements a specific communication abstraction and every actualized *instance of interaction* or communication event consuming resources of the network always has an associated *transport* (communication abstraction), *topic*, a *graphlet* and a set of *roles* for the different node participating in the transport. For example, the roles for the endpoints in datacentric pub/sub transport are a *data source* and a *subscriber*. The topic is identified with an Rid and is used to match the end nodes in correct interaction instances by the scope in which the communication takes place. For example, for datacentric communication, the topic identifies the requested publication(s).

A graphlet defines the network resources used for the payload communication and it can be anything from the route of an IP packet to private virtual circuits. Some protocols may require an additional resource allocation phase for the reservation of a delivery tree described by the given graphlet before the actual payload communication over the delivery tree. A graphlet adheres to a set of scopes that are responsible for the policy-compliant matching of nodes to interaction instances, collecting the needed information to build end-to-end paths, and placing constraints on the chosen resources for the graphlet. A graphlet then connects a set of end nodes that each implement a certain role in the transport. When we talk about one-to-many datacentric transport, the graphlet is called a *delivery tree*, which is basically a multicast tree originating from a data source and having the subscribers as leaves.

Communication always happens inside at least a single scope, but the policies can be divided into aspects of communication handled modularly by different scopes implemented by different entities. Scopes are responsible for combining upgraph information from multiple nodes requesting to participate in an interaction instance inside the scope [8] and the scope selects a subset of the given resource that adhere to its policies. It should be noted, that because a scope can act as a neutral 3rd party for the route selection, it can balance the power between endpoints and optimize the path as a whole.

Figure 13.3 shows the basic concepts of this particular instantiation of the PSIRP/PURSUIT architecture. It should be noted, that there is an authorization to both direction between a scope and data sources it accepts. A data source needs to authorize the scope so that it can prevent malicious scopes from building paths towards it in order to launch a DDoS attack. Each subscription specifies both the scope (Sid) and the publication (Rid). The chosen scope can be thought to answer the question "how" the data should be delivered over the network and the Rid tells "what" the subscriber is interested in.

13.4   Internet Interdomain Structure

When designing a network architecture for the whole Internet, it is necessary to understand the complex scalability, security, and deployability problems stemming from the needs of multiple parties involved. Internet consists of approximately 30000 *autonomous systems* (AS) connected in a graph. Each AS encapsulates a network with its own routing policy typically controlled by a single organizational entity that owns the resources. ASes are identified with IANA assigned 16 or 32 bit AS numbers.

Most of the business relationships between ASes can be categorized on a logical level as either *peer-peer*, *customer-provider*, or *sibling-sibling* type. That is, operators typically form bilateral contracts with each other that determine how packets are routed. A peering relationship between two ASes means that they do not financially compensate at all or pay less for the packets exchanged between each other, as long as the balance of traffic is within predetermined limits or some other similar agreement is met. When two ASes are

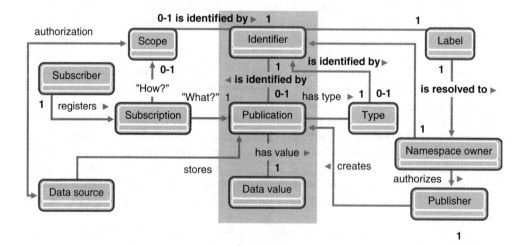

Figure 13.3   PSIRP/PURSUIT concepts [2].

in customer-provider relationship, the customer pays the provider AS for the transit of packets, typically based on some function of traffic volumes over time. A sibling-sibling relationship is between ASes managed by the same organization and, from the policy-routing point of view, can often be thought as a single AS. Even though there is only a single logical business relationship between two networks, the large ASes are typically physically connected to each other via many links that are geographically distributed. Figure 13.4 shows an example subset of ASes and their business relationships.

The connection degree distribution of ASes follows approximately a power law, which means that on the global and national levels there are often few operators, who are connected to hundreds of other networks. The Internet core is formed by about 12 *tier-1* ASes that form together a full mesh and can provide full reachability to the whole Internet via them. A *tier-2* network can peer with other networks but for the full reachability it may need to buy transit service from one or multiple tier-1 networks. A large portion of ASes, so called *stubs*, are only connected to a single provider and are typically operated by enterprises and other organizations. Only about 3000 ASes are managed by Internet service providers (ISPs).

Exact data about the AS structure of the Internet is difficult to gather as not all peering relationships can be seen in the global BGP table and operators are not willing to divulge their business information. However, CAIDA [9] has used BGP information and data gathered from traceroutes to build an inferred map of AS business relationships. In a more recent study [10] more than 200 Exabytes of commercial interdomain traffic was

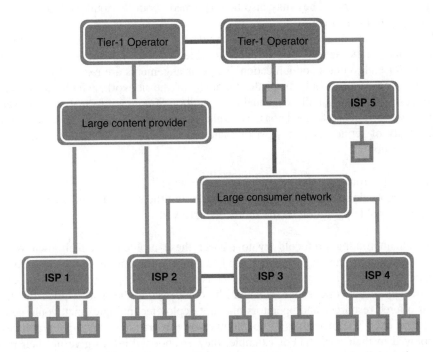

Figure 13.4 AS relationships in the current Internet contain *customer-provider* links and *peer-peer* links. Sibling relationships are not shown in this example.

analyzed from years 2007 to 2009 using more than 3000 peering routers across 110 providers. This is a significant percentage of the total traffic volume of the Internet and from these findings it can be said that the expected hierarchical structure of ASes has somewhat changed recently. The core tier-1 operators have lost some of their importance as content and customers have been consolidated in huge tier-2 networks such as Google, large CDNs (Akamai, LimeLight etc.), or Comcast, that have directly peered with each other around the core. In general, the Internet seems to be more densely connected instead of a hierarchy of backbone operators, regional access, and local access operators. However, the traffic distributions of ASes still approximate a power law distribution, but it is clear that these constant changes in the Internet may cause new business models to emerge.

13.4.1 Policy Routing Problem

In intradomain routing, where the whole network can be thought to be controlled by a single logical entity, routing can be based on optimization of some global metrics such as latency, throughput, fault-tolerance, and/or fairness, but this is not directly possible in the interdomain case. The main stakeholders involved are the operators, their customers, and governments, that can regulate the former two and all of them have their own goals that can be mutually contradictory, forming tussles [11]. In general, the operators want to maximize their long-term profits while end users are interested in the end-to-end quality of service they receive from the network. Governments may impose restrictions on routes for security reasons and, for example, demand that user information is collected and stored by the operators. They may also be concerned about the total welfare the network is producing to their citizens.

The goals of the above-mentioned stakeholders are often contradictory, but by using contracts and compensation, the customers are able to use the resources of their providers' networks for end-to-end communication. These arrangements are external to the network and cannot be deduced only from the topology of the network, which means that the network architecture must allow manual configuration of policy information. For example, the routing decision can be made based on rules that use the source of the packet, protocol used, quality of service and the cost of routes as their input. This creates the technical problems of how to

1. represent the routing policies in an expressive way;
2. distribute topology and routing information in an efficient and scalable way; and
3. guarantee stable operation of the protocols involved.

Interdomain routing protocols try to answer these problems in their own way and network description languages can be used for the representation of policies. However, the game between ISPs introduces some constraints on the possible solutions such as many owners of network resources should be compensated the market price for the use of their offered resources and operators may not deploy technologies that would decrease their profits [12]. The incentives of operators restrict in practice what type of policies can be deployed in their routers. For example, they are not willing to give up their profits, control over their resources, or information without some sort of external pressure or compensation.

From the structure of the current Internet explained in the previous section, it follows that almost all policy compliant paths have the so-called *valley-free* property [13], which means that, on the AS business relationship level, packets always first follow 0-n logical customer-provider "up-hill" links, then 0-1 peer-peer links, and finally 0-n provider-customer "down-hill" links. That is, each AS-level hop needs to be compensated by one of the communicating end nodes.

A typical routing policy of an AS is that incoming packets are first routed to a customer network, then over a peering link, and as the last resort over a transit link, if the destination address is reachable via them. Downward flowing packets from a provider or a peering AS are only routed to customer links. These simple policies follow straightforwardly from the goal of minimizing the expenses of each ISPs providing the transit: every packet sent over a peering link affects the traffic balance that is stipulated in the peering contract and traffic sent over a transit link costs money directly and both of these options need to be compensated by the customer to its provider. It follows, for example, that the shortest path available, defined as the minimum number of AS-level hops, is often not used.

The time granularity of routing policies is typically much more coarse grained than the scales at which traffic engineering, and especially congestion control operate in. However, the policies may affect these two on a general level and guide their operation.

13.4.2 PURSUIT Global Rendezvous

PURSUIT architecture does not dictate a single structure for the global rendezvous function but only specifies the data types and publication naming schemes required for the rendezvous operation. It is assumed that competing implementations can be deployed in parallel as no single solution can satisfy all applications. A possible two-tier architecture for global rendezvous is introduced in [4]. It implements a datacentric pub/sub primitive as a recursive, hierarchical structure, which first joins node local rendezvous implementations into *rendezvous networks* (RN) and then RNs into a global *rendezvous interconnect* (RI) using a hierarchical Chord DHT called Canon [14] as shown in Figure 13.5. The Canon virtual hierarchy is assumed to be formed on top of the Internet AS graph based on the contracts formed by ASes and loosely optimizing the locality of communication as explained in [4].

The RNs can be implemented, for example, using DONA [6] implementations. In another dimension, the rendezvous system is split into common *rendezvous core* and scope-specific implementations of *scope homes* that implement the functionality for a set of scopes as explained in the following sections.

13.4.2.1 Rendezvous Core

At the every level of the hierarchy, the rendezvous core provides an overlay anycast routing to the approximately closest scope home that hosts a given scope. Each subscription message contains the (Sid,Rid) tuple of the publication of which contents the subscriber wishes to receive. Each node in the rendezvous core can cache results of previous rendezvous operations and immediately route the answer back along the reverse path of the rendezvous message to the subscriber.

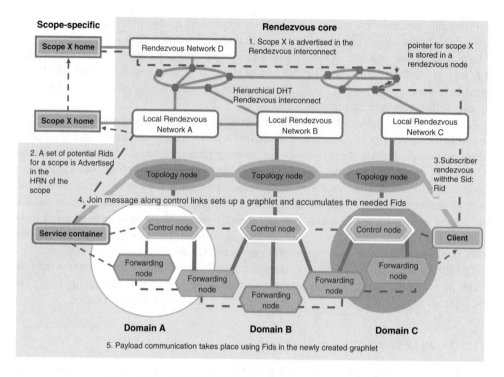

Figure 13.5 PSIRP/PURSUIT phases of communication.

Each rendezvous message is first hierarchically routed towards a rendezvous overlay node storing a *scope pointer* based on the hash of the Sid of the scope in question and then the pointer redirects the request towards the approximately closest scope home that has advertised the scope pointer to the rendezvous core. Publication advertisements by data sources are similarly eventually routed towards the nearest scope home and they contain also the contents of the publication in addition to the (Sid,Rid) pair. The scope then stores the contents of the publication so that it can serve the subscribers requesting it afterwards. In accordance with the fate-sharing principle, both the publication and subscription messages need to be periodically repeated in order to keep the publication data or pending subscription alive. This datacentric request-response anycast routed primitive is the only functionality implemented by the rendezvous core. A detailed description of the security mechanisms used by this architecture is given in [3].

13.4.2.2 Scope Implementation

Scopes can have varying implementations. When a cached result cannot be found in the rendezvous core, the subscription reaches the scope home, which can then dynamically generate the response if it has enough information available. It is possible to avoid caching by including version number in the requested Rid. Each scope implementation may be scaled up by adding more scope home nodes by the scope owner and these nodes can implement their own coordination protocol internally. It should be noted that the *slow-path* control plane overlay rendezvous is not meant for the transfer of large publications, but this type of applications should be supported by adding a datacentric transport to the data plane such as the one described in [5].

13.4.2.3   Phases of Communication

Each node wishing to communicate first requests the description of end-to-end *fast-path* resources used for the graphlet formation. This happens by subscribing to a publication labelled "<Topic_Rid>, <UN>, t" inside the given scope, where UN is the (Sid,Rid) pair naming the ANDL upgraph map of the node and t is the current time. If the scope performing the upgraph joining is access controlled, then the subscribed label also includes the node identity. The upgraph data itself is published by the ISP of the subscriber. Because many nodes share the same upgraph, the rendezvous system caches them orthogonally close to the scope homes. Similarly, the result of the rendezvous is automatically cached and reused if another node in the same domain requests the same service. ANDL maps can also link to other maps and a complete map can be built recursively from smaller publications, which minimizes the amount of communication as only relevant information needs to be transfered.

A typical sequence of operations is shown in Figure 13.5. After a successful rendezvous inside the scope, the client is returned graphlet information about the end-to-end resources it can use for the delivery tree formation. Then a separate resource allocation protocol such as the one described in [5] is used to set up the delivery tree through the domains defined by the graphlet. The resource allocation layer could produce, for example, a set of secure *forwarding identifiers* (Fid), such as zFilter [15], that could be used by the data sources or intermediate in-network caches as an opaque capability to access the created delivery tree securely without affecting the rest of the network. The zFilter concept is based on in-packet Bloom filters that encode multicast trees with sets of link identifiers.

A more detailed example of the delivery tree formation in an intradomain architecture is explained in [5], where the dedicated nodes handling a particular transport abstraction can be scattered in the network. The topology manager node of the domain just routes the packets carrying a given transport through compatible nodes while balancing the load to each node. In the case of datacentric communication, the subscription message that creates the delivery tree in reverse direction on the domain level, is routed in each domain through a *branching node* that implements the needed transport logic for time-decoupled multicast with caching. This means that the transport functionality does not even have to operate at backbone line speeds because of demultiplexing the flows to multiple nodes inside each domain on the path of the delivery tree.

13.5   Summary

In this chapter, we have examined two recent examples of new network architecture proposals that have been inspired by the pub/sub paradigm. CCN proposes a new network architecture in which the naming scheme is based on hierarchical, binary coded, human readable names, aimed to replace IP addresses and DNS names. The names are composed of a number of components so that the top-most is globally routable and the last one

	Layer (OSI)	Underlying transport	API	Namespace	Self-certification	Receiver driven	Key application
DONA	L4-L7	TCP/IP	Anycast	Flat	Yes	No	Content discovery and delivery
CCN	L3	Unreliable data transport, flow control	Name-based (with leases)	Hierarchical	Names can be authenticated	Yes	Content delivery, voice
PSIRP	L2-L7, layerless	Ethernet, TCP/IP and PLA	Pub/sub and metadata-based	Various, recursive	Yes	Yes	Content delivery
Internet Indirection Infrastructure (i3)	L7	TCP/IP	Trigger-based	Flat	Yes	Yes (also sender driven)	Various
Haggle	Layerless	Packet-based (TCP/IP)	Metadata-based	ADU, user level names	Possible	Yes, opportunistic interactions	Mobile dissemination, delay tolerant applications
Siena	L7	TCP/IP	Pub/sub	Content-based	No	Yes	Event dissemination and content delivery
Scribe and other DHT-based systems	L7	TCP/IP	Pub/sub	Topic or content-based	Varies	Yes	Event dissemination and content delivery

Figure 13.6   Comparison of clean-slate and overlay technologies.

is a SHA256 digest of the actual data. The packets are idempotent, self-identifying, and self-authenticating. PSIRP is an FP7-funded collaborative research project with the goal of developing, implementing, and validating an information-centric internetworking architecture based on the publish-subscribe communication paradigm.

Figure 13.6 compares CCN and PSIRP with other similar systems, such as DONA, i3, Siena, and DHT-based pub/sub solutions such as Scribe. These systems can be examined based on the level in the protocol stack, the underlying transport protocol used, for example TCP, the API features, namespace and the addressing model, security properties such as self-certification, how the control over packet delivery is distributed, and the key applications of the systems. We can observe that all of these systems are received driven and they empower the receiver to choose what content is delivered over the network. The systems have been designed for various layers. CCN and PSIRP are examples of so called clean-slate solutions that do not necessarily assume that TCP/IP is used. DONA is an example of a shim layer solution that extends the transport level APIs with anycast and caching support. Siena, i3, and DHT-based solutions are examples of application layer overlays that are typically implemented on top of TCP/IP.

The FP6 Haggle[1]Integrated Project is developing a new autonomic networking architecture for environments with intermittent network connectivity [16]. Haggle exploits autonomic opportunistic communications. A radical departure from the existing TCP/IP protocol suite is proposed that completely eliminates layering above the datalink layer.

---

[1] www.haggleproject.org.

Haggle uses application-driven message forwarding instead of delegating this responsibility to the network layer.

PSIRP shares the vision with Haggle that a new networking stack is needed. The main differences between the projects include PSIRP's focus on publish/subscribe and also that PSIRP focuses on developing architecture for Internet-scale networking rather than environments with intermittent network connectivity.

References

1. Filinski A (1989) *Declarative Continuations and Categorical Duality*. Master's thesis. University of Copenhagen.
2. Jacobson V, Smetters DK, Thornton JD, Plass MF, Briggs NH and Braynard RL (2009) Networking named content *Proceedings of the 5th International Conference on Emerging Networking Experiments and Technologies*, pp. 1–12. CoNEXT '09. ACM, New York, NY.
3. Lagutin D, Visala K, Zahemszky A, Burbridge T and Marias GF (2010) Roles and security in a publish/subscribe network architecture. *ISCC'10*, pp. 68–74.
4. Rajahalme J, Särelä M, Visala K and Riihijärvi J (2010) On name-based inter-domain routing. *Computer Networks Journal: Special Issue on Architectures and Protocols for the Future Internet*, pp. 975–86.
5. Visala K, Lagutin D and Tarkoma S (2009) LANES: An inter-domain data-oriented routing architecture. *ReArch'09*, pp. 55–60. ACM.
6. Koponen T, Chawla M, Chun BG, *et al*. (2007) A data-oriented (and beyond) network architecture. *SIGCOMM Comput. Commun. Rev*. **37**(4): 181–92.
7. Carpenter B (1996) *Architectural Principles of the Internet Internet Engineering Task Force*: RFC 1958.
8. Tarkoma S and Antikainen M (2010) Canopy: publish/subscribe with upgraph combination. *13th IEEE Global Internet Symposium 2010*, pp. 1–6.
9. *The CAIDA AS Relationships Dataset, November 2009* (n.d.) http://www.caida.org/data/active/ asrelationships/.
10. Labovitz C, Iekel-Johnson S, McPherson D, Oberheide J and Jahanian F (2010) Internet inter-domain traffic. *SIGCOMM'10*, pp. 75–86
11. Clark D, Wroclawski J, Sollins K and Braden R (2005) Tussle in cyberspace: Defining tomorrow's Internet. *IEEE/ACM Transactions on Networking* **13**(3): 462–75.
12. Rajahalme J, Särelä M, Nikander P and Tarkoma S (2008) Incentive-compatible caching and peering in data-oriented networks. *ReArch'08*. ACM, 62:1–62:6.
13. Gao L (2001) On inferring autonomous system relationships in the Internet. *IEEE/ACM Transactions on Networking* **9**(6): 733–45.
14. Ganesan P, Gummadi K and Garcia-Molina H (2004) Canon in G major: Designing DHTs with hierarchical structure. *ICDCS'04*, pp. 263–72. IEEE Computer Society.
15. Jokela P, Zahemszky A, Esteve C, Arianfar S and Nikander P (2009) LIPSIN: Line speed publish/subscribe inter-networking. *SIGCOMM'09*, pp. 195–206.
16. Su J, Scott J, Hui P, Crowcroft J, *et al*. (2007) Haggle: Seamless networking for mobile applications. *Proceedings of the 9th International Conference on Ubiquitous Computing*, pp. 391–408 UbiComp '07. Springer-Verlag, Berlin, Heidelberg.

14

Conclusions

There are many ways to classify pub/sub systems, and many possibilities for their use depending on the requirements. In this book, we have examined a number of state of the art solutions for creating pub/sub systems. We observed that one size does not fit all, and a modular approach is needed in building a generic pub/sub solution to be able to meet various application requirements and operating environments.

We examined the key principles and patterns for designing pub/sub systems. The principles included a logically centralized service for decoupling the components, an interest registration service for accepting subscriber interests, and a filtering mechanism for selective information dissemination. The main characteristics include expressiveness and scalability that contrast each other. Additional important characteristics are simplicity, modularity, and interoperability.

There are two general patterns for event notification, namely direct notification and infrastructure-based distributed notification. We examined the key patterns under these two categories. The observer pattern sets a one-to-many dependency between objects where dependent objects are automatically notified/updated whenever the observed object changes state. The event notifier pattern combines the observer and mediator patterns into a logically centralized service that fully decouples the subscribers and publishers and that is suitable for distributed environments. The distributed pub/sub systems follow the event notifier model. Pub/sub is also a vital paradigm for local communications within a single device. The event loop, and a local event broker are the key components of local pub/sub systems. For example, the Java event system presented in Chapter 5 is an example of a local system that was later extended for the distributed case.

We have examined current message-oriented middleware solutions such as JMS, DDS, and various message queue products as well as research oriented solutions for topic and content-based routing. Message-oriented middleware and event notification are becoming more popular in the industry with the advent of the CORBA Notification Service and DSS, the Java Messaging Service, and other related specifications and products from many vendors. Indeed, JMS is frequently used to implement the Enterprise Service Bus component, and DDS is the key standards based solution for embedded and industrial systems that supports datacentric and real-time communications.

Publish/Subscribe Systems: Design and Principles, First Edition. Sasu Tarkoma.
© 2012 John Wiley & Sons, Ltd. Published 2012 by John Wiley & Sons, Ltd.

Many message queue products are now supporting XML-based solutions, such as SOAP, as one of the transport options. MQSeries, MSMQ, and.NET support SOAP, and SIENA has XML bindings as well. XML has many applications in messaging and event-based communication. XML can be used to define the content of messages. For example, JMS facilitates XML-based messages and the routing of XML documents with a flexible header-based system. The YFilter is an example of an efficient XML document matcher that supports XPath and XQuery languages.

The research oriented solutions are currently only partially used by commercial systems and they indicate possible extensions and even new products. For example, at the moment there are no large-scale content-based systems deployed outside specific application domains; however, there are many deployed event services, for example Facebook Messages and Chat, Twitter, various alert services, and Pubsubhubbub. Many of these offer only simple matching, their implementation is cluster-based and they are proprietary.

These limitations are addressed by the research solutions, such as DHT-based systems and content-based routing, that can be realized as overlays and they can scale to wide-area environments. There are many specific solutions for pub/sub related features and problems, and we have given an outline of the selected important solutions pertaining to basic functionality in topic, content, and keyword based systems as well as a number of advanced topics. Many research projects have addressed and are addressing issues of scalability, composite event detection, mobility, and fault tolerance, to name a few topics.

Traditional messaging systems are being influenced by pub/sub systems. For instance, JMS supports both queues and pub/sub style communication with filtering. However, these systems usually lack support for distributed coordination in notification delivery, and they employ topic-based routing. Current event systems are evolving towards content-based routing, which uses the whole notification as an address. In content-based systems clients can change their interests without changing the addressing scheme (adding a new topic).

The observation that one-size does not fit all cases has led to the development of modular and reconfigurable pub/sub systems. Certainly reconfiguration is necessary in order to support runtime modifications to the system. Runtime upgrades and modifications are necessary for today's highly scalable and available systems as we discussed with the Facebook Messages backend system.

Figure 14.1 provides a summary of the components of a modular pub/sub router that would be suitable for various environments and capable of meeting differing application requirements. This design is based on the various systems and solutions presented in the book. The key ideas are the core router that provides the basic pub/sub API, and an interface and system that allows pluggable router components. Various components can then be discovered and utilized either at runtime or at configuration time. Thus the pub/sub router can be tailored and adapted for different environments and to meet the given application requirements. Runtime configuration and adaptation of pub/sub systems is still a research topic, and typically not supported by products with the exception of load balancing and fault-tolerance features.

The figure highlights the modular router core that is deployed over an overlay network with specific configurable modules that together form the runtime implementation. The core and its components expose certain API methods for developers and users of the system.

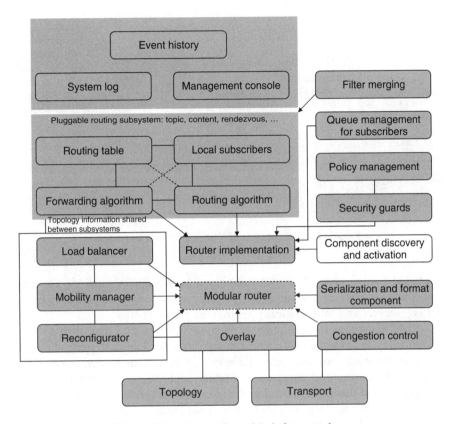

Figure 14.1   A modular pub/sub framework.

The internal communication of the components is not presented by the figure, but it could be simply local method calls or based on a local system bus following the pub/sub model. It is interesting to observe that pub/sub, as well as the blackboard pattern, can be used to realize the internal organization of a pub/sub router engine. For example, the routing algorithm component could subscribe updates from the router and overlay network regarding the status of neighbouring routers. This internal communication mechanism should be flexible, and allow the discovery of components as well as support information brokering inside the router.

The basic components include congestion control, message processing with the serialization and format component, and the reconfigurator. The core of the router consists of the routing and forwarding algorithms, and the modules that they need to function, namely the routing table, local subscriber table, and queue management component. The separation of routing and forwarding allows the development of different fast forwarding algorithms, and to treat routing table updates and published messages differently. Additional security guards are needed to meet the given security requirements.

Some of the components work in the distributed environment. For example, the lower overlay level needs to maintain the overlay network routing tables and monitor the overlay network. The pub/sub routing table needs to keep track of the neighbouring routers in the

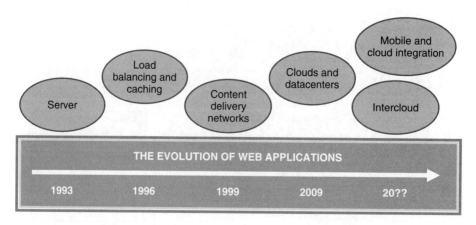

Figure 14.2 Evolution of Web application provision.

content-based routing layer. The configurator and load balancing components may need to know about a subset of pub/sub routers. Moreover, security components may need to contact key servers and other logically centralized points of trust.

Figure 14.2 illustrates the evolution of Web applications and their provisioning. The cloud computing environment offers new possibilities for low-cost application development and deployment, but also new challenges in how data is stored and distributed across different devices and datacenters. Mobile applications are also integrating with cloud resources and there is a general problem of how to synchronize applications across devices and servers and keep data up-to-date. Thus there is a need for an asynchronous communications substrate that connects all the components together. Indeed, such message queue and topic-based pub/sub solutions are used today to manage systems. We examined several solutions in Chapter 12.

The future appears to be promising for pub/sub solutions. We presented a number of applications in Chapter 12 that highlight current uses of the technology. Classical applications include stock market and banking, detection of complex events, logistics, industrial environments and systems, and SOA. Social networking sites such as Facebook and Twitter use pub/sub technology extensively in order to connect people and applications. Moreover, cloud computing systems and Internet of Things rely on message delivery and pub/sub. It should also be noted that pub/sub engines can be used for content brokering. Indeed, online advertisement is an emerging area for pub/sub systems. Online advertisement requires content-based routing solutions were presented in Chapters 7 and 8. The advertisement engine relies on the efficient matching of user profiles with active advertisement campaigns. Pub/sub systems can offer this matching engine and also distribute the matching process across the Internet.

On the other hand, there is no existing global pub/sub technology or service for the Web. Facebook and Twitter provide open APIs as well as other Web sites, and they are at least partly offering this service. With the emergence of HTML5 and WebSockets, the Web technology offers a more efficient basis for implementing pub/sub solutions. The question remains open how a global pub/sub hub would form over the Internet. Pubsubhubbub provides an interesting example of such a design, but it is not yet attained such a status.

We briefly examined a number of security solutions for pub/sub including the Event-Guard system that is based on the modular guard components. Basic security can be supported through modular security solutions that build on key generation and distribution, encryption, and digital signatures. Subscriber and publisher privacy are new security challenges for pub/sub systems and most pub/sub systems do not consider privacy issues.

Another open issue is whether the event service should reside at the network level or at the application level. For Internet-scale routing, as proposed in SIENA, it might be beneficial to have some support at the network level. This has been investigated recently by the PSIRP and CCN projects presented in Chapter 13. These proposals require large changes to the way routers and end systems operate, and thus they are research oriented ideas and prototypes at the moment. The idea of building a network on top of pub/sub is a very interesting one.

Index

Publish/Subscribe Systems: Design and Principles, First Edition. Sasu Tarkoma.
© 2012 John Wiley & Sons, Ltd. Published 2012 by John Wiley & Sons, Ltd.

Centennial
College
Libraries